SCHOLAR Study Guide

CfE Advanced Higher Chemistry

Unit 2: Organic Chemistry and Instrumental Analysis

and

Unit 3: Researching Chemistry

Authored by:

Diane Oldershaw (previously Menzieshill High School)

Reviewed by:

Helen McGeer (Firrhill High School)

Nikki Penman (The High School of Glasgow)

Previously authored by:

Peter Johnson

Brian T McKerchar

Arthur A Sandison

Heriot-Watt University

Edinburgh EH14 4AS, United Kingdom.

Distributed by the SCHOLAR Forum.

SCHOLAR Study Guide Unit 2 and Unit 3: CfE Advanced Higher Chemistry

1. CfE Advanced Higher Chemistry Course Code: C713 77

   ISBN 978-1-911057-20-8

Print Production and Fulfilment in UK by Print Trail www.printtrail.com

Acknowledgements

Thanks are due to the members of Heriot-Watt University's SCHOLAR team who planned and created these materials, and to the many colleagues who reviewed the content.

We would like to acknowledge the assistance of the education authorities, colleges, teachers and students who contributed to the SCHOLAR programme and who evaluated these materials.

Grateful acknowledgement is made for permission to use the following material in the SCHOLAR programme:

The Scottish Qualifications Authority for permission to use Past Papers assessments.

The Scottish Government for financial support.

The content of this Study Guide is aligned to the Scottish Qualifications Authority (SQA) curriculum.

Contents

Topic 1

Molecular orbitals (Unit 2)

Contents

Prerequisite knowledge

Before you begin this topic, you should know:

- *what an orbital is and about the different types of orbital (s,p,d and f); (Advanced Higher Chemistry Unit 1)*

- *the different types of bonding that occur within compounds and elements and how they fit into the bonding continuum. (Advanced Higher Chemistry Unit 1, Higher Chemistry Unit 1, National 5 Chemistry Unit 1)*

Learning objectives

By the end of this topic, you should know about:

- *how molecular bonding orbitals are formed;*

- *how sigma and pi bonds are formed through hybridisation;*

- *the bonding continuum.*

1.1   Formation of molecular bonding orbitals

When atoms approach each other, their separate sets of atomic orbitals merge to form a single set of molecular orbitals. Some of the molecular orbitals, known as 'bonding molecular orbitals', occupy the region between two nuclei. The attraction of positive nuclei to negative electrons occupying bonding molecular orbitals is the basis of bonding between atoms. Each molecular orbital can accommodate a maximum of two electrons.

If we look at the example of $H_2(g)$ one of the molecular orbitals is formed by adding the mathematical functions for the two 1s orbitals that come together to form this molecule. A molecular orbital is a mathematical function describing the wave-like behaviour of an electron in a molecule. This function can be used to calculate chemical and physical properties such as the probability of finding an electron in any specific region. This can be described simply as the region of space in which the function has a significant amplitude. Molecular orbitals are usually constructed by combining atomic orbitals or hybrid orbitals from each atom of the molecule, or other molecular orbitals from groups of atoms. The shape of orbitals are determined through quantum mechanics (see Unit 1 Topic 2.4).

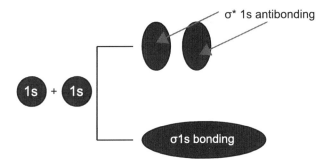

S-orbitals have a spherical symmetry surrounding a single nucleus, whereas σ-orbitals have a cylindrical symmetry and encompass two nuclei.

The bonding orbital is where the electrons spend most of their time between the two nuclei. Electrons in the other antibonding orbital spend most of their time away from the region between the two nuclei.

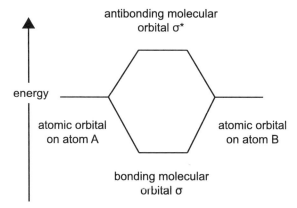

Electrons are added to orbitals in order of increasing energy. The two electrons associated with H_2 bonding are placed in the bonding molecular orbital suggesting that the energy of an H_2 molecule is lower than 2 separate H atoms. This results in a molecule of H_2 being more stable than two separate H atoms.

We can use this to explain why molecules of helium do not exist. Combining two helium atoms with the $1s^2$ electronic configuration would result in two electrons in the bonding orbital and two electrons in the antibonding orbital. The total energy of a He molecule would be exactly the same as two isolated He atoms with nothing to hold the atoms together in a molecule. For further atoms it is only the valence shell electrons used in combining atoms that will be considered as core electrons make no contribution to the stability of molecules.

1.2   Hybridisation and the role in formation of σ and π bonds

It is difficult to explain the shapes of even the simplest molecules with atomic orbitals. A solution to this problem was proposed by Linus Pauling, who argued that the valence orbitals on an atom could be combined to form hybrid atomic orbitals.

A covalent bond is formed when two half-filled atomic orbitals overlap. If they overlap along the axis of the bond (end on) a covalent bond is known as a sigma (σ) bond.

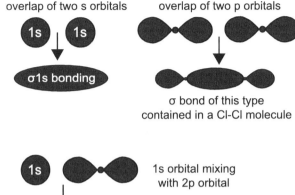

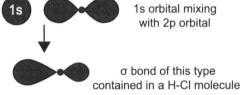

Pi (π) bonds arise where atoms make multiple bonds, for example the double bond in a molecule of oxygen O_2 is made up of one sigma and one pi bond. The triple bond in a molecule of nitrogen N_2 is made up of one sigma and two pi bonds. Pi bonds are formed when atomic orbitals lie perpendicular to the bond and overlap side on. End to end overlap is more efficient than side on overlap and therefore σ bonds are stronger than π bonds.

side-on overlap of two p orbitals in a π bond

If we look at carbon in its ground state it has the electronic configuration $1s^2\ 2s^2\ 2p^2$. This means it has two half-filled orbitals in the 2p subshell which may lead us to believe it would form two bonds instead of the four we know it forms. A simple explanation might involve promotion of an electron from the 2s orbital to the empty 2p orbital, producing four unpaired electrons which could then form four bonds with hydrogen.

A more satisfactory explanation involves hybridisation. The theory assumes that the 2s and three 2p orbitals combine during bonding to form four new identical hybrid orbitals.

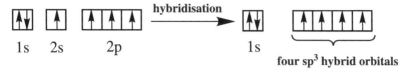

hybridisation

1s   2s   2p   1s

four sp³ hybrid orbitals

The hybrid orbitals are known as sp³ orbitals because they are formed by combining one s and three p orbitals.

The energy required to promote the electron would be more than offset by the formation of two extra covalent bonds. However, whereas the others would involve 2p orbitals. Spectroscopic measurements show that all four bonds in methane are identical.

Let's look at an alkane, ethane for example. Each carbon has three 2p orbitals and one 2s orbital which mix to form four degenerate (equal energy) hybrid orbitals. These are known as sp³ hybrid orbitals and point towards the corners of a tetrahedron in order to minimise repulsion from each other. The four sp³ orbitals on each carbon atom overlap end to end with one sp³ orbital on the other carbon atom and the three hydrogen 1s orbitals. This forms 4 σ bonds.

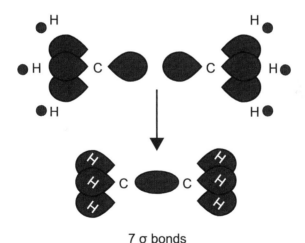

7 σ bonds

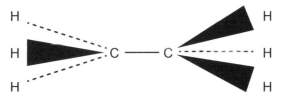

The bonding in ethane can be described as sp^3 hybridisation and sigma bonds.

If we take a look at the bonding in the corresponding alkene ethene we can see that the 2s orbital and two of the three 2p orbitals mix on each carbon atom to form three sp^2 hybrid orbitals. To minimise repulsion these orbitals form a trigonal planar arrangement. The carbon atoms use the three sp^2 hybrid orbitals to form sigma bonds with two hydrogen atoms and with the other carbon atom. The unhybridised 2p orbitals left on the carbon can overlap side-on to form a pi bond.

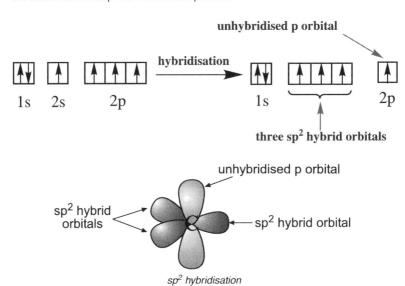

sp^2 hybridisation

unhybridised p orbitals

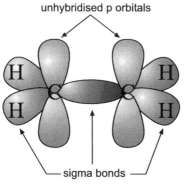

sigma bonds

C to C sigma bond forming

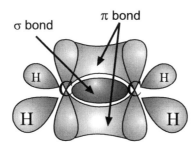

In alkynes for example ethyne (C_2H_2) the 2s orbital and one of the three 2p orbitals mix on each carbon to form two sp hybrid orbitals. This will form 3 sigma bonds and 2 pi bonds. The alkyne adopts a linear structure to minimise repulsion.

Bonding in hydrocarbons

It may help to draw Lewis dot structures for the molecule to identify the bonding and non-bonding pairs of electrons before answering the question.

Go online

For the next three questions, consider the hydrocarbon ethyne, HC≡CH.

Q1: What is the hybridisation of the carbon atoms required to form a C≡C bond?

a) sp^3
b) sp^2
c) sp

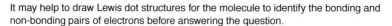

Q2: Which of the following statements is true about an ethyne molecule?

a) There are three π bonds and two σ bonds.
b) There are two π bonds and two σ bonds.
c) There are two π bonds and three σ bonds.

d) There are three π bonds and three σ bonds.

....

Q3: What shape will the ethyne molecule be?

....

1.3   The bonding continuum

The symmetry and position of the bonding orbitals between atoms will determine the type of bonding: ionic, polar covalent or pure covalent.

For pure covalent bonds where there is no or very little difference in electronegativity between the atoms involved in the bond the bonding orbital will be symmetrical around a point where the bonding electrons are predicted to be found. A fluorine molecule F_2 would be an example of where this would be found.

In polar covalent bonds (H^{δ^+}-Cl^{δ^-}) where there is a difference in electronegativity between the two atoms involved in the bond the bonding orbital will be asymmetrical with the bonding electrons more likely to found around chlorine due to it having a larger electronegativity.

In ionic bonds (Na^+Cl^-) where the difference in electronegativity between the two atoms involved in the bond is large, the bonding orbital will be extremely asymmetrical and almost entirely around one of the atoms. In this case it would be around chlorine due to it having a larger electronegativity compared to sodium.

1.4   Summary

Summary

You should now be able to state that:

- molecular bonding orbitals are formed by the overlap of orbitals from individual atoms when bonding.

- hybridisation occurs when different types of atomic orbitals are mixed.

- sigma bonds are formed when two half filled atomic orbitals overlap end-on.

- pi bonds are formed when two half filled atomic orbitals overlap side-on.

- where compounds fit onto the bonding continuum is determined by the difference in electronegativity between bonded atoms.

1.5   Resources

- LearnChemistry, Chemistry Vignettes: Molecular Orbitals (http://rsc li/2a1zkg0)
- Molecular Orbital Theory (http://bit.ly/2abKNqV)

1.6 End of topic test

End of Topic 1 test

The end of topic test for *Molecular orbitals.*

Go online

Q4: Which is the correct description of the numbers of sigma and pi bonds in propene?

a) 1 sigma bond, 8 pi bonds
b) 2 sigma bonds, 8 pi bonds
c) 7 sigma bonds, 1 pi bonds
d) 8 sigma bonds, 1 pi bond

....

Q5: The bonds in 2-bromobutane ($CH_3CH_2CHBrCH_3$) are sp^3 hybridised. Explain what is meant by sp^3 hybridisation.

....

Q6: How many sp^2 hybridised carbon atoms does the following molecule contain?

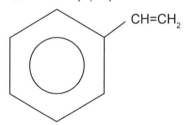

$$CH=CH_2$$

a) 0
b) 2
c) 6
d) 8

....

Q7: Ethene molecules contain:

a) sp^2 hybridised carbon atoms but no sp^3 hybridised carbon atoms.
b) sigma bonds but no pi bonds.
c) sp^3 hybridised carbon atoms but no sp^2 hybridised carbon atoms.
d) pi bonds but no sigma bonds.

....

Q8: Predict the bonding in $TiCl_4$ and where the bonding orbitals will be found.

....

Topic 2

Molecular structure (Unit 2)

Contents

Prerequisite knowledge

Before you begin this topic, you should know how to:

- *determine the molecular formula of a compound. (National 5 Chemistry unit 1 and 2)*

- *draw the structural formula of a compound. (National 5 Chemistry unit 2)*

Learning objectives

By the end of this topic, you should know:

- *how to draw the skeletal structure of a compound from the molecular formula or the structural formula and vice versa.*

2.1   Practising with molecular structure

There are many representations of organic molecules and you need to be able to convert easily between the different forms. These include molecular formulae and structural formulae of which you should be familiar with but we can also represent molecules in skeletal formulae.

The molecular formula tells us the number of each type of atom within an organic molecule. For example the molecular formula of butane is C_4H_{10} telling us we have a molecule with four carbon atoms and ten hydrogen atoms. The table below shows some molecular formulae for a few selected hydrocarbon molecules.

Hydrocarbon molecule	Molecular formula
Ethane	C_2H_6
Propane	C_3H_8
Butene	C_4H_8
Pentene	C_5H_{10}
Cyclobutane	C_4H_8
Cyclohexane	C_6H_{12}
Ethyne	C_2H_2

For molecules with functional groups including hydroxyl (OH), amine (NH_2), carboxyl (COOH) and carbonyl (CO) these are shown as part of the molecular formula of the molecule. Some examples are given in the table below.

Organic molecule	Molecular formula
Ethanol	C_2H_5OH
Ethanoic acid	CH_3COOH
Ethylamine	$C_2H_5NH_2$
Ethanal	CH_3CHO

If we go back to the example of butane we can represent this molecule through its structural formula. This gives a representation of how the elements are bonded within the molecule.

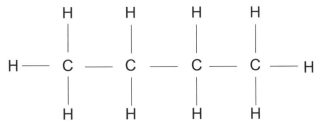

This shows us that the four carbons are singly bonded in a straight chain with the appropriate number of hydrogen atoms bonded to each carbon.

The structure of butane can also be represented as a skeletal formula.   In organic chemistry, skeletal formulae are the most abbreviated diagrammatic descriptions of moloculoc in common uoc.   Thcy look vcry barc because in skeletal formulae the hydrogen atoms (attached directly to carbons) are removed, leaving just a "carbon skeleton" with functional groups attached to it. You must remember that hydrogen atoms that are not shown in skeletal formula are assumed to be there.

Some points to note with skeletal formulae:

a) There is a carbon atom at each junction between bonds in a chain and at the end of a bond (unless there is something else there like a functional group);

b) There are enough hydrogen atoms attached to each carbon atom to make the total number of bonds of each carbon four.

So looking at butane again, the skeletal formula would be:

The four carbon atoms are positioned in the following places in the above skeletal formula. Attached to these carbons would be the appropriate number of hydrogen atoms to give each carbon four bonds.

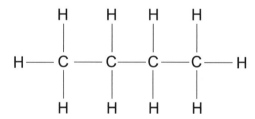

If we look at the molecule 2-methylbutane the skeletal structure would look like this:

skeletal formula   structural formula

If we look at the skeletal formula of ethanol, for example, the functional group OH is included in the formula.

the two carbon atoms are positioned here with the appropriate number of hydrogen atoms attached

Some other molecules drawn as skeletal formulae are given below.

amino acid asparagine

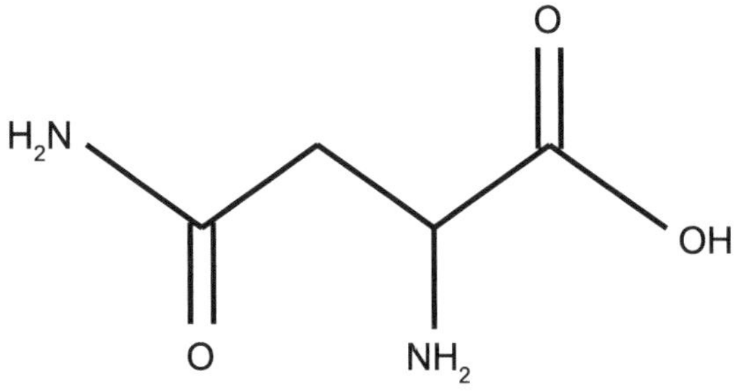

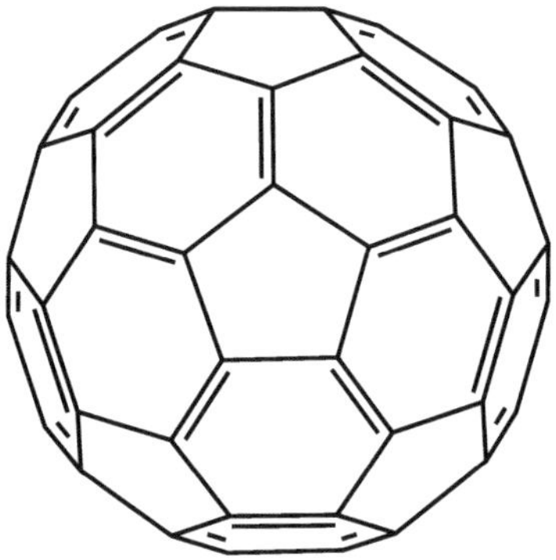

Buckminsterfullerene C60 (http://commons.wikimedia.org/wiki/File:C60a.png, by http://en.wikipedia.org/wiki/User:Mstroeck, licensed under http://creativecommons.org/licenses/by-sa/3.0/deed.en)

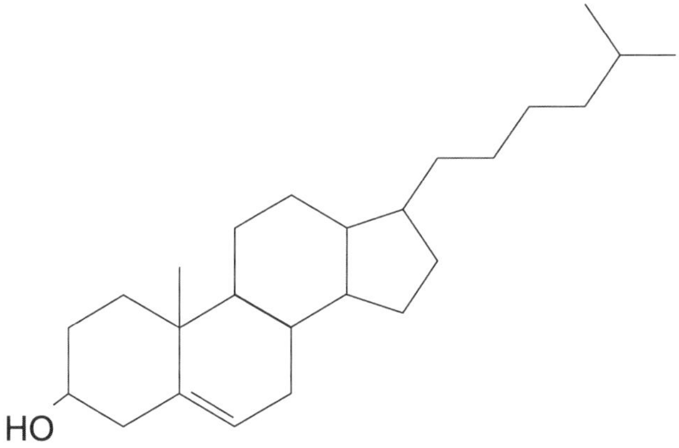

Cholesterol

You should practise converting between the molecular, structural and skeletal formulae for various molecules in particular the different families of hydrocarbons studied at National 5, Higher and Advanced Higher level. This should be done for molecules with up to 10 carbon atoms in the longest chain.

2.2 Summary

Summary

You should now be able to:

- interchange between and recognise the molecular, structural and skeletal formulae of various compounds.

2.3 Resources

- LearnChemistry, Organic formulae (http://rsc.li/2az02xA)

Topic 3

Stereochemistry (Unit 2)

Contents

Prerequisite knowledge

Before you begin this topic, you should know:

- *that isomers are compounds that have the same molecular formula but differ in structural formulae. (National 5 chemistry unit 2, Higher Chemistry unit 2)*

Learning objectives

By the end of this topic, you should know:

- *stereoisomers are isomers that have the same molecular formula but differ in structural formulae (a different spatial arrangement of their atoms);*

- *geometric isomers are stereoiosmers where there is a lack of rotation around one of the bonds mostly a C=C;*

- *these isomers are labelled cis and trans dependent on whether the substitutes are on the same or different sides of the C=C;*

- *optical isomers are non-superimposable mirror images of asymmetric molecules and are referred to as chiral molecules or enantiomers;*

- *isomers can often have very different physical or chemical properties from each other.*

3.1 Isomers

Molecules which have the same molecular formula but differ in the way their atoms are arranged (structural formula) are called isomers. They often have very different physical and chemical properties from each other. There are two ways that atoms can be arranged in each isomer, one being that the atoms are bonded together in a different order. These are called structural isomers.

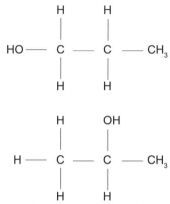

An example of structural isomers: propan-1-ol and propan-2-ol respectively

The second way involves the atoms being bonded in the same order but the arrangement of the atoms in space is different for each isomer. These are called stereoisomers and there are two different types we will study in this unit, geometrical and optical isomers.

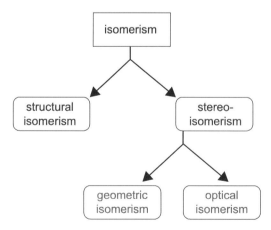

3.2 Geometric isomers

Geometric isomers can be illustrated by looking at the alkene but-2-ene.

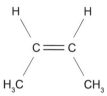

It contains a carbon to carbon double bond and the arrangement of atoms and bonds is planar with all the bond angles at 120°. The bonds are fixed in relation to each other meaning it is impossible to rotate one end of the alkene molecule around the carbon carbon double bond while the other end is fixed. This is why alkenes can exhibit geometric isomerism.

It would be advantageous at this point to make a model of but-2-ene using molymods to show the lack of rotation around the carbon to carbon bond.

There are two possible geometric isomers of but-2-ene and they are referred to as the **cis** isomer and the **trans** isomer.

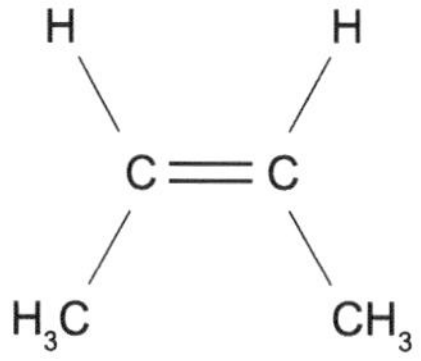

cis-but-2-ene

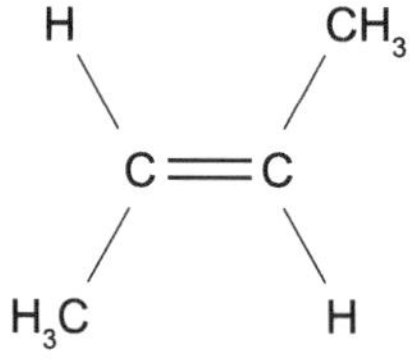

trans-but-2-ene

For geometric isomers cis means "on the same side" and you can see from the diagram that the two methyl groups are on the same side of the carbon to carbon double bond. Trans means "on different sides" and you can again see from the diagram that the methyl groups are on different sides of the carbon to carbon double bond. These two isomers have distinct physical properties; the melting point of the cis isomer is -139°C while the melting point of the trans isomer is -105°C.

There are also geometric isomers found in but-2-enedioic acid.

H   H
 \   /
  C ═══ C   cis-but-2-enedioic acid
 /   \
HOOC   COOH

H   COOH
 \   /
  C ═══ C   trans-but-2-enedioic acid
 /   \
HOOC   H

These isomers have differing physical properties but also have a chemical difference too in that the cis form is readily dehydrated. This is not possible in the trans isomer due to the carboxyl groups being on opposite sides of the carbon to carbon bond which is not a suitable orientation to undergo such a reaction. The melting point of the cis isomer is 135°C whereas the trans isomer has a melting point of 287°C again showing the effect on physical properties of isomers.

Geometric isomers can only occur in organic molecules which have two different groups attached to each of the carbon atoms of the double bond, in addition to the carbon to carbon double bond. For example propene would not exhibit geometric isomerism.

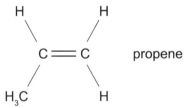

propene

propene has two identical H atoms attached to one of the carbon atoms from the carbon to carbon bond

Geometric isomerism is most commonly found in organic molecules containing a C=C bond but it can also be found in saturated rings where rotation about the C-C single bonds is restricted.

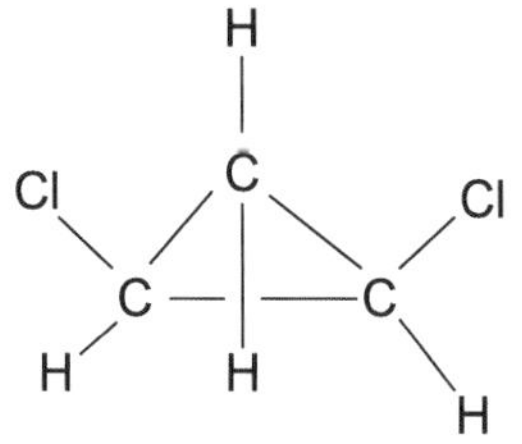

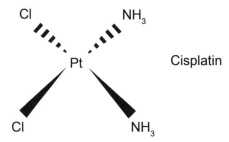

cis isomer

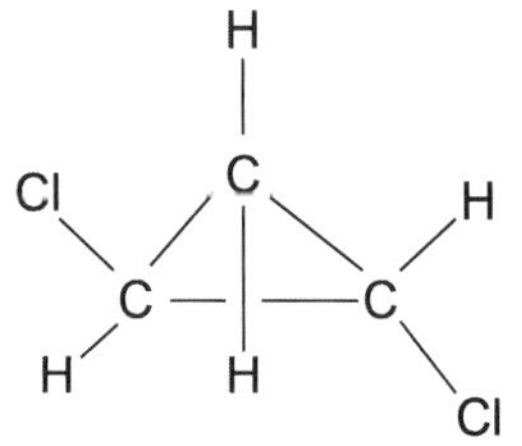

trans isomer

1,2-dichlorocyclopropane

Non-organic compounds can also exhibit geometric isomerism. For example the anti-cancer drug cisplatin.

Cisplatin

The trans-isomer has no comparable useful pharmacological effect.

3.3 Optical isomers

In the same way as your left hand and your right-hand are mirror images of each other, many chemical compounds can exist in two mirror image forms. Right and left hands cannot be superimposed on top of each other so that all the fingers coincide and are therefore not identical. A right hand glove does not fit a left hand and is said to be **chiral**.

The hands and the compounds have no centre of symmetry, plane of symmetry or axis of symmetry. Chirality arises from a lack of symmetry. Lack of symmetry is called asymmetry.

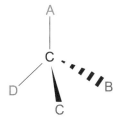

tetrahedral arrangement of an isomer showing the four different groups attached

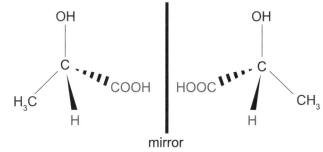

mirror

two optical isomers of lactic acid: the mirror images are non-superimposable

Optical isomers are identical in all physical properties except their effect on plane polarised light, i.e. they exhibit optical activity. When plane polarised light is passed through a solution containing one optical isomer, the plane is rotated through a certain angle. If you have a solution of the other optical isomer at the same concentration the plane of polarised light is rotated by exactly the same angle but in the opposite direction. So if one isomer rotated the plane of polarised light by +60° (clockwise direction) the other isomer would rotate it by -60° (anticlockwise direction).

If you have a mixture of equimolar optical isomers it would have no effect on plane-polarised light since the rotational effect of one isomer would be cancelled out by the opposite rotational effect of the other. This mixture is optically inactive and is known as a **racemic mixture** .

Optical isomers of substances can be labelled R or S. The symbol R comes from the Latin rectus for right, and S from the Latin sinister for left. The 4 different groups are numbered in order of priority and if the order 1,2,3,4 is in a clockwise direction we call the enantiomer R. If they are in that order in an anticlockwise direction then we call the enantiomer S. The order of priority is determined by the element directly bonded to the chiral centre, the one with the highest atomic mass being numbered 1. If two groups have the same element directly bonded to the chiral centre then the next element bonded to that is considered and so on.

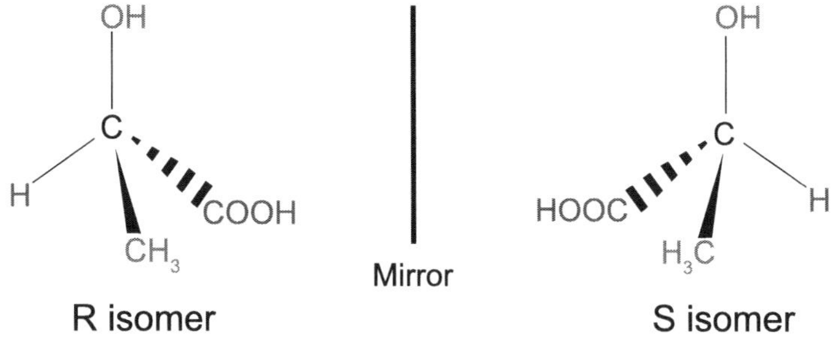

Priority of groups

1 = OH
2 = COOH
3 = CH$_3$
4 = H

Optical isomerism is immensely important in biological systems. In most biological systems only one optical isomer of each organic compound is usually present. For example, when the amino acid alanine is synthesised in the laboratory, a mixture of the two possible isomers (a racemic mixture) is produced. When alanine is isolated from living cells, only one of the two forms is seen.

The proteins in our bodies are built up using only one of the enantiomeric forms of amino acids. Only the isomer of alanine with structure 2 (Figure 7.19) occurs naturally in organisms such as humans. Since amino acids are the monomers in proteins such as enzymes, enzymes themselves will be chiral. The active site of an enzyme will only be able to operate on one type of optical isomer. An interesting example of this is the action of penicillin, which functions by preventing the formation of peptide links involving the enantiomer of alanine present in the cell walls of bacteria. This enantiomer is not found in humans. Penicillin can therefore attack and kill bacteria but not harm the human host, as this enantiomer is not present in human cells.

3.4 Biological systems

Optical isomerism is immensely important in biological systems. In most biological systems only one optical isomer of each organic compound is usually present. For example, when the amino acid alanine is synthesised in the laboratory, a mixture of the two possible isomers (a racemic mixture) is produced. When alanine is isolated from living cells, only one of the two forms is seen.

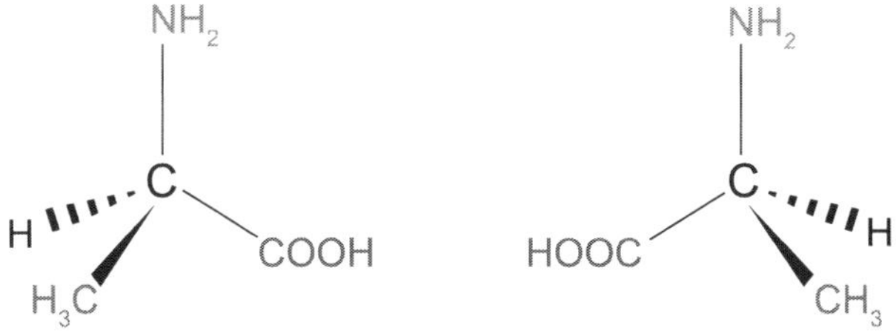

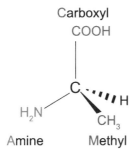

Structure 1   Structure 2

Alanine mixture

Optical isomers of alanine

Go online

This illustration shows one of the **enantiomers** of alanine, drawn in a different way.

Carboxyl
COOH

The carbon atom surrounded by four different groups is the **chiral centre**. As well as the hydrogen atom it has an amine group, a carboxyl group and a methyl group.

Amine   Methyl

Q1: Compare above picture with the structures shown below. Which structure is it?

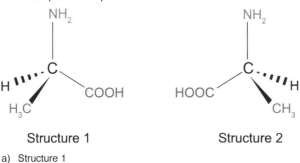

Structure 1   Structure 2

a) Structure 1
b) Structure 2

. . . .

If you have access to molecular models, you should build the molecule on the left of Picture 1 by starting with a carbon and using different coloured spheres for the four groups. (The suggested colours would be red for carboxyl, blue for amine, yellow for methyl and white for hydrogen.) Try to build the other enantiomer as the exercise progresses.

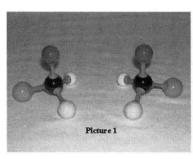 

Picture 1 shows two isomers side by side and when one of them is taken away and replaced with a mirror, the two images in Picture 2 look the same as in Picture 1.

In Picture 3, the isomer that was removed is placed behind the mirror, showing that it is identical to the image. Picture 4 points out how the two isomers are non-superimposable because even with the two substituent groups on the left of each isomer lined up, the two arrowed groups do not match.

Q2: The molecules shown in Picture 4 are

a) superimposable
b) non-superimposable

. . . .

Q3: The two molecules are said to be

a) symmetric
b) asymmetric

....

Q4: The two molecules can be said to be

a) identical
b) chiral

....

....

3.5 Optical isomerism in medicines and other substances

Many medicines are produced as a mixture of enantiomers, only one of which is pharmacologically active, as it can prove costly to separate the isomers. One of the enantiomers of salbutamol used in the treatment of asthma is 68 times more effective than the other. Great care has to be taken when using drugs with enantiomeric forms as this has led to tragedy in the past, specifically in the case of thalidomide. A mixture of the isomers was used to treat nausea during pregnancy, and one enantiomer, which was thought to be inactive, turned out to cause damage to the unborn child. Many handicapped babies were born before the drug was recognised as being responsible. Screening of pharmaceuticals has to be very thorough. Regulations were tightened significantly after the thalidomide tragedy to ensure that both enantiomeric forms of chiral drugs are tested.

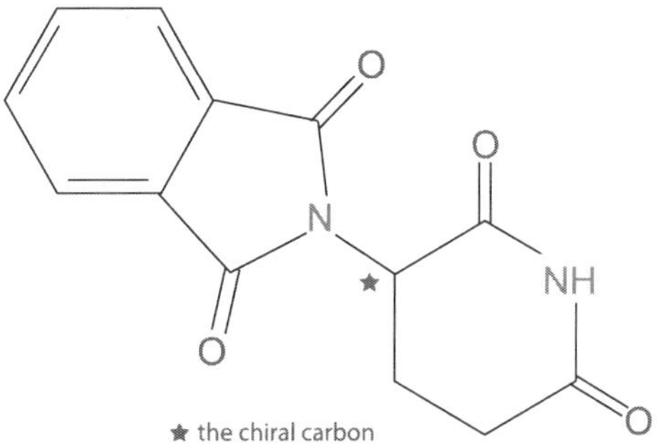

★ the chiral carbon
structure of thalidomide

Another example where optical isomers are different is in limonene. One isomer of limonene smells of oranges, the other of lemons. Note the placement of the chiral carbon in the structure of limonene diagrams below.

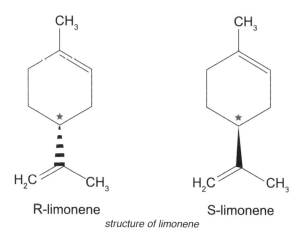

R-limonene   S-limonene

structure of limonene

3.6 Summary

Summary

You should now be able to state that:

- stereoisomers are isomers that have the same molecular formula but differ in structural formulae.

- geometric isomers are stereoiosmers where there is a lack of rotation around one of the bonds mostly a C=C.

- these isomers are labelled cis and trans dependent on whether the substitutes are on the same or different sides of the C=C.

- optical isomers are non-superimposable mirror images of asymmetric molecules and are referred to as chiral molecules or enantiomers.

- isomers can often have very different physical or chemical properties from each other.

3.7 Resources

- LearnChemistry, Introduction to Chirality (http://rsc.li/2aiQ3ub)

3.8 End of topic test

Go online

End of Topic 3 test

Q5: Which of the following could **not** exist in isomeric forms?

a) $C_2H_4Cl_2$
b) C_3H_6
c) C_3H_7Br
d) C_2F_4

....

Q6: Which of the following pairs represent the same chemical substance?

a) $H_2C{=}CH{-}CH_2{-}CH_3$ and $H_3C{-}CH{=}CH{-}CH_3$

b)

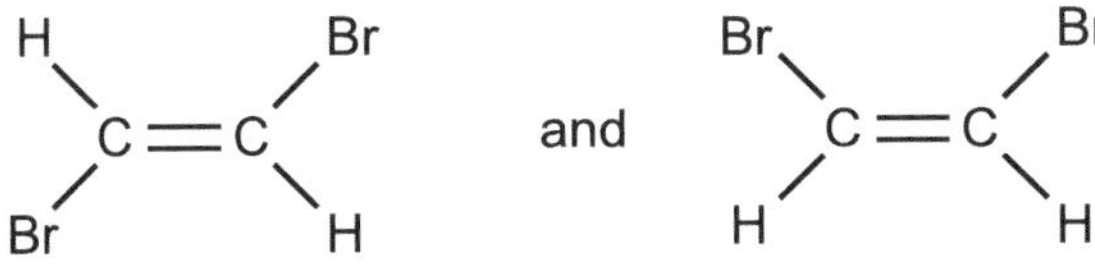

$$\underset{Br}{\overset{H}{\diagdown}}C{=}C\underset{H}{\overset{Br}{\diagup}} \quad \text{and} \quad \underset{H}{\overset{Br}{\diagdown}}C{=}C\underset{H}{\overset{Br}{\diagup}}$$

c)

$$H_3C{-}CH_2{-}\underset{\underset{CH_3}{|}}{CH}{-}CH_3 \quad \text{and} \quad CH_3{-}\underset{\underset{CH_3}{|}}{\overset{\overset{CH_3}{|}}{C}}{-}CH_3$$

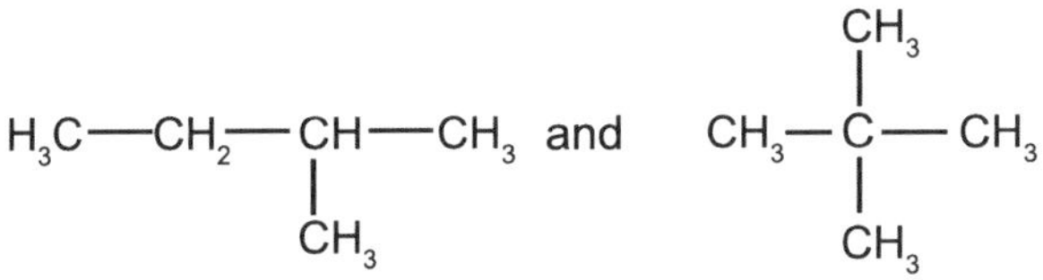

d)

$$Cl{-}\underset{\underset{H}{|}}{\overset{\overset{H}{|}}{C}}{-}\underset{\underset{Cl}{|}}{\overset{\overset{H}{|}}{C}}{-}Cl \quad \text{and} \quad H{-}\underset{\underset{Cl}{|}}{\overset{\overset{Cl}{|}}{C}}{-}\underset{\underset{Cl}{|}}{\overset{\overset{H}{|}}{C}}{-}H$$

....

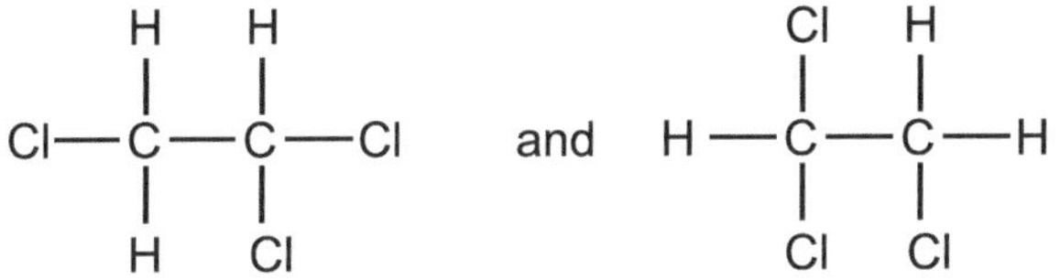

Q7: Which of these alcohols exists as optical isomers?

a) Propan-1-ol
b) Propan-2-ol
c) Butan-2-ol
d) Butan-1-ol

....

Q8: Which of these will have enantiomeric forms?

a) 1,2-dibromoethane
b) 1-chloroethanol
c) 1,2-bromochloroethene
d) 1-chloroethene

....

Take a look at these brominated alkenes:

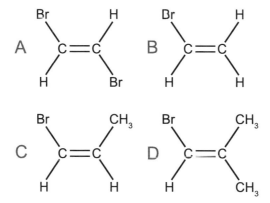

Q9: Which one contains an isomer of bromocyclopropane?

a) A
b) B
c) C
d) D

. . . .

Q10: Which two brominated alkenes contain a molecule which is one of a cis/trans pair?

a) A and B
b) B and D
c) A and C
d) C and D

. . . .

Q11: Which molecule will be the **least** polar?

a) A
b) B
c) C
d) D

. . . .

Look at the structure of these cycloalkanes:

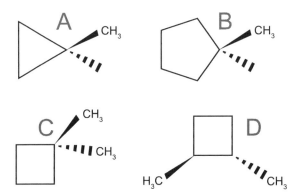

Q12: Which of the cycloalkanes shows cis/trans isomerism?

a) A
b) B
c) C
d) D

. . . .

Q13: Which of the cycloalkanes is one of an enantiomeric pair?

a) A
b) B
c) C
d) D

. . . .

Q14: Which of these amino acids does **not** have a chiral centre?

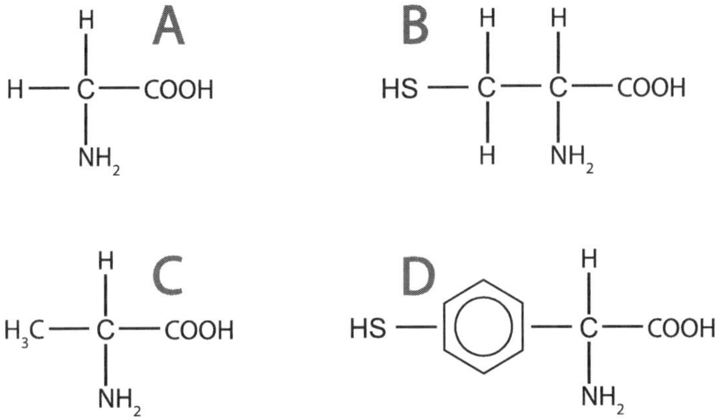

a) A
b) B
c) C
d) D

....

Q15: The part of a polarimeter capable of rotating a beam of polarised light is called the:

a) polariser.
b) analyser.
c) sample tube of chiral molecules.
d) light source.

....

Q16: Thalidomide is a drug which has optical isomers. Which atom is the chiral centre?

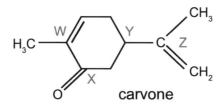

★ the chiral carbon

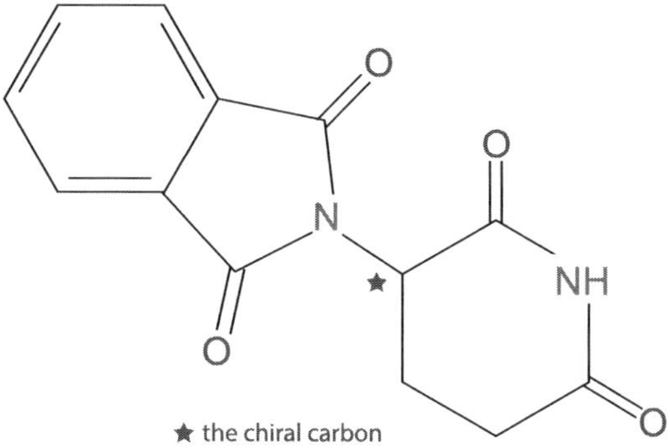

a) W
b) X
c) Y
d) Z

....

Q17: Carvone has a chiral centre. One form smells like spearmint and the other form smells like caraway seeds. Which atom is the chiral centre?

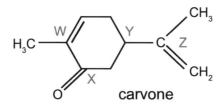

carvone

a) W
b) X
c) Y
d) Z

....

Q18: Carvone can undergo addition reactions with hydrogen molecules. How many moles of hydrogen would add onto one mole of carvone?

a) One
b) Two
c) Three
d) Four

....

Q19: What other word could be used to replace the phrase **optical isomers** of carvone?

....

This grid shows four compounds which all occur naturally in the fragrances of particular plants. A is geraniol (roses), B is nerol (bergamot), C is linalool (lavender) and D is citronellol (geraniums).

Geraniol (roses)
Nerol (bergamot)
Linalool (lavender)
Citronellol (geraniums)

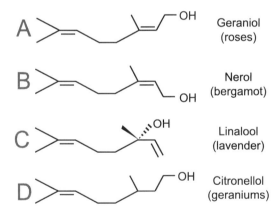

Q20: What is the relationship between A and B?

a) Structural isomers
b) Geometric isomers
c) Optical isomers
d) Not isomers

....

Q21: What is the relationship between A and C?

a) Structural isomers
b) Geometric isomers
c) Optical isomers
d) Not isomers

. . . .

Q22: Which compound (or compounds) could produce compound D on hydrogenation?

a) A only
b) B only
c) A and B
d) A, B and C

. . . .

Q23: Which of the following contains a chiral entity?

a)

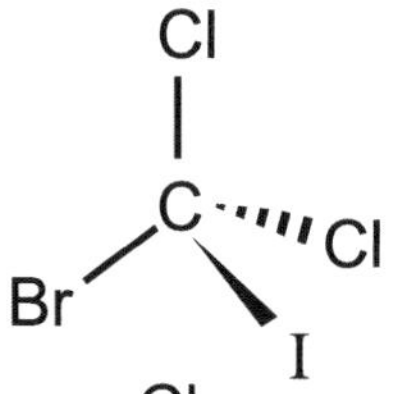

b)

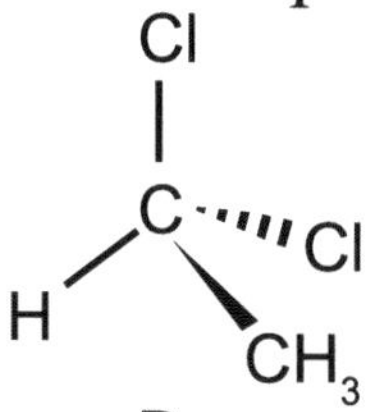

c)

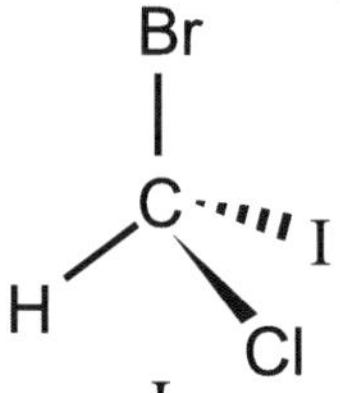

d)

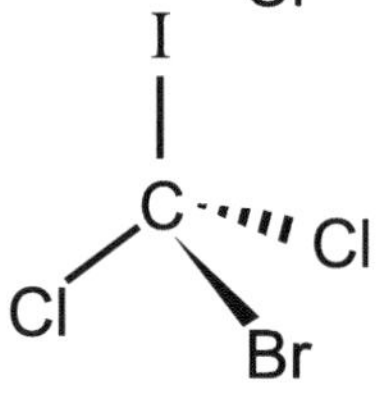

. . . .

Q24: Which of the following contain identical entities?

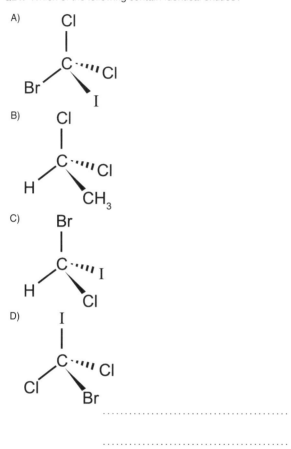

A)

B)

C)

D)

. . . .

. . . .

Topic 4

Synthesis (Unit 2)

Contents

Learning objectives

By the end of this topic, you should be able to:

- *recognise and use different types of reaction in organic synthesis including substitution, addition, elimination, condensation, hydrolysis, oxidation, reduction;*

- *devise synthetic routes, with no more than three steps, from a given reactant to a final product;*

- *look at molecular structures and deduce the reactions they can undergo.*

4.1   Homo- and heterolytic bond fission

Alkanes are not particularly reactive due to the non-polar nature of their bonds. They can however, react with halogens in the presence of sunlight or UV light where halogenoalkanes are produced along with steamy fumes of the corresponding hydrogen halide. In this reaction an atom of hydrogen has been replaced with an atom of a halogen and is an example of a substitution reaction. This substitution reaction is thought to occur by a chain reaction which has three main steps: propagation, initiation and termination (chain reactions have already been covered briefly in the CfE Higher course and you will be familiar with these steps).

Bond breaking: if we look at the reaction between methane and chlorine:

$$CH_4 + Cl_2 \rightarrow CH_3Cl + HCl$$

This reaction does not take place in the dark, it requires UV light to provide the energy to break the chlorine-chlorine bond. This splits the chlorine molecules into chlorine atoms.

$$Cl - Cl \rightarrow Cl \cdot + Cl \cdot \text{ (the dot represents an unpaired electron)}$$

This type of bond breaking is known as homolytic fission and usually occurs when the bond is non polar or very slightly polar. In homolytic fission one electron from the bond goes to one atom while the other electron goes to the other atom. Atoms with unpaired electrons are known as radicals which are incredibly unstable and therefore extremely reactive. The initiation step in a chain reaction produces radicals.

Note: If a bond were to split unevenly (one atom getting both electrons, and the other none), ions would be formed. The atom that got both electrons would become negatively charged, while the other one would become positive. This is called heterolytic fission and will be favoured when the bond is polar. Reactions proceeding via heterolytic fission tend to produce far fewer products and are therefore better suited for synthesis.

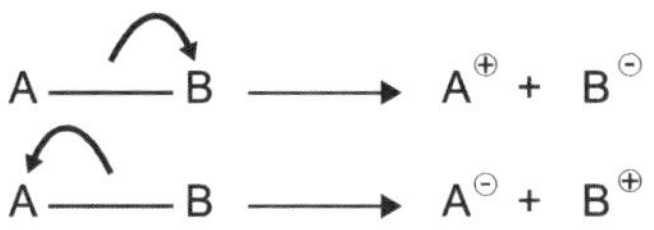

$$A \curvearrowright B \longrightarrow A^{\oplus} + B^{\ominus}$$
$$A \curvearrowright B \longrightarrow A^{\ominus} + B^{\oplus}$$

Heterolytic fission

Initiation steps are followed by propagation and termination steps in chain reactions. Looking at the reaction between methane and bromine the propagation and termination steps can be seen below.

Chain initiation
The chain is initiated by UV light breaking some bromine molecules into free radicals by homolytic fission.

$$Br_2 \rightarrow 2Br \cdot$$

Chain propagation reactions
These are the reactions which keep the chain going. In each of these propagation steps one radical enters the reaction and another is formed.

$$CH_4 + Br\cdot \rightarrow CH_3\cdot + HBr$$
$$CH_3\cdot + Br_2 \rightarrow CH_3Br + Br\cdot$$

Chain termination reactions
These are reactions which remove free radicals from the system without replacing them by new ones. This brings the chain reaction to an end.

$$Br\cdot + Br\cdot \rightarrow Br_2$$
$$CH_3\cdot + Br\cdot \rightarrow CH_3Br$$
$$CH_3\cdot + CH_3\cdot \rightarrow CH_3CH_3$$

4.2   Electrophiles and nucleophiles

In reactions involving heterolytic bond fission, attacking groups are classified as "nucleophiles" or "electrophiles".

Electrophiles are chemical species that are electron deficient and therefore are "electron loving" species. Electrophiles are molecules or positively charged ions which are capable of accepting an electron pair. They will seek out electron rich sites in organic molecules; examples include NO_2^+ and SO_3H^+.

Nucleophiles are chemical species that are electron rich and are "electron donating" species. Nucleophiles are molecules or negatively charged ions which have at least one lone pair of electrons that they can donate and form dative bonds. They will seek out electron-deficient sites in organic molecules; examples include H_2O, NH_3 and the halide ions.

4.3   Curly arrow notation

Double headed curly arrows are used to indicate the movement of electron pairs in a reaction. The tail of the arrow shows where the electrons originate from and the head shows where they end up. An arrow starting at the middle of a covalent bond indicates that heterolytic bond fission is occurring. When an arrow is drawn with the head pointing to the space between two atoms, this indicates that a covalent bond will be formed between the two atoms.

A single headed curly arrow indicates the movement of a single electron. These are useful in discussions about radical chemistry mechanisms.

4.4 Haloalkanes

Haloalkanes can be regarded as substituted alkanes where one or more of the hydrogen atoms have been replaced with a halogen atom. In naming of haloalkanes the halogen atoms are treated as branches and naming is done in the same way as for branched alkanes.

This haloalkane is named 2-bromo-2-chloro-1,1,1-trifluoroethane (remember branches are named in alphabetical order)

Monohaloalkanes (where one H atom is substituted for a halogen atom) have three different structural types which are primary, secondary and tertiary.   These are determined by the number of alkyl groups attached to the carbon atom that is directly attached to the halogen atom. In these diagrams, X represents a halogen atom:

Primary monohaloalkane, one alkyl group attached to the carbon atom directly attached to the halogen atom

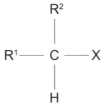

Secondary monohaloalkane, two alkyl groups attached to the carbon atom directly attached to the halogen atom

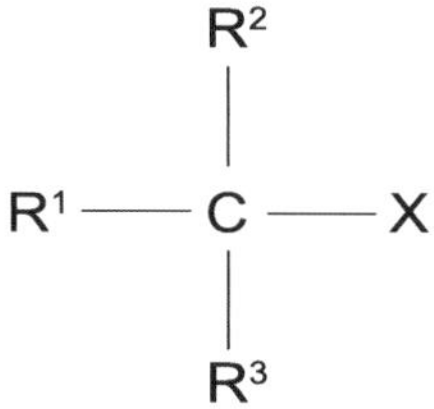

$$R^1 - \underset{\underset{R^3}{|}}{\overset{\overset{R^2}{|}}{C}} - X$$

Tertiary monohaloalkane, three alkyl groups attached to the carbon atom directly attached to the halogen atom

Due to the polar nature of the carbon-halogen bond haloalkanes are susceptible to nucleophilic attack.

$$> C^{\delta+} \longrightarrow X^{\delta-}$$

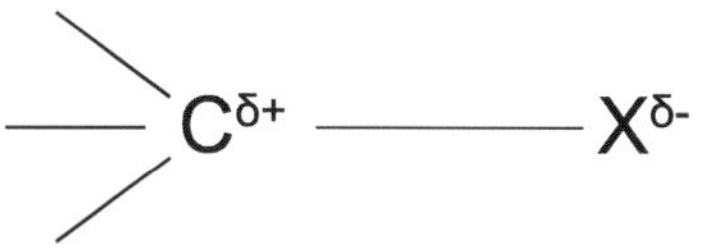

The presence of the slight positive charge on the carbon atom makes haloalkanes susceptible to nucleophilic attack. The nucleophile donates a pair of electrons forming a bond with the carbon atom of the C-X bond. The halogen atom is "thrown out" and substituted by the nucleophile. The mechanism for this is included in the Haloalkanes section of this topic under S_N1 and S_N2 reactions.

Nucleophilic substitutions

Reaction of monohaloalkanes with alkalis produces alcohols. A solution of aqueous potassium hydroxide (KOH) or sodium hydroxide (NaOH) is used.

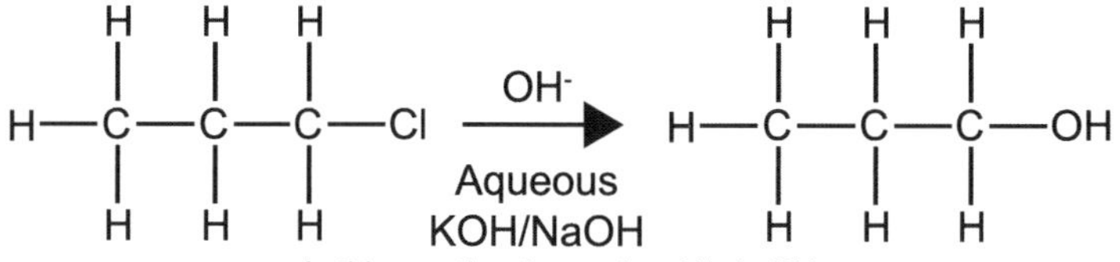

In this reaction the nucleophile is OH⁻

Reaction with alcoholic potassium alkoxide (potassium methoxide in methanol CH_3OK) produces ethers.

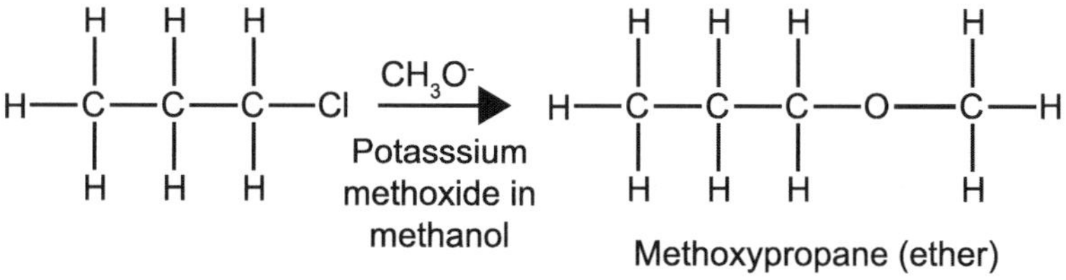

Methoxypropane (ether)

In this reaction the nucleophile is CH_3O^-

Reaction with ethanolic potassium cyanide or sodium cyanide (KCN or NaCN in ethanol) produces nitriles.

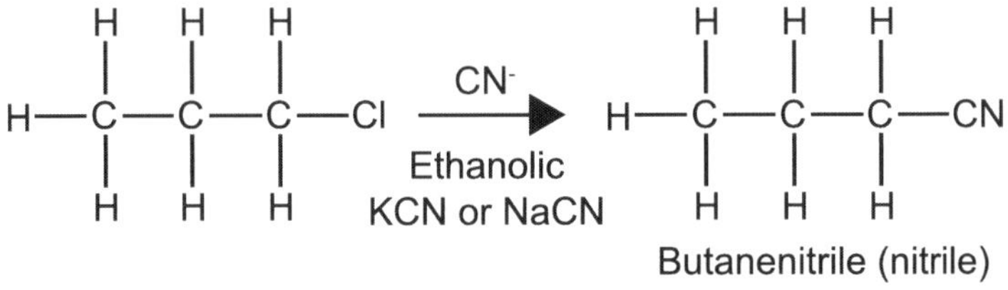

Butanenitrile (nitrile)

The nucleophile in this reaction is CN⁻

The end nitrile contains one more carbon than the original haloalkane. This is very useful in synthetic organic chemistry as a way of increasing the chain length of an organic compound. The nitrile can be converted into the corresponding carboxylic acid through acid hydrolysis.

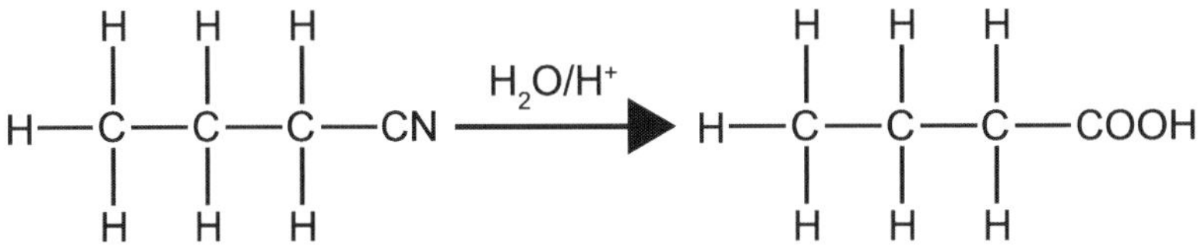

Monohaloalkanes can also undergo elimination reactions to form alkenes. This is achieved by heating the monohaloalkane under reflux with ethanolic potassium or sodium hydroxide.

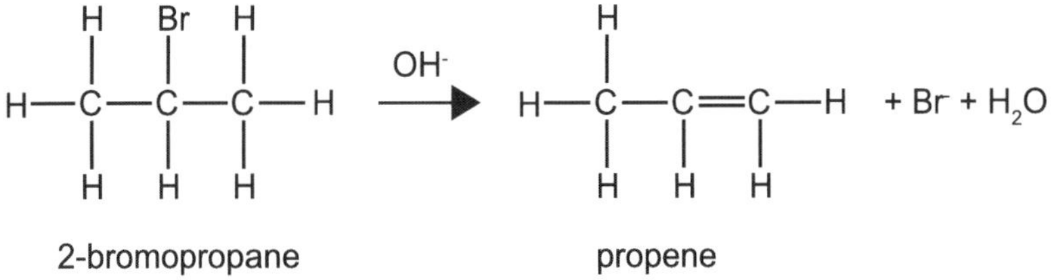

2-bromopropane   propene

In this reaction a hydrogen halide is removed from the original monohaloalkane and for some it can result in two different alkenes being produced. This is due to the availability of more than one H atom that can be removed in the formation of the hydrogen halide. For example 2-chlorobutane can result in but-1-ene and but-2-ene, of which but-2-ene is the major product.

The resulting alkene can be tested by the addition of bromine water which will be decolourised in the presence of an alkene due to the bromine being added across the double bond in an addition reaction.

The presence of halogen ions can be tested by reacting the substance with silver nitrate solution. The silver halide is precipitated due to them being insoluble in water. The colour of the precipitate can tell you which halide ion was originally present in the substance.

Silver halide	Colour of precipitate
Sliver chloride	White
Silver bromide	Cream
Silver Iodide	Yellow

It is quite difficult to tell the colours of these precipitates apart especially if very little precipitate is produced. You can add the precipitate to ammonia solution and confirm which halide is present.

Original precipitate	Observation
AgCl	Precipitate dissolves to give a colourless solution.
AgBr	Precipitate is almost unchanged using dilute ammonia solution, but dissolves in concentrated ammonia solution to give a colourless solution.
AgI	Precipitate is insoluble in ammonia solution of any concentration.

Haloalkanes have been used as anticancer agents however are known for their toxic side effects. They are also often used as alkylating agents.

S_N1 and S_N2 reactions

Haloalkanes will undergo **nucleophilic substitution** by one of two different reaction mechanisms which are called S_N1 and S_N2.

S_N1 reaction mechanism

A kinetic study of the reaction between 2-bromo-methylpropane (tertiary haloalkane) and the nucleophile OH^- shows it has the rate equation rate=$k[(CH_3)_3CBr]$. This means it is first order with respect to the haloalkane implying the rate determining step can only involve the haloalkane. It is a two-step process. The polarity of the C-Br bond is shown in green in Step 1 below.

Step 1

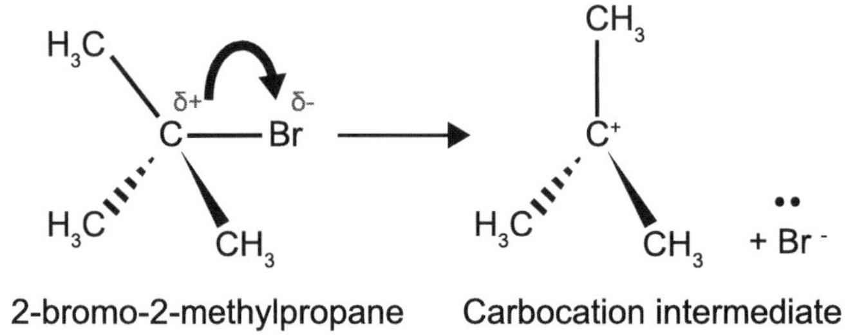

2-bromo-2-methylpropane   Carbocation intermediate

The first step produces a carbocation intermediate

Step 2

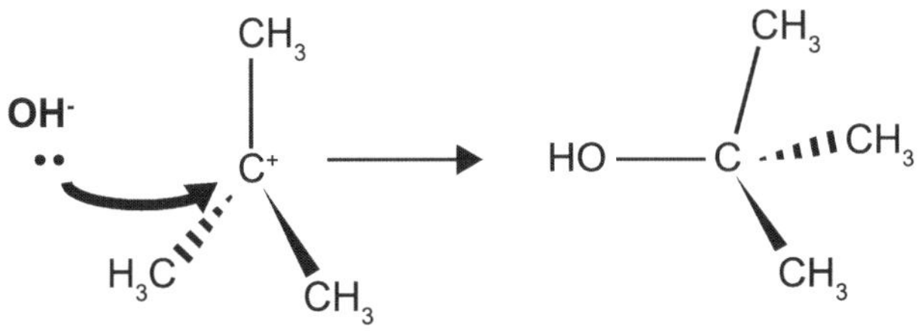

S$_N$2 reaction mechanism

A kinetic study of the reaction of bromoethane (primary haloalkane) and the nucleophile OH$^-$ has the rate equation rate=k[CH$_3$CH$_2$Br][OH$^-$]. This means it is first order with respect to both the haloalkane and the hydroxide ion implying the rate determining step should involve both these species. This is a **one-step process**.

Step 1

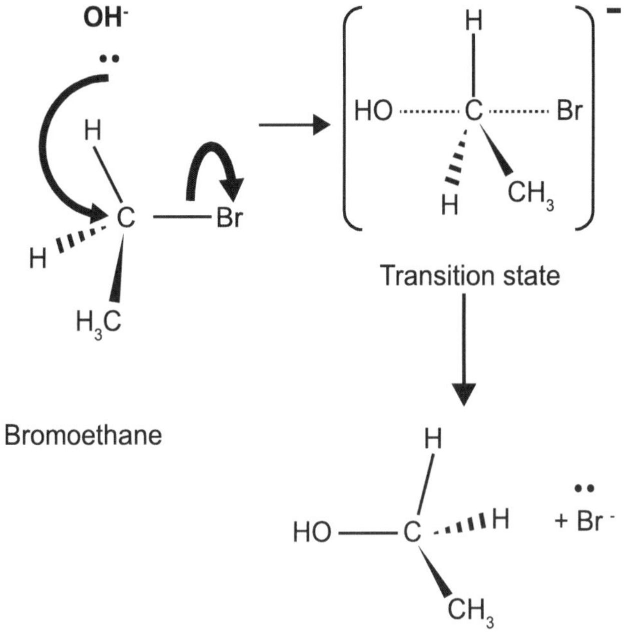

Transition state

Bromoethane

The negative hydroxide ion in the bromoethane is a nucleophile so attacks the slightly positive carbon atom. A negatively charged intermediate is formed.

How do you know which reaction mechanism (S$_N$1 or S$_N$2) a haloalkane will undergo? You need to look at what type of haloalkane primary, secondary or tertiary you are dealing with. In the S$_N$1 reaction a **carbocation** intermediate is formed which could be a primary, secondary or tertiary carbocation but since alkyl groups are electron donating the tertiary carbocation will be the most stable. Tertiary haloalkanes are the most likely and primary haloalkanes the least likely to proceed by a S_N1 reaction mechanism.

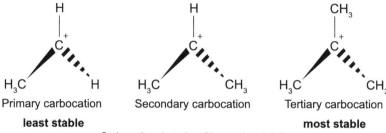

Primary carbocation	Secondary carbocation	Tertiary carbocation
least stable		**most stable**

Carbocations in order of increasing stability

In S_N1 reactions a positively charged intermediate is formed, whereas in S_N2 reactions a negatively charged intermediate is formed.

In a S_N2 reaction the OH⁻ nucleophile attacks the carbon atom of the carbon-halogen bond from the side opposite to the halogen atom. In the case of tertiary haloalkanes that position is most likely to be hindered by three bulky alkyl groups. Tertiary haloalkanes are least likely and primary haloalkanes most likely to proceed by a S_N2 reaction mechanism.

4.5 Alcohols

Prior knowledge from National 5 and Higher Chemistry is required for this section on alcohols.

This includes:

a) Basic structure of alcohols including the functional group hydroxyl (OH);

b) Structural types of alcohols including primary, secondary and tertiary;

c) Oxidation reactions of alcohols.

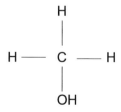

Methanol: first member of the alcohol family containing the hydroxyl (OH) functional group

Properties of alcohols

As the chain length of alcohols increase with the addition of a CH_2 unit between each progressive member their boiling points show a progressive increase. However, if we compare the boiling point of an alcohol to the boiling point of an alkane of similar relative formula mass and shape we can see that they are considerably higher. This is due to the presence of the polar hydroxyl (OH) group in the alcohol molecule allowing hydrogen bonding to be set up between the individual molecules. This is shown in the diagram below.

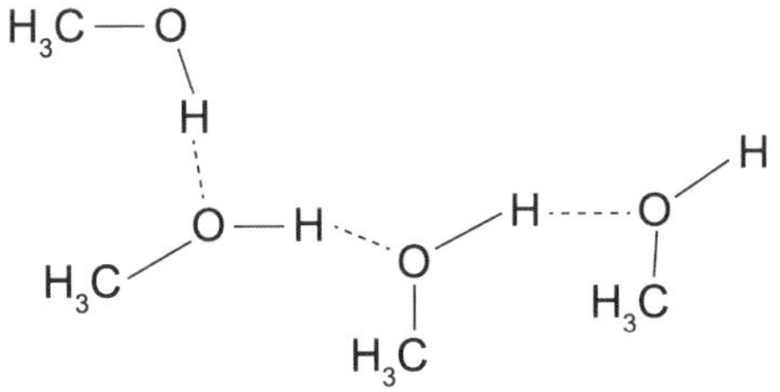

Hydrogen bonding between methanol molecules (shown as dashed lines)

Between the alkane molecule of similar relative formula mass and shape only London dispersion forces are found and since hydrogen bonds are stronger extra energy is required to break them giving a reason for the higher boiling point of alcohols.

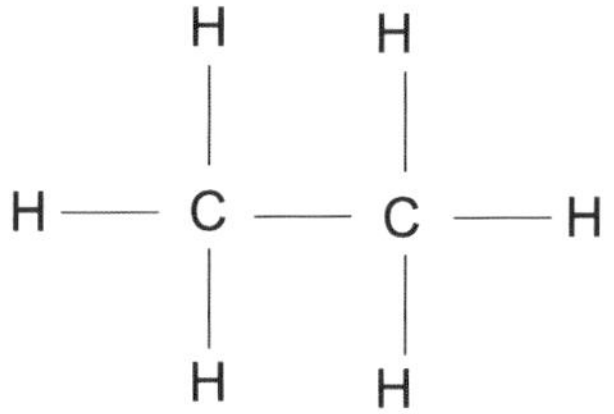

Ethane is the alkane closest to methanol in terms of relative formula mass (methanol 32 g and ethane 30 g). Only London dispersion forces are found between individual molecules of ethane. The boiling point of methanol is 64.7° C and the boiling point of ethane is -89° C showing the increased effect that hydrogen bonding has on the boiling point of methanol.

There is also a graduated decrease in the solubilites of alcohols in water as the chain length of the alcohol increases. Lower chain length alcohols (methanol, ethanol and propan-1-ol) are completely soluble in water (miscible with water) but alcohols such as heptan-1-ol and other higher chain length alcohols are insoluble in water. The smaller chain alcohols are soluble in water as the energy released in forming hydrogen bonds between the alcohol and water molecules is enough to break the hydrogen bonds between the water molecules. By the time you reach heptan-1-ol the large non-polar hydrocarbon part of the molecule disrupts the hydrogen bonding ability of the water with the hydroxyl, hence reducing solubility in water.

Preparation of alcohols

Alcohols can be prepared by two different reactions:

a) Heating haloalkanes under reflux with aqueous sodium/potassium hydroxide by nucleophilic substitution (see the Haloalkane section of this topic under nucleophilic substitution reactions);

b) Acid catalysed hydration of alkenes described below.

Alkenes undergo addition reactions with water to form alcohols. This reaction is an acid catalysed hydration proceeding through a carbocation intermediate.

Step 1

The hydrogen ion of the acid catalyst is an electrophile and the electrons of the double bond in the alkene (electron rich) attack the hydrogen ion forming a carbocation.

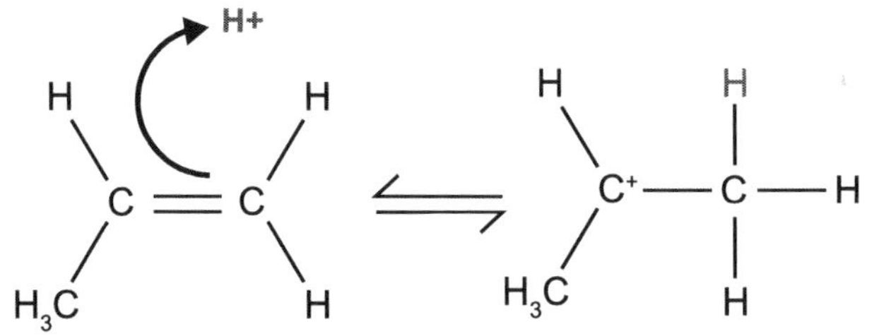

See Markovnikov's rule, discussed in the 'Electrophilic addition' section later in this topic, to find out why the hydrogen is more likely to attach to that particular carbon in step 1 (above).

Step 2

The carbocation undergoes rapid nucleophilic attack by a water molecule to give a protonated alcohol (alcohol with a hydrogen ion attached).

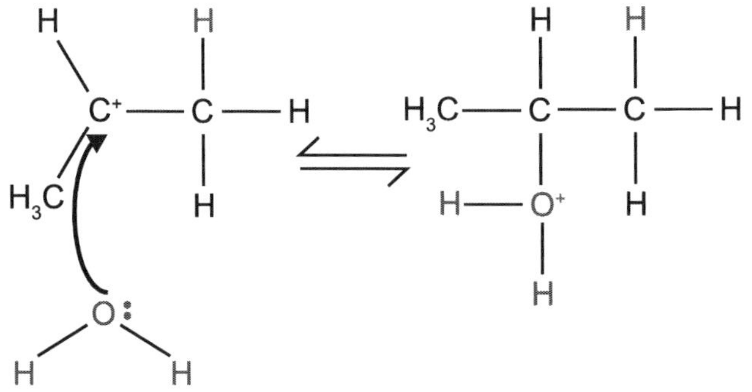

Step 3

Formation of the alcohol propan-2-ol. The protonated propan-2-ol is a strong acid and readily loses a proton to give propan-2-ol the final product.

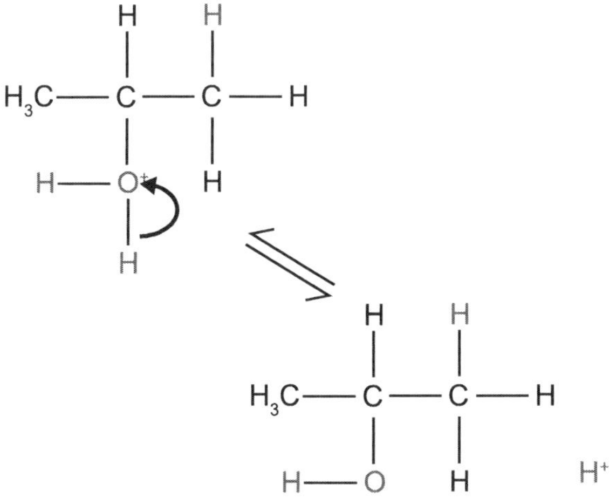

Alkoxides

Mentioned briefly in the Haloalkanes section, alkoxides are formed by adding an alkali metal such as potassium or sodium with an alcohol. For example when potassium is added to methanol, potassium methoxide is formed.

$$2K + 2CH_3OH \rightarrow H_2 + 2CH_3O^-K^+$$

Dehydration of alcohols

Dehydration of alcohols forms alkenes. This can be done in two ways either by passing the vapour of the alcohol over hot aluminium oxide or by treating the alcohol with concentrated sulfuric acid or phosphoric acid (orthophosphoric acid).

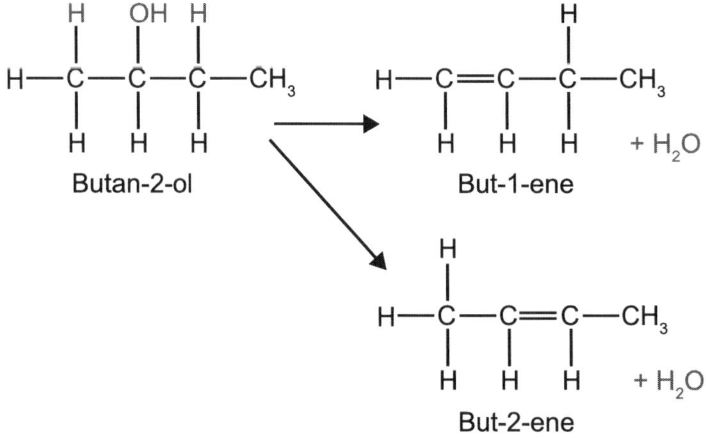

Butan-2-ol → But-1-ene + H_2O

But-2-ene + H_2O

During dehydration the **OH** is removed along with an **H** atom on an adjacent carbon. This forms 2 alkenes with but-2-ene being the major product. With some alcohols such as propan-2-ol and butan-1-ol only one alkene is produced.

Esters

Alcohols can be reacted with carboxylic acids or acid chlorides to form esters. This is a condensation or esterification reaction carried out with a catalyst of concentrated sulfuric acid if using the carboxylic acid. When using an acid chloride the reaction is much faster and a catalyst is not required. Esters have been covered at National 5 and Higher level.

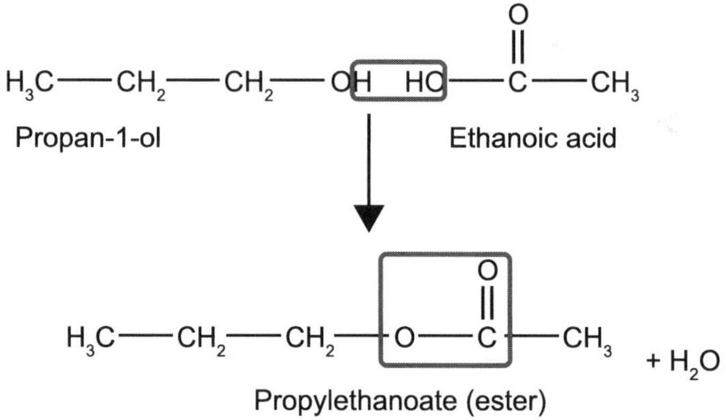

Propan-1-ol   Ethanoic acid

Propylethanoate (ester)   + H_2O

Ester linkage formed from the loss of water between the alcohol and carboxylic acid

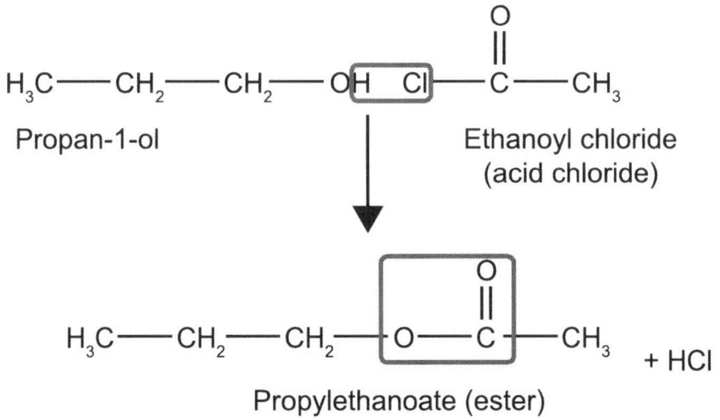

Propan-1-ol

Ethanoyl chloride
(acid chloride)

Propylethanoate (ester)

Ester linkage formed from the loss of HCl between the alcohol and acid chloride

Reduction of aldehydes and ketones

Aldehydes formed from the mild oxidation of primary alcohols using hot copper (II) oxide or acidified potassium dichromate can be reduced back to primary alcohols by reacting with lithium aluminium hydride ($LiAlH_4$) dissolved in ether. Sodium borohydride ($NaBH_4$) can also be used as a reducing agent. Similarly ketones formed from the mild oxidation of secondary alcohols can be reduced back to secondary alcohols.

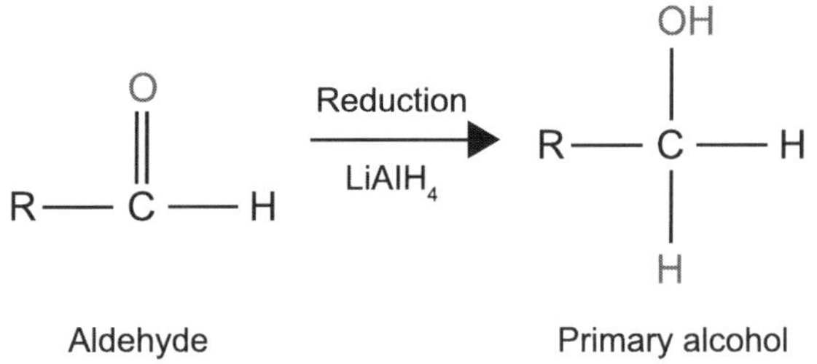

Aldehyde

Primary alcohol

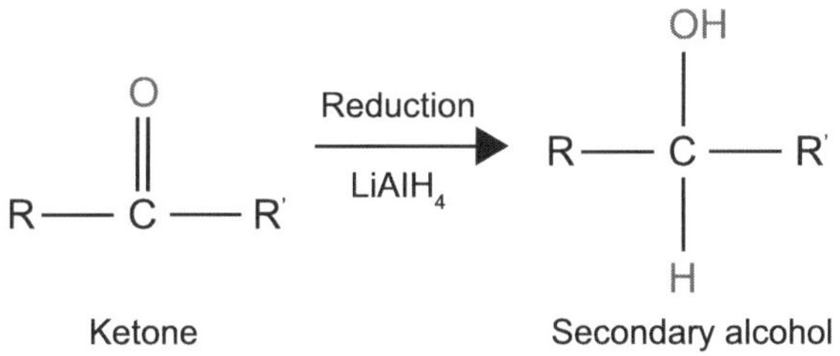

Ketone

Secondary alcohol

The reactions are usually carried out in solution in a carefully dried ether such as ethoxyethane (diethyl ether).

Alcohol hydroxyl groups are present in a lot of pharmaceutical drugs as they are involved in hydrogen bonding with protein binding sites particularly beta blockers and anti-asthmatics.

4.6   Ethers

Ethers are synthesised from haloalkanes and alkoxides. They were the first anaesthetics and have the general formula R-O-R' where R and R' are alkyl groups. If R and R' are different then the ether is unsymmetrical and when identical they are symmetrical.

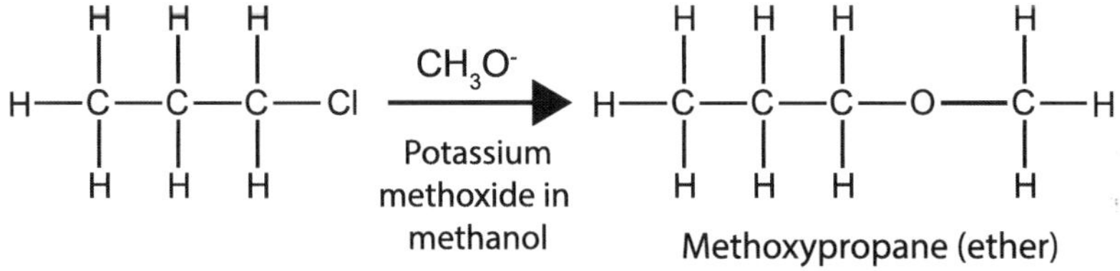

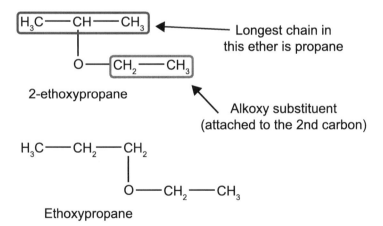

Ethers are named by assigning the longest carbon chain as the parent name. This is prefixed by the alkoxy substituent which has been named by removing the 'yl' from the name of the alkyl substituent and adding 'oxy'. CH_3CH_2O- is named ethoxy and CH_3O-is named methoxy, for example.

Properties of ethers

The boiling points of ethers are much lower than that of their isomeric alcohols due to the fact that hydrogen bonding does not occur between ether molecules. This is due to the highly electronegative oxygen atom not being directly bonded to a hydrogen atom. They can however form hydrogen bonds with water molecules as shown below.

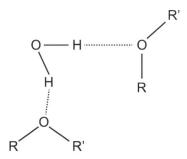

Hydrogen bonding can occur between water and ether molecules but not between ether molecules themselves

This also explains why low relative formula mass ethers are soluble in water for example methoxymethane and methoxyethane. The larger ethers are insoluble in water and therefore are useful in extracting organic compounds from aqueous solutions.

Ethers are highly flammable and when exposed to air slowly form peroxides which are unstable and can be explosive. Ethers are used as solvents as most organic compounds dissolve in them and they are relatively chemically inert. They are easily removed by distillation due to being volatile.

4.7 Alkenes

Alkenes can be prepared in the laboratory by:

   a) dehydration of alcohols using aluminium oxide, concentrated sulfuric acid or orthophosphoric acid (see the 'Alcohols' section of this topic).

   b) base-induced elimination of hydrogen halides from monohaloalkanes (see the 'Haloalkanes' section of this topic).

4.8 Electrophilic addition

Alkenes undergo **electrophilic addition** reactions with a variety of substances to form different products.

Catalytic addition of hydrogen

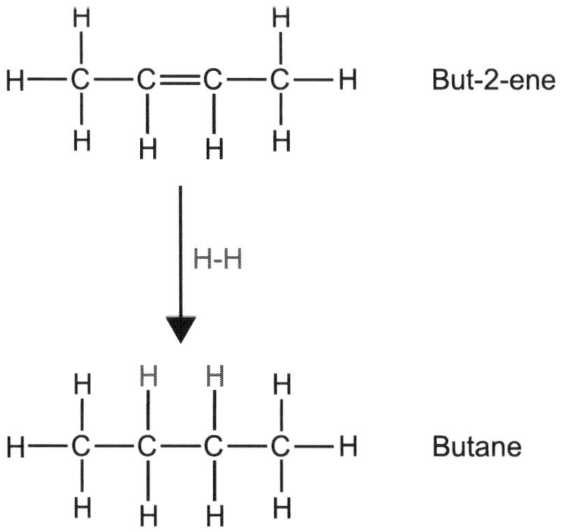
But-2-ene
H-H
Butane

This reaction is also known as hydrogenation and is catalysed by nickel or palladium.

Addition of halogens (halogenation)

The mechanism for the addition of halogens to alkenes involves two steps.

Step 1

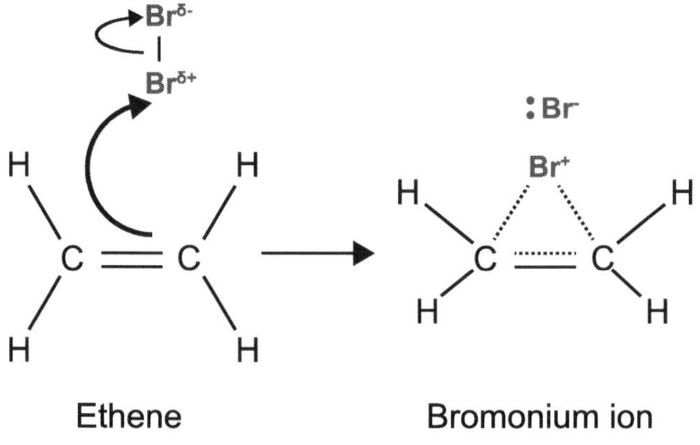
Br$^{\delta-}$
Br$^{\delta+}$
:Br^{-}
Br^{+}
Ethene
Bromonium ion

The bromine molecule is the electrophile in this reaction and undergoes heterolytic fission. It approaches the double bond in ethene and becomes polarised. The electron rich double bond pushes the electrons in the bromine molecule towards the bromine atom which is furthest away from the double bond which gains a slight negative charge. The other bromine atom gains a slight positive charge and the Br-Br bond breaks heterolytically creating a cyclic intermediate and a bromide ion.

Step 2

The bromide ion attacks the cyclic intermediate ion from the opposite side to the Br atom which prevents access to the side where it is located. The bromide ion is acting as a nucleophile seeking out a centre of positive charge.

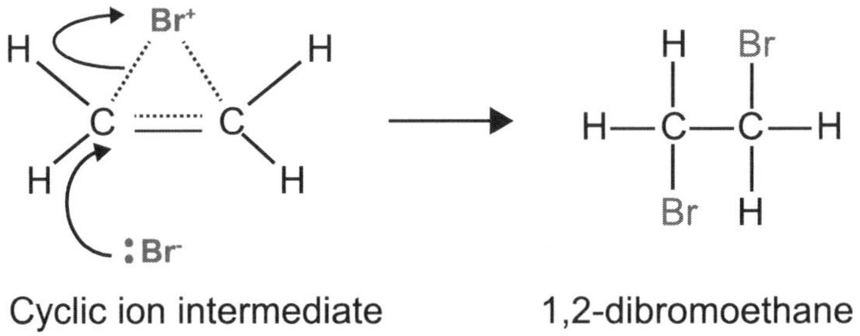

Cyclic ion intermediate   1,2-dibromoethane

Addition of hydrogen halides (hydrohalogenation)

The mechanism for this addition is also a two-step process.

Step 1

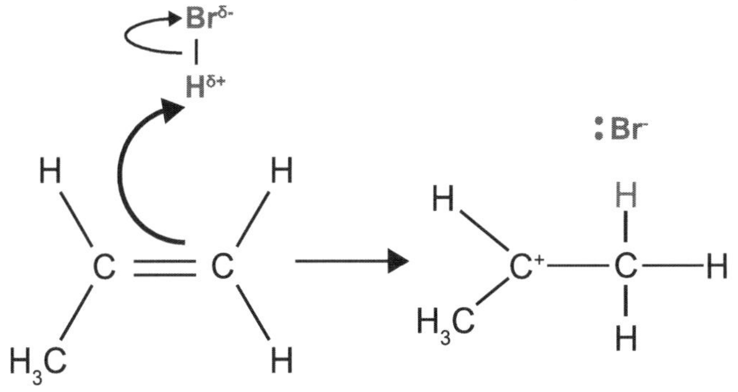

The H-Br molecule is already polarised and the electrons of the double bond attack the hydrogen. The H-Br bond breaks heterolytically and a bromide ion is formed at the same time. A carbocation is formed after the double bond breaks to form a new bond to the hydrogen.

Step 2

The second step involves the bromide ion attacking the carbocation intermediate which can be done from either side of the carbocation.

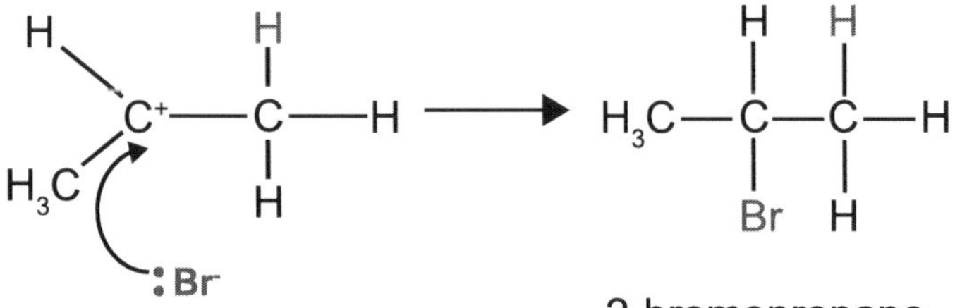

2-bromopropane

In the above diagram the product is 2-bromopropane but as propane is an unsymmetrical alkane 1-bromopropane will also be formed.

Markovnikov's rule

When a hydrogen halide is added to an unsymmetrical alkene (one where the groups attached to one carbon of the double bond are different from the groups attached to the other carbon of the double bond) two products are formed. Markovnikov's rule states that when H-X is added onto an unsymmetrical alkene the major product is the one where the hydrogen bonds to the carbon atom of the double bond that has already the greatest number of hydrogen atoms attached to it.

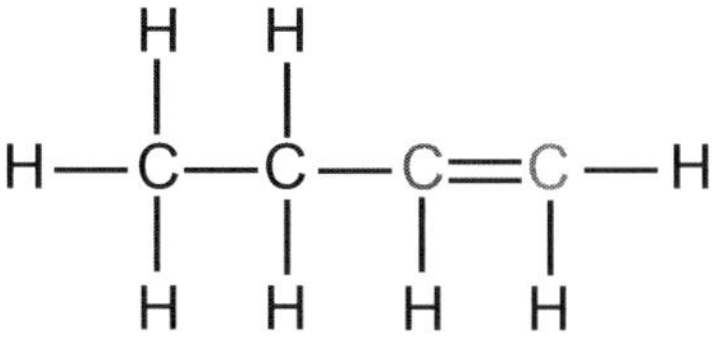

If we add for example H-Cl onto but-1-ene the hydrogen atom should attach itself to the carbon atom shown in blue as it has the greatest number of hydrogen atoms already attached to it. This would form the product 2-chlorobutane as the major product.

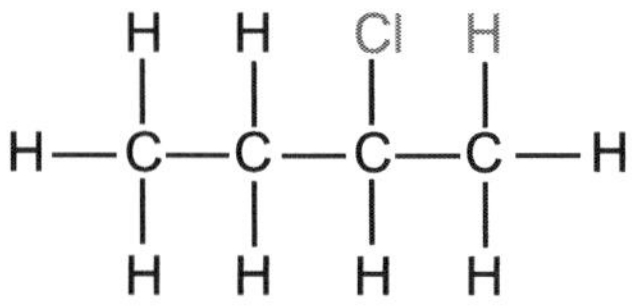

2-chlorobutane

The minor product would be 1-chlorobutane:

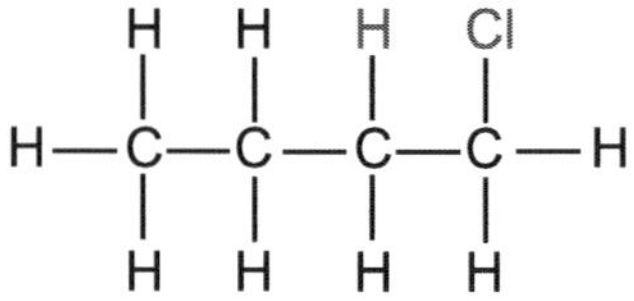

1-chlorobutane

Addition of water (hydration)

The addition of water which is catalysed by an acid has a very similar mechanism for the addition of hydrogen halides proceeding through a carbocation intermediate. This has already been covered in the Alcohols section of this topic but has been included again in this section.

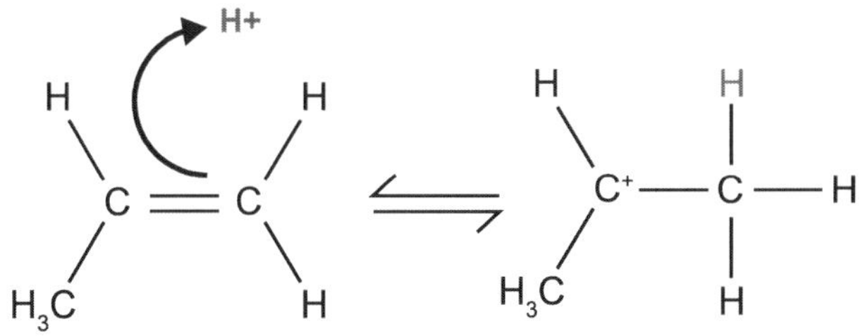

Step 1: formation of a carbocation

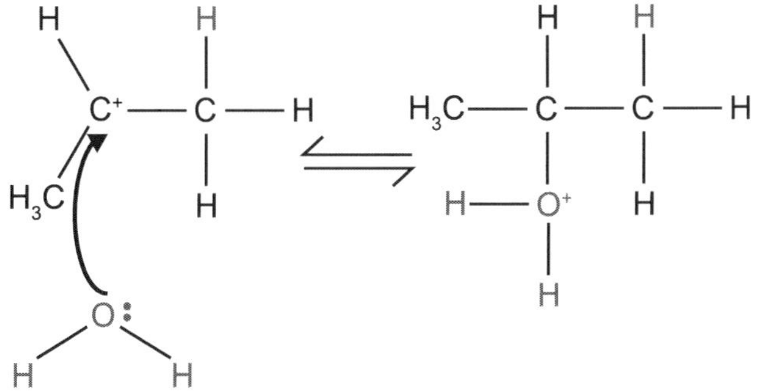

Step 2: the carbocation undergoes nucleophilic attack by a water molecule to give a protonated alcohol

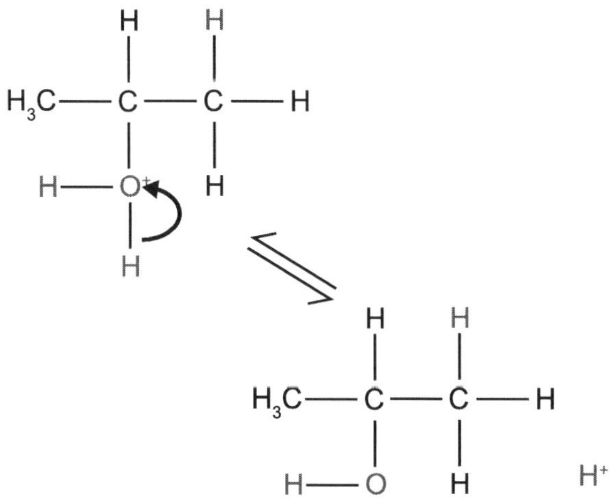

Step 3: the protonated alcohol which is a strong acid loses a proton to the give the final alcohol

This reaction also follows Markovnikov's rule in determining the major product when water is added to an unsymmetrical alkene. The major product in this reaction is propan-2-ol with propan-1-ol formed as the minor product also. This is also due to the stability of the carbocation formed in the first step.

4.9 Carboxylic acids

Preparation of carboxylic acids

Carboxylic acids can be prepared by:

a) oxidising primary alcohols or aldehydes by heating them with acidified potassium dichromate (this has been covered at higher chemistry and should be revised).

b) hydrolysis of nitriles, esters or amides by heating them in the presence of a catalyst which can be either an acid or an alkali. The hydrolysis of nitriles was covered earlier in the section 'Haloalkanes' under the heading 'Nucleophilic substitutions'.

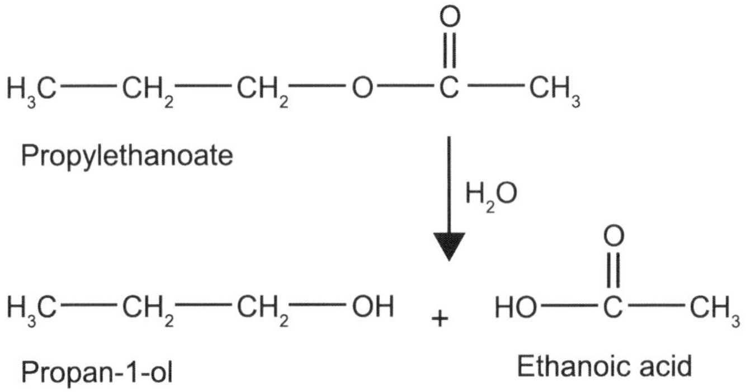

$$H_3C\text{——}CH_2\text{——}CH_2\text{——}O\text{——}\overset{\displaystyle O}{\overset{\displaystyle \|}{C}}\text{——}CH_3$$

Propylethanoate

$$\downarrow H_2O$$

$$H_3C\text{——}CH_2\text{——}CH_2\text{——}OH \quad + \quad HO\text{——}\overset{\displaystyle O}{\overset{\displaystyle \|}{C}}\text{——}CH_3$$

Propan-1-ol   Ethanoic acid

Ester hydrolysis (catalyst concentrated sulfuric acid)

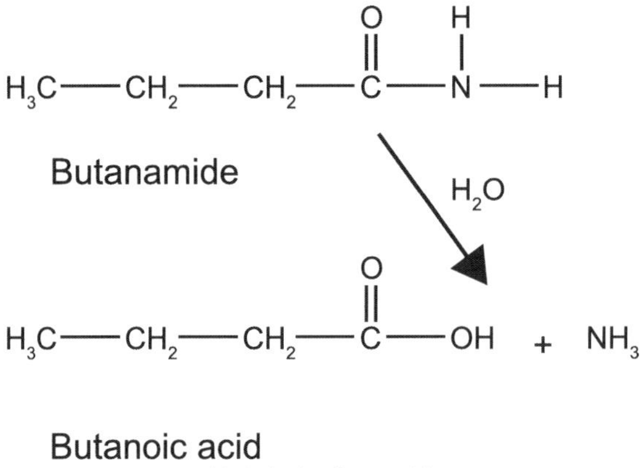

$$H_3C\text{——}CH_2\text{——}CH_2\text{——}\overset{\displaystyle O}{\overset{\displaystyle \|}{C}}\text{——}\overset{\displaystyle H}{\overset{\displaystyle |}{N}}\text{——}H$$

Butanamide H_2O

$$H_3C\text{——}CH_2\text{——}CH_2\text{——}\overset{\displaystyle O}{\overset{\displaystyle \|}{C}}\text{——}OH \quad + \quad NH_3$$

Butanoic acid

Hydrolysis of an amide

Properties of carboxylic acids

Aqueous solutions of carboxylic acids have a pH less than 7. Since they are weak acids, the pH is dependent on the concentration of the acid and the pKa value (see the 'Strong/weak acids and bases' section in the Unit 1 topic 'Chemical equilibrium').

Carboxylic acids are capable of forming hydrogen bonds. This means that they have higher boiling points than alkanes of a similar molecular mass. In a concentrated carboxylic acid, dimers can form between two molecules due to the hydrogen bonding.

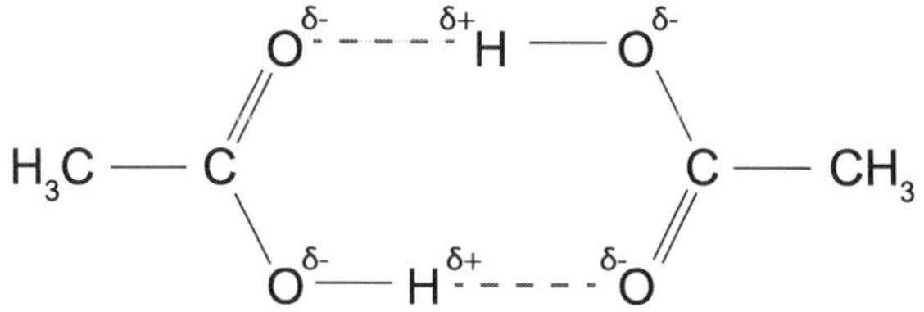

Dimers forming in a concentrated carboxylic acid

Small carboxylic acids are soluble in water since they can form hydrogen bonds with water molecules.

Reactions of carboxylic acids

Carboxylic acids behave as typical acids in aqueous solution and form salts by reacting with metals and bases including alkalis (soluble bases). They also undergo condensation reactions with alcohols to form esters (covered in the Alcohols section of this topic). Carboxylic acids react with amines to form amides and they can be reduced by using lithium aluminium hydride ($LiAlH_4$) to directly form primary alcohols due to the $LiAlH_4$ being such a powerful reducing agent. Sodium borohydride ($NaBH_4$) can also be used as a reducing agent.

Metal and carboxylic acid

An example of this reaction is that of magnesium and ethanoic acid to form the salt magnesium ethanoate and hydrogen gas.

$$Mg(s) + 2CH_3COO^-H^+(aq) \rightarrow H_2(g) + Mg^{2+}(CH_3COO^-)_2(aq)$$

Base and carboxylic acid

Sodium carbonate reacts with ethanoic acid to form the salt sodium ethanoate, water and carbon dioxide gas.

$$Na_2CO_3(s) + 2CH_3COOH(aq) \rightarrow CO_2(g) + H_2O(l) + 2Na^+CH_3COO^-(aq)$$

Condensation reactions to form esters

Please see the Alcohols section of this topic and Higher Chemistry.

Condensation reactions to form amides

Also covered at Higher Chemistry.

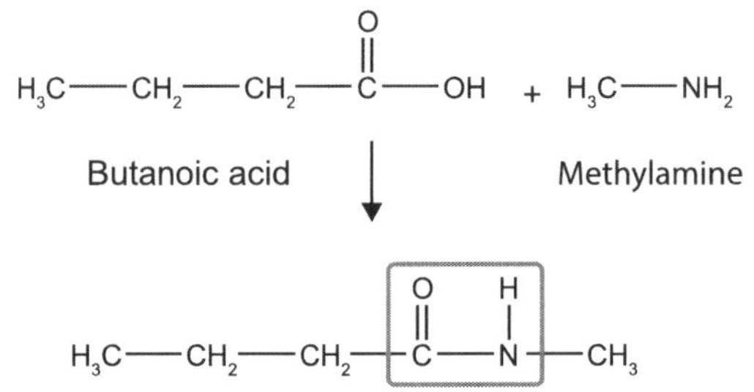

$$H_3C-CH_2-CH_2-\overset{\overset{\displaystyle O}{\|}}{C}-OH \quad + \quad H_3C-NH_2$$

Butanoic acid   Methylamine

$$H_3C-CH_2-CH_2-\boxed{\overset{\overset{\displaystyle O}{\|}}{C}-\overset{\overset{\displaystyle H}{|}}{N}}-CH_3$$

An amide with the amide linkage highlighted,
this can also be called a peptide link

Reaction with LiAlH₄ (or NaBH₄) to form primary alcohols

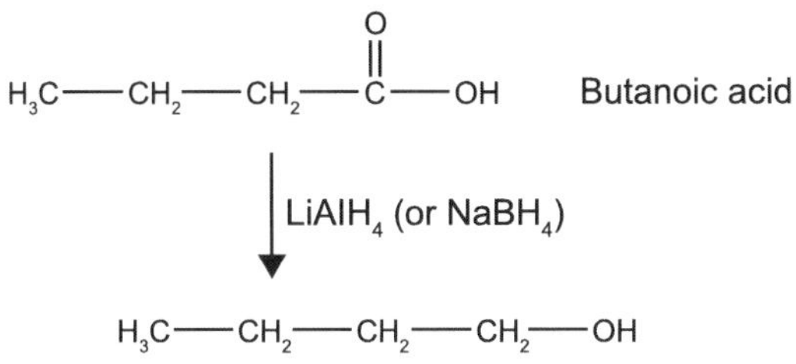

$$H_3C-CH_2-CH_2-\overset{\overset{\displaystyle O}{\|}}{C}-OH \qquad \text{Butanoic acid}$$

LiAlH₄ (or NaBH₄)

$$H_3C-CH_2-CH_2-CH_2-OH$$

Butan-1-ol (primary alcohol)

4.10 Amines

Amines are derived from ammonia where one or more of the hydrogen atoms have been replaced with an alkyl group. Amines are classified according to the number of alkyl groups attached to the nitrogen atom.

Primary amine where one of the hydrogens of ammonia has been replaced with an alkyl group.

$$H_3C-N\overset{\displaystyle H}{\underset{\displaystyle H}{<}}$$

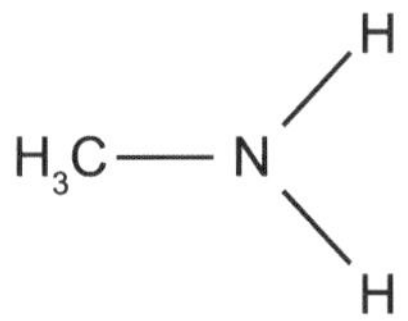

Primary amine

Secondary amine where two of the hydrogen atoms of ammonia have been replaced with an alkyl group. These do not necessarily have to be the same alkyl group.

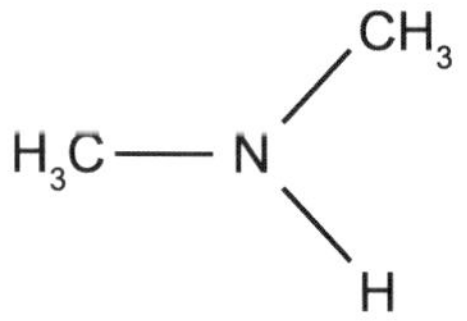

$$H_3C - N \overset{\displaystyle CH_3}{\underset{\displaystyle H}{<}}$$

Secondary amine

Tertiary amine where all three hydrogen atoms of ammonia have been replaced with alkyl groups. Again these alkyl groups are not necessarily all the same.

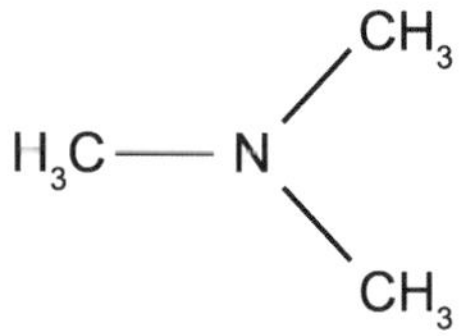

$$H_3C - N \overset{\displaystyle CH_3}{\underset{\displaystyle CH_3}{<}}$$

Tertiary amine

Naming amines

Amines are named by prefixing the word amine with the names of the alkyl groups attached to the nitrogen atom arranged in alphabetical order.

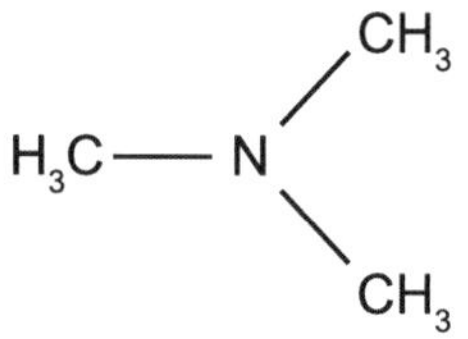

$$H_3C - N \overset{\displaystyle CH_3}{\underset{\displaystyle CH_3}{<}}$$

Trimethylamine

$$H_3C - N \overset{\displaystyle CH_2CH_3}{\underset{\displaystyle CH_3}{<}}$$

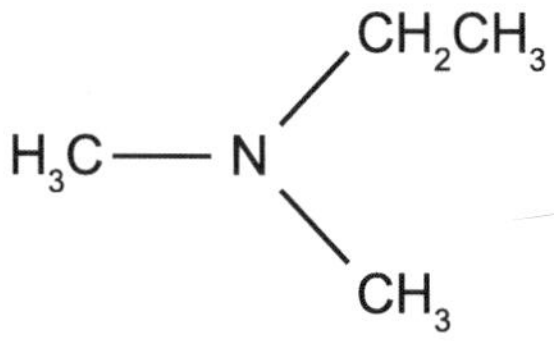

Ethyldimethylamine

Properties of amines

A polar N-H bond is found in primary and secondary amines therefore they have hydrogen bonding between their molecules. These do not occur in tertiary amines due to the lack of a hydrogen atom bonded directly to the highly electronegative nitrogen atom. This causes primary and secondary amines to have higher boiling points compared to their isomeric tertiary amines.

All types of amine can form hydrogen bonds with water and hence they are soluble in water. This is shown in the diagram below.

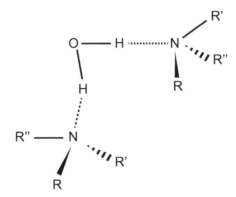

Amines are weak bases like ammonia and will dissociate slightly in aqueous solutions. The lone pair of electrons on the nitrogen in the amine molecule accepts a proton from the water molecule forming a alkylammonium ions and hydroxide ions. The hydroxide ions make the solution alkaline.

$$CH_3NH_2(aq) + H_2O(l) \rightleftharpoons CH_3NH_3^+(aq) + OH^-(aq)$$

Reactions of amines

Amines react with hydrochloric acid, sulfuric acid and nitric acid to form salts.

$$CH_3CH_2NH_2 + HNO_3 \rightarrow CH_3CH_2NH_3^+NO_3^- \text{ (ethylammonium nitrate)}$$

They also react with carboxylic acids to form salts. Amides are formed if these salts are heated losing water.

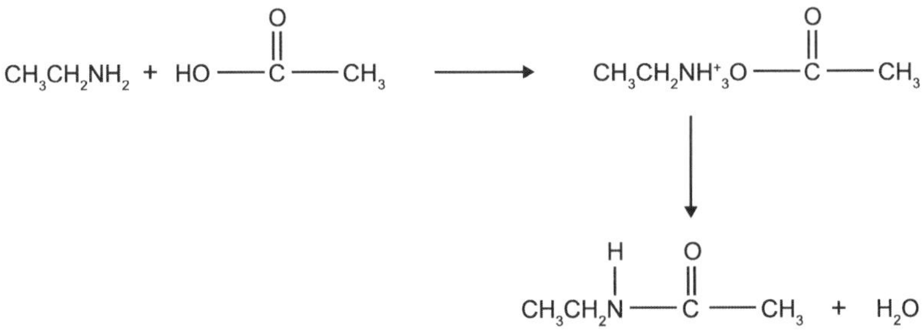

Reaction of an amine with a carboxylic acid to form a salt

4.11 Aromatic compounds

The simplest member of the class of compounds known as **aromatic** compounds is benzene. It has the molecular formula C_6H_6. In the planar benzene molecule each carbon atom is sp^2 hybridised and the three filled sp^2 hybrid orbitals form sigma bonds with a hydrogen atom and two neighbouring carbon atoms. This leaves an electron occupying a p orbital on each carbon atom. Each of these p orbitals overlaps side on with the two p orbitals on either side and a pi molecular orbital forms.

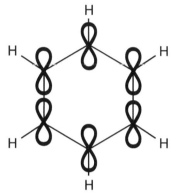

P orbitals above and below each carbon atom

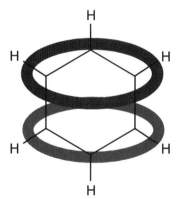

Overlap of p orbitals to form a pi molecular orbital which extends above and below the benzene ring

The six electrons that occupy the pi orbital are not tied to any one of the carbon atoms and are shared by all six. They are known as delocalised electrons. The structure is represented as follows:

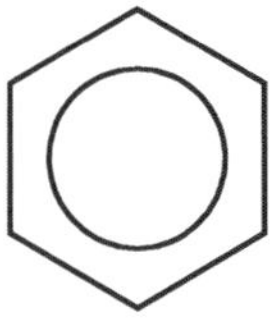

The structure below, called the Kekulé structure, wrongly illustrates that benzene has alternating single and double bonds. Historically, this was an accepted structure for benzene but it is not correct as benzene does not react with Br_2(aq) proving that it doesn't have C=C bonds. Analysis of the bonds shows that all six C-C bonds in the ring are equal length and energy, between those of C-C single bonds and C=C double bonds.

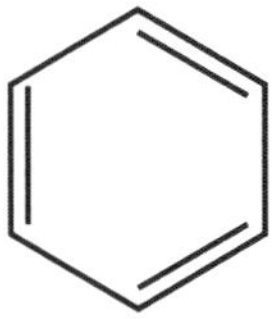

4.12   Reactions of benzene

The delocalised electrons give benzene an unusual stability and enable it to undergo substitution reactions rather than addition reactions which would disrupt the stability of the ring. Benzene is readily attacked by electrophiles, due to the high electron density of the delocalised ring system, through reactions such as alkylation, chlorination, nitration and sulfonation, which are all examples of **electrophilic substitution**.

When writing reaction mechanisms for aromatic systems, it is often useful to represent them as Kekulé structures (the localised arrangements).   It should always be remembered however that this is not a true picture of the benzene molecule.

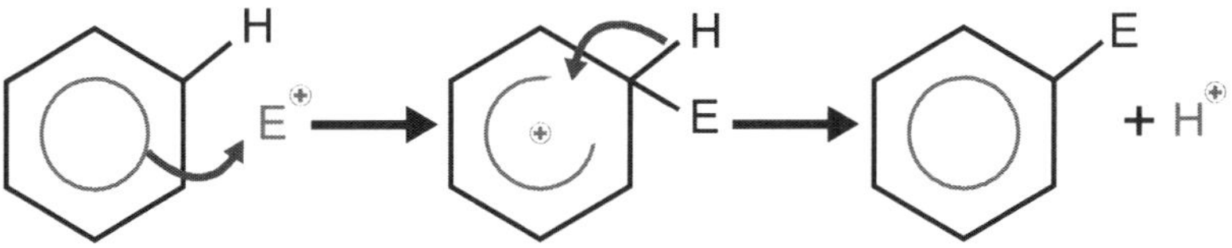

The electrophile attacks one of the carbons atoms in the ring, forming an intermediate ion with a positive charge that can be stabilised by delocalisation around the ring. The intermediate ion then loses H^+ to regain aromatic character and restore the system.

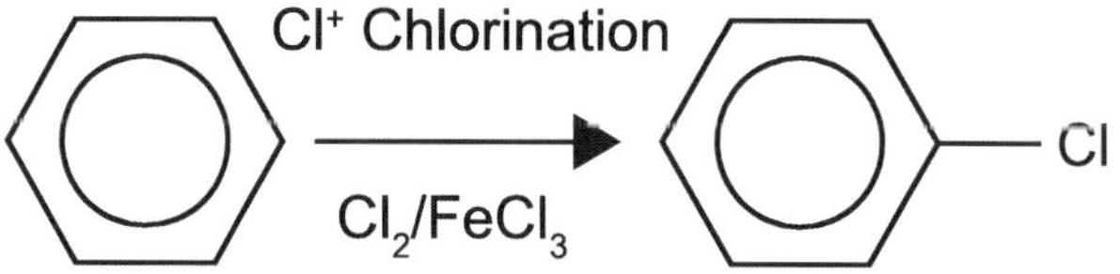

Cl⁺ Chlorination

$$\text{benzene} \xrightarrow[\text{Cl}_2/\text{FeCl}_3]{} \text{C}_6\text{H}_5\text{-Cl}$$

As well as chlorination using aluminium chloride or iron (III) chloride, benzene can undergo bromination using iron (III) bromide.

The catalyst polarises the halogen molecule (here it is bromine) by accepting a pair of electrons from one atom and creating an electrophilic centre on the other atom.

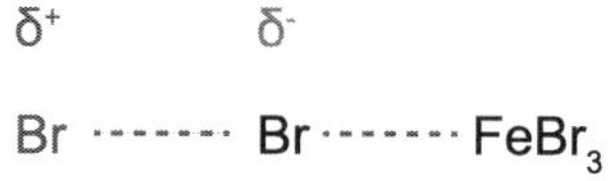

$$\overset{\delta^+}{\text{Br}} \cdots\cdots \overset{\delta^-}{\text{Br}} \cdots\cdots \text{FeBr}_3$$

Electrophilic centre on bromine

This partially positive bromine atom attacks one of the carbons on the benzene and creates a carbocation which is stabilised by delocalisation on the ring. The intermediate then loses a hydrogen ion by heterolytic fission from the benzene, regaining the aromatic character and forming bromobenzene. The iron(III) bromide is regenerated. The same mechanism applies if chlorine is used with a suitable catalyst.

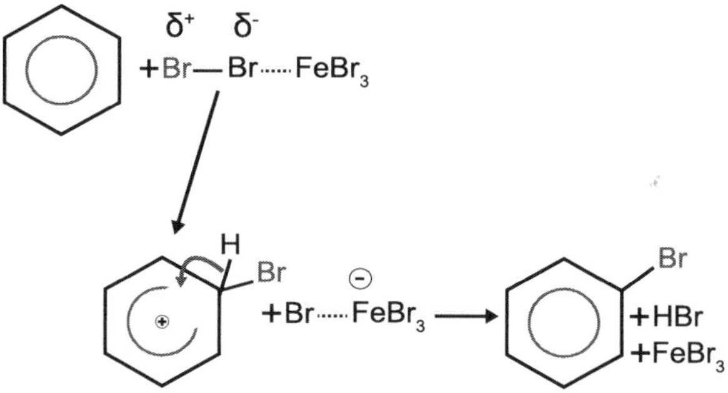

$$\text{benzene} + \overset{\delta^+}{\text{Br}} - \overset{\delta^-}{\text{Br}} \cdots\cdots \text{FeBr}_3$$

$$\downarrow$$

intermediate with H, Br $+ \text{Br} \cdots\cdots \text{FeBr}_3^{\ominus} \longrightarrow$ bromobenzene $+ \text{HBr} + \text{FeBr}_3$

The reaction of chlorine with benzene

Answer these questions concerning the similar reaction of chlorine with benzene in the presence of aluminium(III) chloride.

Q1: What kind of reaction would take place?

a) Electrophilic addition
b) Nucleophilic substitution
c) Electrophilic substitution
d) Nucleophilic addition

. . . .

Q2: What is generally the best name for the product?

a) Chlorophenyl
b) Chlorobenzene
c) Benzene chloride
d) Phenylchloride

....

Q3: What type of fission does the chlorine molecule undergo in this reaction?

a) Nuclear
b) Electrolytic
c) Homolytic
d) Heterolytic

....

Q4: What type of organic ion is created as an intermediate in this reaction?

....

Benzene will react with nitric acid when a mixture of concentrated nitric acid and concentrated sulfuric acid (known as a nitrating mixture) is used and the temperature is kept below 55°C. The nitrating mixture generates the nitronium ion, NO_2^+.

$$HNO_3 + 2H_2SO_4 \longrightarrow H_3O^+ + 2HSO_4^- + NO_2^+$$

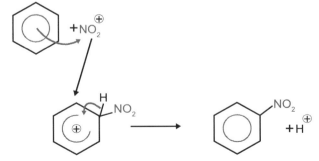

If the temperature is allowed to rise above 55°C, further substitution of the ring can take place and small amounts of the di- and tri- substituted compounds can result.

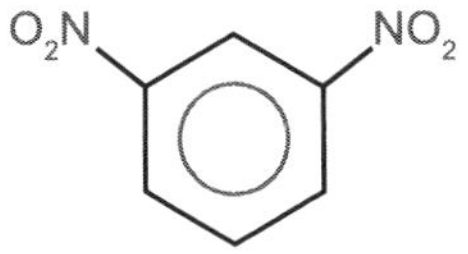

Structure x

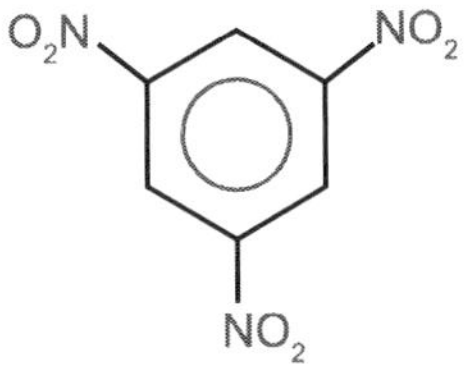

Di- and trinitrobenzene structures

Di- and trinitrobenzene structures

Q5: What type of reactant is the nitronium ion?

a) Carbanion
b) Nucleophile
c) Carbocation
d) Electrophile

....

Q6: What type of intermediate is formed?

a) Free radical
b) Carbocation
c) Carbanion
d) Nitronium anion

....

Q7: Describe how the intermediate is stabilised. Try writing a sentence before checking your answer.

....

Q8: What name would you suggest for structure (x)?

a) 1,1-dinitrobenzene
b) 1,2-dinitrobenzene
c) 1,3-dinitrobenzene
d) 1,4-dinitrobenzene

....

Benzene will react with concentrated sulfuric acid if the reactants are heated together under reflux for several hours. If fuming sulfuric acid is used (sulfuric acid enriched with sulfur trioxide) under cold conditions $^+SO_3H$ substitutes onto the ring. This suggests that the active species is sulfur trioxide. The sulfur atom in the sulfur trioxide molecule carries a partial positive charge and can attack the benzene ring. The mechanism is the same as that shown for nitration.

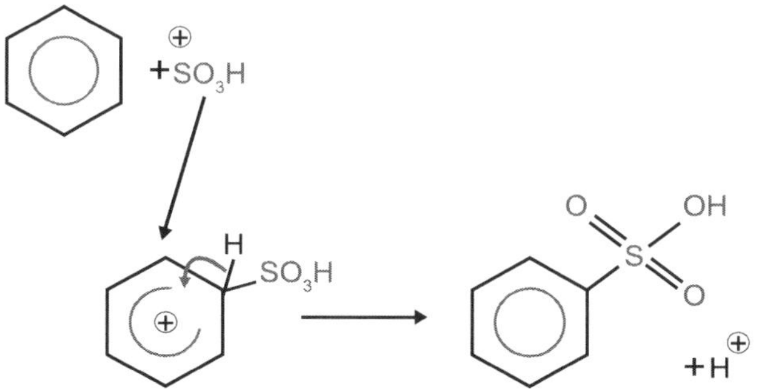

Sulfonation of benzene

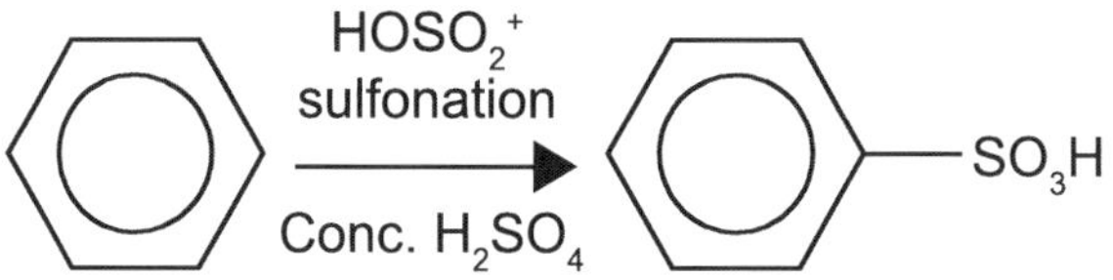

Benzenesulfonic acid

Benzene

Q9: In this reaction is the sulfur trioxide acting as a nucleophile or a electrophile?

....

Q10: Does it react with the benzene in an addition or a substitution reaction?

....

The aluminium chloride catalyst mentioned in the halogenation reactions can be used to increase the polarisation of halogen containing organic molecules like halogenoalkanes. This allows an electrophilic carbon atom to attack the benzene ring and builds up side-chains. The reaction is called a Friedel-Crafts reaction, after the scientists who discovered it.

$$\delta^+ \qquad \delta^-$$
$$R \cdots\cdots Cl \cdots\cdots AlCl_3$$

Electrophilic centre on alkyl group

The catalyst increases the polarity of the halogenoalkane producing the electrophilic centre that can attack the benzene ring . The carbocation formed is stabilised by delocalisation and the intermediate so formed regains its stability by loss of a hydrogen ion forming an alkylbenzene.

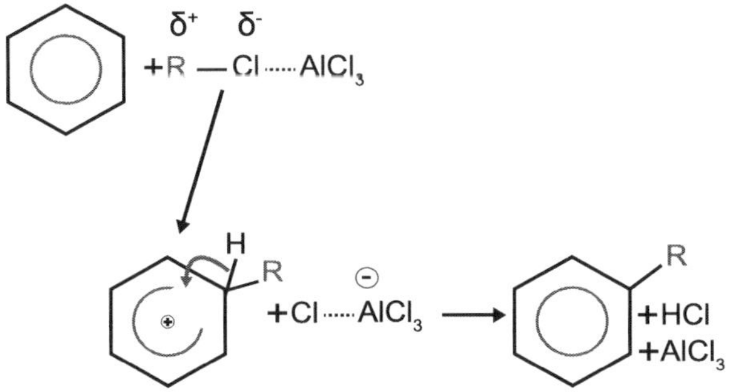

$$\delta^+ \quad \delta^-$$
$$+R - Cl \cdots AlCl_3$$

$$+Cl \cdots AlCl_3^{\ominus} \longrightarrow \quad +HCl$$
$$+AlCl_3$$

Where R- is an alkyl group such as methyl $-CH_3$

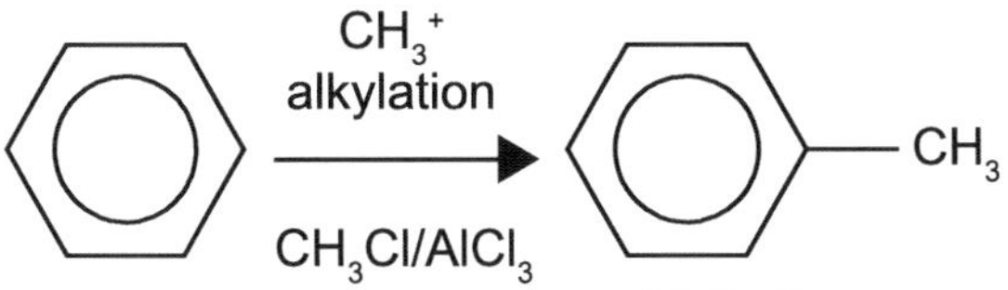

$$\xrightarrow[CH_3Cl/AlCl_3]{\substack{CH_3^+ \\ \text{alkylation}}} \quad -CH_3$$

Methylbenzene

Methylbenzene

Q11: What word describes the role of the aluminium(III) chloride?

....

Q12: What type of fission does the halogenoalkane undergo in this reaction?

a) nuclear
b) homolytic
c) heterolytic
d) electrolytic

....

Q13: Name the chloroalkane necessary to produce ethylbenzene in a Friedel-Crafts reaction.

....

One or more H atoms can be substituted on the benzene ring which leads to a wide range of consumer products including many pharmaceutical drugs. (The benzene ring is usually drawn in a different manner in these structures).

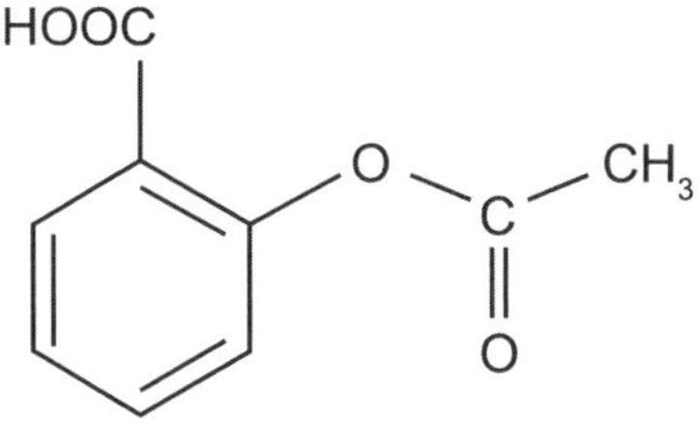

HOOC

Structure of
aspirin

Structure of
paracetamol

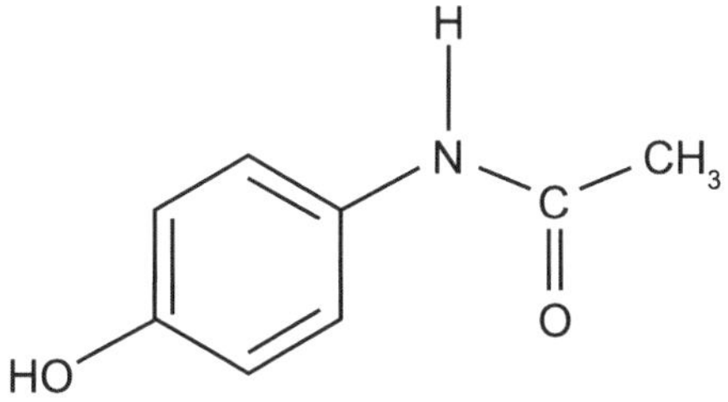

Phenyl

Where one of the hydrogen atoms has been substituted in a benzene ring it is known as
the phenyl group (C_6H_5).

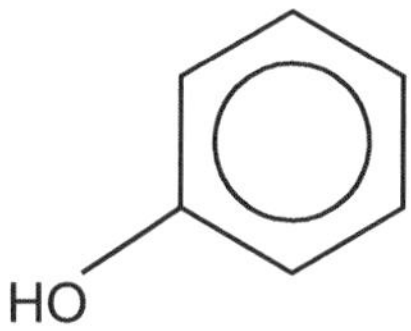

This molecule is
known as phenol

HO

4.13   Extra practice

Extra practice questions

Some possible reactions of propenal are shown below

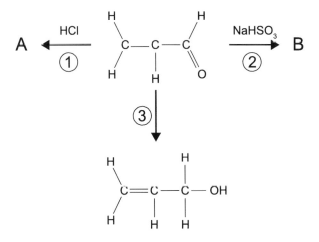

Q14: Draw a structural formula for compound **A** which is the major product formed during the reaction.

. . . .

Q15: Which reagent could be used to carry out reaction **3**?

. . . .

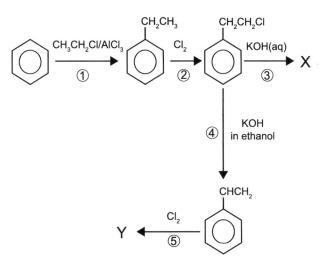

Q16: What type of reaction is taking place in step **1**?

. . . .

Q17: What experimental condition would be required in step **2**?

....

Q18: Draw a structural formula for product **X**.

....

Q19: What type of reaction is taking place in step **4**?

....

Q20: Draw a structural formula for product **Y**.

....

Consider the structure of lactic acid:

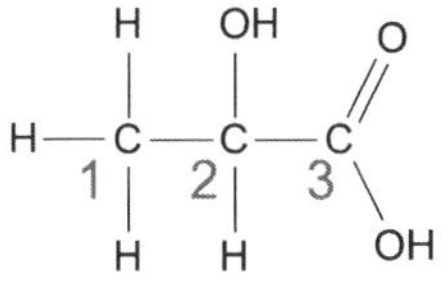

Q21: What is the systematic name of lactic acid?

....

Q22: Lactic acid contains an aysmmetric carbon atom. Identify, and **explain**, which of the numbered carbon atoms is asymmetric.

....

Q23: Lactic acid can be produced from ethanal by the reaction sequence below.

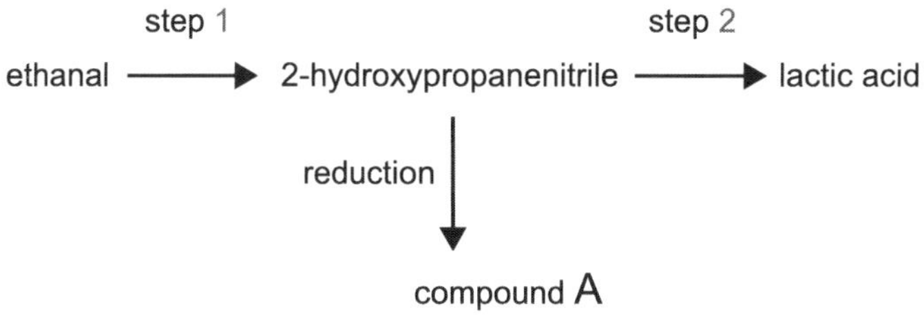

Which reagent could be used in **step 1**?

....

Q24: What type of reaction takes place in **step 2**?

....

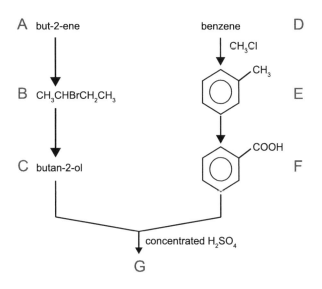

Q25: Explain why but-2-ene exhibits geometric isomerism yet its structural isomer but-1-ene does not.

. . . .

Q26: But-2-ene undergoes electrophilic addition to form **B**. Draw a structure for the carbocation intermediate formed in this electrophilic addition reaction.

. . . .

Q27: Name a reagent used to convert **B** to **C**.

. . . .

Q28: Name a catalyst required in converting **D** to **E**.

. . . .

Q29: Draw a structural formula for ester **G**.

. . . .

. . . .

4.14   Summary

Summary

You should now be able to:

- recognise and use different types of reaction in organic synthesis including substitution, addition, elimination, condensation, hydrolysis, oxidation, reduction.

- devise synthetic routes, with no more than three steps, from a given reactant to a final product.

- look at molecular structures and deduce the reactions it can undergo.

4.15   Resources

- Chemguide, Free radical substitution (http://bit.ly/2azRAex)

- Avogadro (http://www.avogadro.co.uk/chemist.htm)

- The 8 Types of Arrows In Organic Chemistry, Explained (http://bit.ly/2awwQWo)

- Chemistry in Context Laboratory Manual, fifth edition, Graham Hill and John Holman ISBN 0-17-448276-0

- Chemguide, Nucleophilic substitution (http://bit.ly/2ajukQb)

- Comparing the SN1 and SN2 Reactions (http://bit.ly/1Nwjiml)

- The Williamson Ether Synthesis (http://bit.ly/2aMcW7A)

- Chemguide, Hydrolysing Nitriles (http://bit.ly/2az2O5Q)

4.16 End of topic test

End of Topic 4 test

Q30:

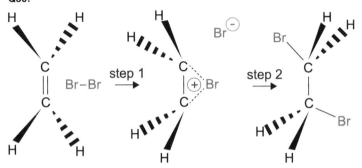

Go online

The two steps in the mechanism shown could be described as:

a) ethene acting as an nucleophile and Br⁻ acting as an nucleophile.
b) homolytic fission of the Br_2 followed by Br⁻ acting as a nucleophile.
c) ethene acting as an electrophile and Br⁻ acting as an electrophile.
d) ethene acting as an electrophile and Br⁻ acting as an nucleophile.

....

Q31: Name the missing compound in this reaction of propene.

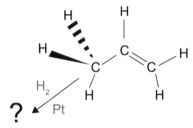

....

Q32: Name the missing compound in this reaction of propene.

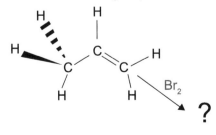

....

Q33: What type of halogenoalkane is shown below?

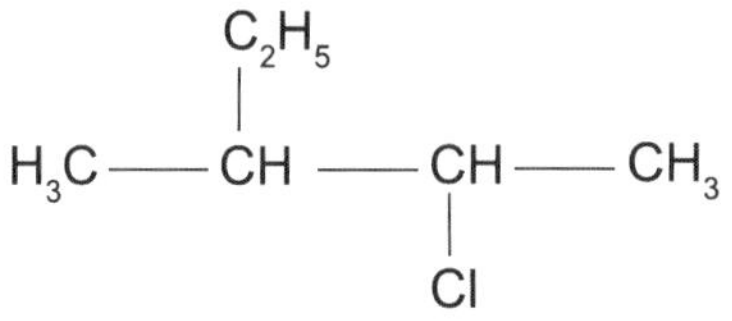

$$H_3C \!-\! CH \!-\! CH \!-\! CH_3$$

with C_2H_5 attached above the first CH and Cl attached below the second CH.

a) Primary
b) Secondary
c) Tertiary
d) None of the above

....

Q34: Give the correct name for the compound shown.

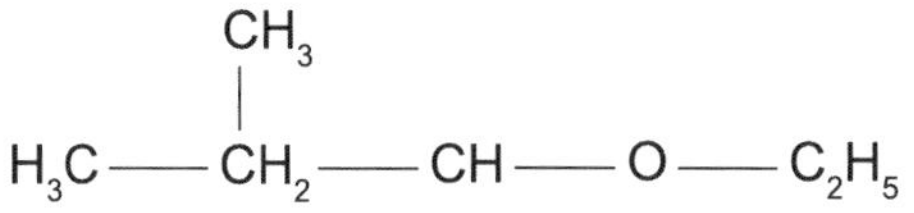

$$H_3C \!-\! CH_2 \!-\! CH \!-\! O \!-\! C_2H_5$$

with CH_3 attached above the CH.

....

Q35: The diagram shows the structure of an alcohol. What type of alcohol is it?

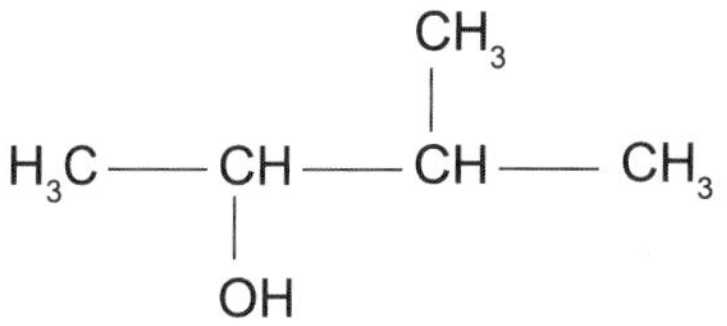

$$H_3C \!-\! CH \!-\! CH \!-\! CH_3$$

with CH_3 attached above the second CH and OH attached below the first CH.

....

Q36: Which of the following could not be used to prepare this alcohol?

a) 2-methylbut-2-ene
b) 2-methylbut-1-ene
c) 3-methylbut-1-ene
d) 2-bromo-3-methylbutane

....

Esters can be prepared from alcohols by two alternative routes.

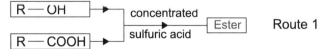

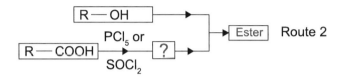

Q37: What type of compound is missing in Route 2?

....

Q38: What is the disadvantage in using Route 1?

....

Q39: Which of the following shows the structure of the ester formed when propan-2-ol and ethanoic acid react?

1   H_3C —— CH_2 —— $\overset{\overset{\textstyle O}{\|}}{C}$ —— O —— CH_2 —— CH_3

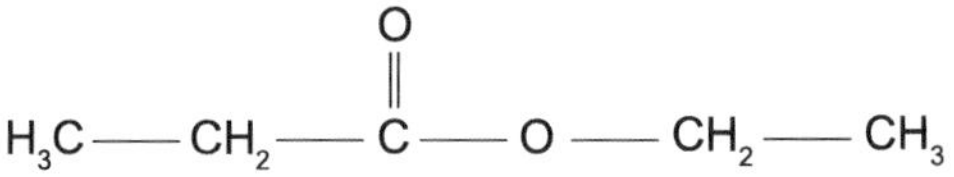

2   H_3C —— $\overset{\overset{\textstyle CH_3}{|}}{CH}$ —— O —— $\overset{\overset{\textstyle O}{\|}}{C}$ —— CH_2 —— CH_3

3   H_3C —— $\overset{\overset{\textstyle CH_3}{|}}{CH}$ —— O —— $\overset{\overset{\textstyle O}{\|}}{C}$ —— CH_3

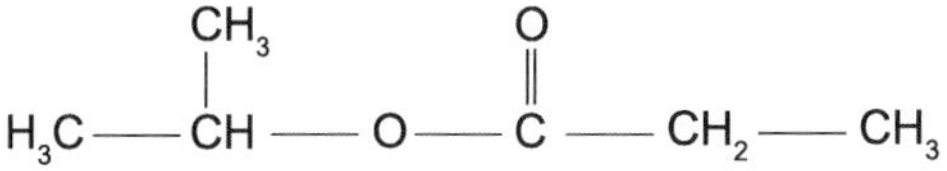

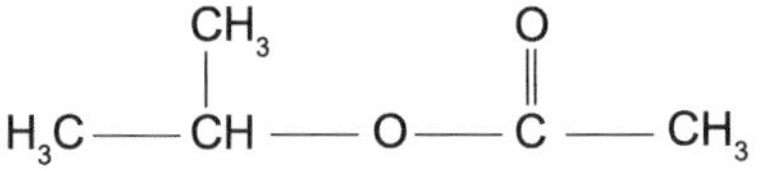

4   H_3C —— CH_2 —— CH_2 —— O —— $\overset{\overset{\textstyle O}{\|}}{C}$ —— CH_3

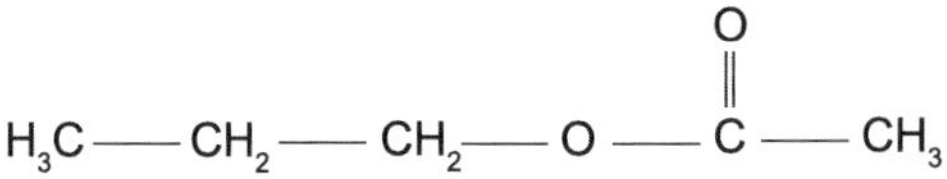

a) 1
b) 2
c) 3
d) 4

....

Q40: Carboxylic acids are weak acids. Which of the following statements is true?

a) Hexanoic acid will be more soluble than propanoic acid.
b) Ethanol will have a lower value for pK_a than ethanoic acid.
c) The ethanoate ion is more stable than the ethoxide ion due to electron delocalisation.
d) In dilute aqueous ethanoic acid, hydrogen bonding between acid molecules produces dimers.

. . . .

Propanoic acid can react with a number of different reagents as show in the scheme below.

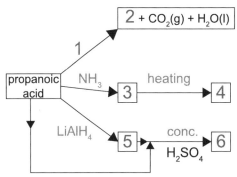

Q41: What could reagent **1** be?

a) $MgCO_3$
b) NaOH
c) CuO
d) H_2SO_4

. . . .

Q42: What type of compound are **2** and **3**?

. . . .

Q43: What type of compound is **4**?

. . . .

Q44: Name compound **5**.

. . . .

Q45: Name compound 6

. . . .

Q46: Which of these statements describes the reactivity of the benzene ring? It:

a) resists electrophilic attack.
b) rapidly decolourises bromine water.
c) reacts rapidly with nucleophiles.
d) resists addition reactions.

. . . .

Q47: Which of these is a major product of the reaction between 2-chloropropane and benzene in the presence of aluminium(III) chloride?

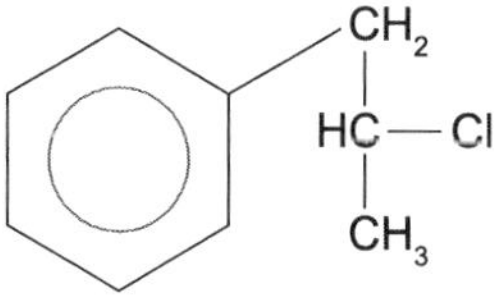

a)
$$CH_2$$
$$HC - Cl$$
$$CH_3$$

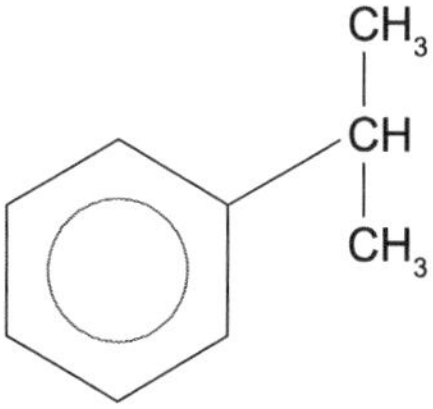

b)
$$CH_3$$
$$CH$$
$$CH_3$$

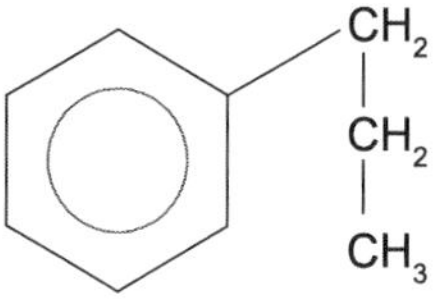

c)
$$CH_2$$
$$CH_2$$
$$CH_3$$

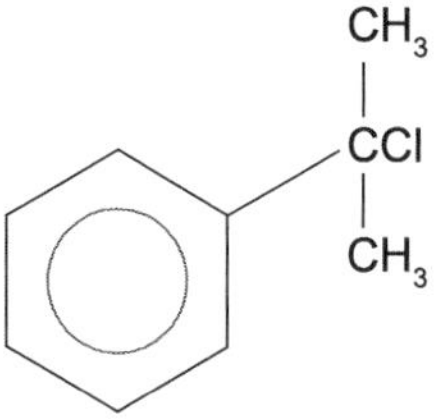

d)
$$CH_3$$
$$CCl$$
$$CH_3$$

. . . .

Derivatives of benzene with more than one substituted side group are named by numbering the ring and defining the substituted group position. Example:

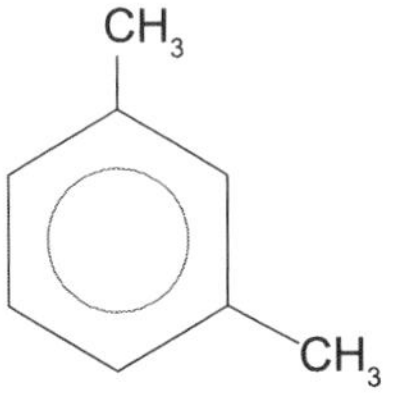

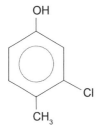

1,3-dimethylbenzene 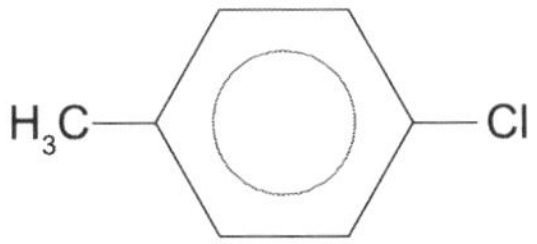   1-chloro-4-methylbenzene

Q48: Name this molecule:

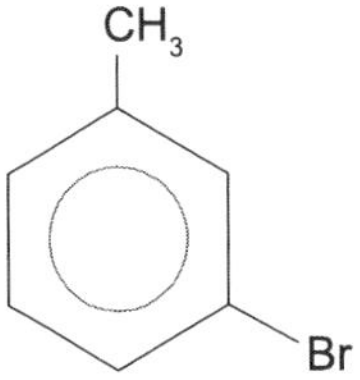

Q49: Name this molecule:

....

Q50: Suggest a name for a possible aromatic isomer of 1,3-dimethylbenzene

....

Q51: Which of the following reaction types are likely to be involved in the conversion of benzene to a typical detergent?

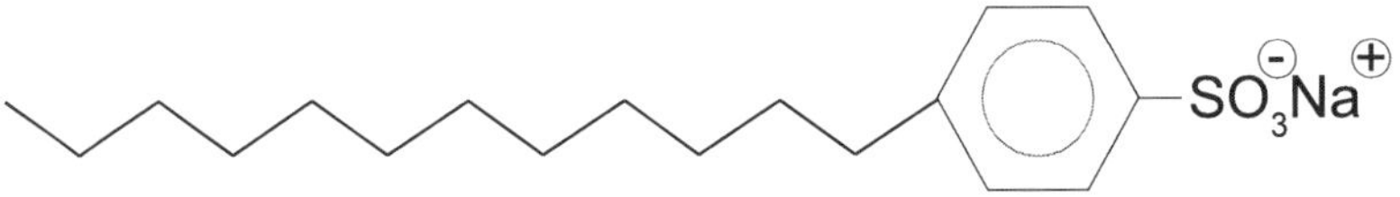

1. Addition
2. Alkylation
3. Condensation
4. Nitration
5. Polymerisation
6. Sulfonation

....

Q52: Which of these reactions would occur between benzene and a mixture of concentrated nitric acid and concentrated sulfuric acid?

a) Addition
b) Alkylation
c) Condensation
d) Nitration
e) Polymerisation
f) Sulfonation

....

Look at this molecule:

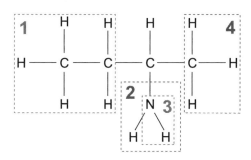

Q53: Which area is the functional group in this molecule?

a) 1
b) 2
c) 3
d) 4

....

Q54: To which class of amine does it belong?

a) Primary
b) Secondary
c) Tertiary
d) Aromatic

. . . .

Q55: What name would you give the molecule?

a) Butan-2-amine
b) 2-methylpropanamine
c) 2-aminobutane
d) Ethylmethylamine

. . . .

Ethylamine reacts with nitric acid with both in the gaseous state and produces tiny white crystals.

Q56: What name would you give to the white crystals?

. . . .

Q57: What type of reaction has taken place?

. . . .

Q58: Which element is the electrophilic centre in this reaction?

. . . .

Q59: Which element is the nucleophilic centre in this reaction?

. . . .

Q60: Name a primary amine which should have a higher solubility in water than ethylamine.

. . . .

. . . .

Topic 5

Experimental determination of structure (Unit 2)

Contents

Learning objectives

By the end of this topic, you should be able to:

- *explain that elemental microanalysis can be used to determine the masses of C, H, O, S and N in a sample of an organic compound in order to determine its empirical formula;*

- *work out the empirical formula from data given;*

- *explain that mass spectrometry can be used to determine the accurate molecular mass and structural features of an organic compound;*

- *explain infra-red spectroscopy can be used to identify certain functional groups in an organic compound and work out which compound is responsible for a spectra by identifying which functional groups are responsible for peaks;*

- *proton nuclear magnetic resonance spectroscopy (proton NMR) can give information about the different environments of hydrogen atoms in an organic molecule, and about how many hydrogen atoms there are in each of these environments;*

- *be able again to identify which compound is responsible for a spectra and be able to draw low resolution NMR spectra;*

- *explain how absorption of visible light by organic molecules occurs;*
- *state that the chromophore is the group of atoms within a molecule which is responsible for the absorption of light in the visible region of the spectrum.*

5.1   Elemental microanalysis

Elemental microanalysic (combustion analysis) can be used to determine the empirical formula of an organic compound.   The empirical formula shows the simplest whole number ratio of the different atoms that are present in a compound. If we look at propene for example then the molecular formula is C_3H_6 and hence its C:H ratio is 3:6.   This simplifies to 1:2 and hence the empirical formula would be CH_2.

To carry out elemental microanalysis in modern combustion analysers a tiny sample (2 mg approximately) is accurately weighed and oxidised at a high temperature in an atmosphere of oxygen. This produces a mixture of gases SO_2, N_2, CO_2 and H_2O which are separated by gas chromatography and the mass of each component measured using a thermal conductivity detector. By converting the masses of each product gas into the masses of the original element (see below for example) and subtracting each of these from the original mass of the sample, the mass of oxygen in the compound can be obtained. The empirical formula is then determined from the calculated element masses using the method below.

Calculating the empirical formula

An antibiotic contains C, H, N, S and O. Combustion of 0.3442 g of the compound in excess oxygen yielded 0.528 g CO_2, 0.144 g H_2O, 0.128 g SO_2 and 0.056 g N_2. What is the empirical formula of the antibiotic?

1 mol CO_2 (44.0 g) contains 1 mole of C (12.0 g)
Mass of C in sample = 0.528 × 12 / 44 = 0.144 g

1 mol H_2O (18.0 g) contains 1 mole of H_2 (2 g)
Mass of H in sample = 0.144 × 2 / 18 = 0.016 g

1 mol SO_2 (64.1 g) contains 1 mole of S (32.1 g)
Mass of S in sample = 0.128 × 32.1 / 64.1 = 0.064 g

1 mol N_2 (28.0 g) contains 1 mole of N_2 (28.0 g)
Mass of N in sample = 0.056 × 28 / 28 = 0.056 g

Mass of O in sample = 0.3442 - (0.144 + 0.016 + 0.064 + 0.056) = 0.064 g

Element	C	H	S	N	O
Mass (g)	0.144	0.016	0.064	0.056	0.064
Number of moles	0.144 / 12 = 0.012	0.016 / 1 = 0.016	0.064 / 32.1 = 0.002	0.056 / 14 = 0.004	0.064 / 16 = 0.004
Mole ratio	0.012 / 0.002 = 6	0.016 / 0.002 = 8	0.002 /0.002 = 1	0.004 / 0.002 = 2	0.004 / 0.002 = 2

Empirical formula is therefore $C_6H_8SN_2O_2$

You can also work out the empirical formula by knowing the percentage of each element in the compound.

A white solid is found to contain 66.67% carbon, 7.407% hydrogen and 25.936% nitrogen. What is the empirical formula of the solid?

Element	C	H	N
Percentage %	66.670	7.407	25.936
Number of moles	66.670 / 12 = 5.555	7.407 / 1 = 7.401	25.936 / 14 = 1.853
Mole ratio	5.555 / 1.853 = 3	7.407 / 1.853 = 4	1.853 / 1.853 = 1

Empirical formula is therefore C_3H_4N. The formula mass is reported to be 108 g and the molecular formula can be worked out from this.

Empirical formula = C_3H_4N = $(3 \times 12) + (4 \times 1) + 14 = 54$ g
Therefore the molecular formula is twice the empirical formula to make the mass 108 g = $C_6H_8N_2$.

Go online

Calculate the empirical formula

Q1: A liquid is found to contain 52.17% carbon, 13.04% hydrogen and the remainder is made up of oxygen. Work out the empirical formula of the liquid.

. . . .

5.2   Mass spectrometry

Mass spectroscopy is a technique used in determining the accurate molecular mass of an organic compound. It can also determine structural features. A tiny sample (1×10^{-4} g approximately) of the unknown compound is vaporised and injected into the mass spectrometer. High-energy electrons bombard it with enough energy to knock electrons out of the molecules which are ionised and broken into smaller ion fragments.

The instrument is divided into four main sections:

1. *The inlet system.* This is maintained at a high temperature, so that any sample introduced will be vaporised rapidly. The interior of a mass spectrometer must be maintained at high vacuum to minimise collisions, so the injector and the rest of the instrument are connected to vacuum pumps.

2. *The ion source.* Some of the vaporised sample enters the ionisation chamber where it is bombarded with electrons. Provided the energy of the collision is greater than the molecular ionisation energy, some positive ions are produced from molecules in the sample material.
   Some of these molecular ions (sometimes called **parent ions**) will contain sufficient energy to break bonds, producing a range of fragments, some of which may also be positive ions (daughter ions).

$$M + e^- \rightarrow M^{\cdot +} + 2e^-$$

The parent ions and ion fragments from the source are accelerated and focussed into a thin, fast moving ion beam by passing through a series of focusing slits.

3. *The analyser.* This stream of ions is then passed through a powerful magnetic field, where the ions experience forces that cause them to adopt curved trajectories.

    Ions with the same charge (say, 1+) will experience the same force perpendicular to their motion, but the actual path of the ion will depend on its mass. Heavier ions will be deflected less than lighter ones.

4. *The detector.* Ions with the same mass/charge (m/z) ratio will have the same trajectory and can be counted by a detector. As the magnetic field strength is altered ions with different m/z ratios will enter the detector so that a graph of abundance (the ion count) against m/z values can be constructed. This is the mass spectrum of the sample compound(s).

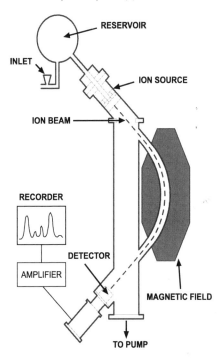

You can try a simple experiment at home. Get someone to roll golf balls and table tennis balls (heavy and light ions!) across a table. As the balls pass you, blow gently across their path. The two types of ball are about the same size, will experience about the same deflecting force from your breath, but will have different amounts of deflection owing to their different masses.

The following mass spectrum is for benzoic acid. The height of the vertical lines represents the relative abundance of ions of a particular m/z (mass/charge) ratio. The most abundant peak (the tallest one) is called the base peak and is given the abundance of 100%. The percentages of the other peaks are assigned relative to the base peak.

The peak with the highest m/z value:

- is often the molecular ion (heaviest ion), but it might be too small or undetectable in cases where it is unstable;

- will give the molecular mass and therefore the molecular formula of the compound.

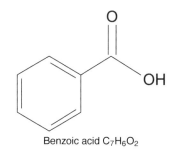

Benzoic acid $C_7H_6O_2$

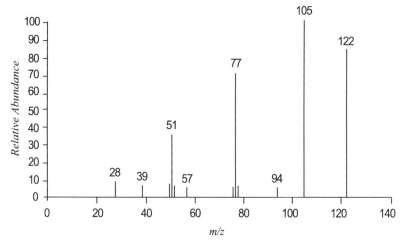

The base peak in the above example has a m/z ratio of 105 and the molecular ion peak has a m/z ratio of 122 $[C_7H_6O_2]^+$ (molecule with one electron removed).

Mass difference	Suggested group
15	$[CH_3]^+$
17	$[OH]^+$
28	$[C{=}O]^+$ or $[C_2H_4]^+$
29	$[C_2H_5]^+$
31	$[CH_3O]^+$
45	$[COOH]^+$
77	$[C_6H_5]^+$

This table shows common mass differences between peaks in a mass spectrum and suggested groups that correspond to the difference.

Example : Structure from fragmentation pattern

Compounds **A** and **B** both have the same molecular formula, C_3H_6O, but their molecular structures are different. The mass spectra of **A** and **B** are shown below.

Mass spectrum of Compound A

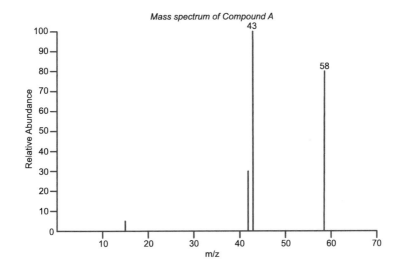

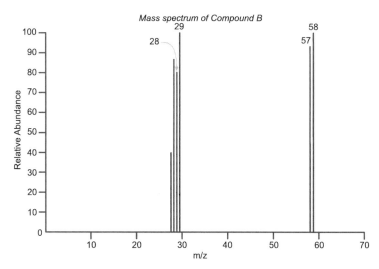

A1. For compound **A**, which group of atoms could be lost when the ion of m/z 43 forms from the ion of m/z 58?
→The mass difference from 58 to 43 is 15. This indicates that a **methyl group** is likely to be responsible for this.

A2. Suggest a formula for the ion of m/z 43 in the spectrum of compound **A**.
→The ion of m/z 43 is probably due to loss of a methyl group from the molecule, so that a fragment of formula $[C_2H_3O]^+$ would form the m/z 43 ion. A likely structure for this is $[CH_3CO]^+$.

A3. Identify **A**.
→**A** is propanone, $[CH_3COCH_3]^+$, which has major fragments of $[CH_3CO$ and $CH_3]^+$.

B1. For compound **B**, what atom or groups of atoms could be lost when

i) the molecular ion changes into the ion of m/z 57?
→A loss of 1 mass unit from 58 to 57 can only be due to loss of a **hydrogen atom**. The molecule must have a single, 'loose' H.

ii) the ion of m/z 57 changes into the ion of m/z 29?
→The mass difference from 57 to 29 is 28, probably due to loss of **C=O**.

B2. Suggest formulae for the ions of m/z 28,29, and 57 in the spectrum of compound **B**.
→The ion of m/z 28 is the $[C=O]^+$ fragment; addition of 1 mass unit (an **H**) to this would form a $[CHO]^+$ (aldehyde) as the m/z 29 peak.

B3. Identify **B**.
→The m/z 57 is loss of hydrogen from the molecule, leaving a fragment of formula $[C_3H_5O]^+$. We know this contains a C=O group, so is likely to be $[C_2H_5CO]^+$.

. . . .

Questions about fragmentation

Q2: Draw the structures for the six isomers of butanol - four alcohols and two ethers.

Go online

....

The mass spectra X, Y and Z below are for three isomers of butanol, excluding 2-methylpropan-1-ol. There is one example of a primary alcohol, a secondary alcohol and a tertiary one, but not necessarily in that order.

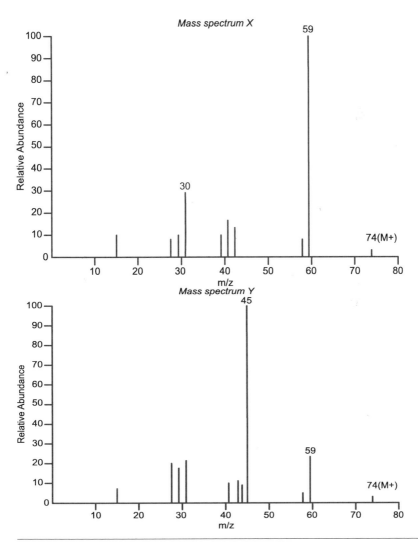

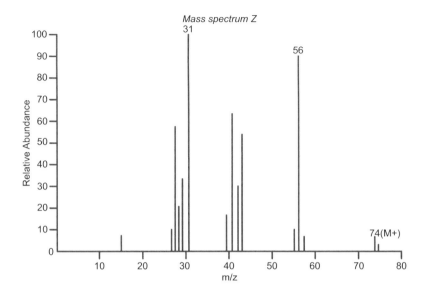

Q3: What is the m/z value for the molecular (M+) ion of all three butanols? Answer with one integral value.

....

If you would like to try to link the mass spectra with the appropriate molecular structures, have a go now, by working out the main fragments lost from the M+ ion. Otherwise, the remaining questions will guide you through to an answer, with the conclusions summarised at the end.

Q4: Spectrum X has a base peak at m/z 59. What fragment has been lost from the molecular ion to produce this m/z 59 ion?

....

Q5: Using the fragmentation information, which of the three butanols would be likely to produce a CH_3 fragment most readily?

....

Q6: Closer examination of spectrum Y shows a base peak at m/z 45. What could be the structure of this ion?

....

Q7: When the C-C bond adjacent to the the C-OH bond breaks in butan-1-ol what are the structures and m/z values of the fragments?

....

Q8: In which spectrum are these most prominent?

....

Q9: Given that alcohols tend to fragment at the C-C bond adjacent to the C-OH group (i.e. H-CH$_2$OH would produce R$^+$ and [CH$_2$OH]$^+$ fragments), can you explain, with reasons, which spectrum is for which isomer?

....

....

The following is the mass spectrum of naphthalene an aromatic molecule.

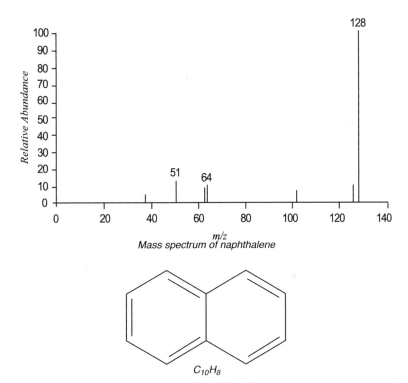

Mass spectrum of naphthalene

$C_{10}H_8$

Naphthalene has a stable molecular ion [C$_{10}$H$_8$]$^+$ which gives a very large response at m/z = 128 with minor fragmentation peaks in the rest of the spectrum. In this case the base peak is the molecular ion and naphthalene has a molecular mass of 128. The peak at m/z = 64 indicates the stability of the molecular ion being assigned to a doubly charge ion M^{2+} with m = 128 and z = 2 so that m/z = 64. This is fairly typical of aromatic compounds giving further evidence for their stability.

5.3   Infrared spectroscopy

Infrared is found on the electromagnetic spectrum between microwaves and visible light (see Unit 1 topic on 'Electromagnetic radiation and atomic spectra'). When it is absorbed by organic compounds there is sufficient energy to cause the bonds in the molecules to vibrate but not break them.

The wavelength of the infrared radiation that is absorbed (quoted as a wavenumber cm^{-1}) by a vibrating bond depends on the type of atoms which make up the bond along with the strength of the bond (wavenumber = 1/wavelength). Light atoms that are joined by stronger bonds absorb radiation of shorter wavelengths (higher energy) in general than heavier atoms joined by weaker bonds. Infrared can be used to identify certain bonds and functional groups in organic molecules.

Simple diatomic molecules, for example halogens, have only one bond, which can vibrate only by stretching and compressing. The two halogen atoms can pull apart and then push together.

Intermediate	Slower (heavier spheres)	Faster (stronger spring)
	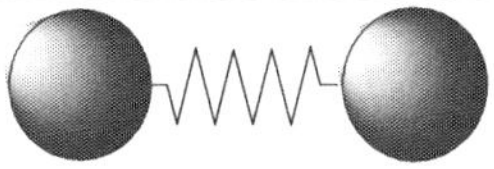	

Compound	Bond enthalpy/kJ mol^{-1}
HCl	432
HBr	366
HI	298

The table above shows the bond enthalpies for various hydrogen halides.

Go online

Questions on bond vibration

Q10: Which molecule has the strongest bond?

a) HCl
b) HBr
c) HI

. . . .

Q11: Which molecule has the largest molecular mass?

a) HCl
b) HBr
c) HI

. . . .

Q12: Which molecule will have the largest vibration frequency?

a) HCl
b) HBr
c) HI

. . . .

Some organic molecules with typically more than two atoms can give rise to modes of vibration other than simple stretching of bonds. Consider water (H_2O) each of the O-H bonds will have a characteristic stretching frequency, but these can interact and stretch either symmetrically or asymmetrically, giving rise to two slightly different IR absorption frequencies for the O - H bonds in a water molecule.

See Figure 5.1

Figure 5.1: Stretching modes in water molecules

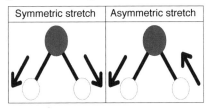

. . . .

Figure 5.2: Bending modes

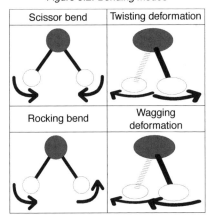

. . . .

Infrared spectrometer

The source beam of infrared radiation gets split into two with one part of the beam passing through the sample and the other through a reference cell (may contain solvent).

The monochromator grating scans the wavelengths prior to a detector which compares the intensity of the two beams. The amplified signal is plotted as % transmission or absorbance.

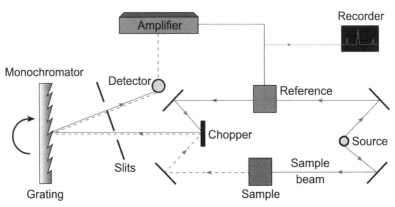

Interpreting infrared spectra

Below is an IR spectrum of ethanol C_2H_5OH. An infrared correlation table can be found on page 14 of the CfE Higher and Advanced Higher Chemistry data booklet. Tables of values have been built up by observation of a large number of compounds.

The region of an IR spectrum from 4000 to 1400 cm^{-1} contains many absorbance wavenumbers for specific bond types. Below 1400 cm^{-1} IR spectra typically have a number of absorbances not assigned to a particular bond, and this is called the fingerprint region. In fact, these relatively low energy vibrations are due to complex vibrations which are unique to that molecule. Comparison of this region for an unknown material with a set of standards run under identical conditions will allow identification.

There is a range of values for a particular group, because the value is slightly altered by the surrounding groups.

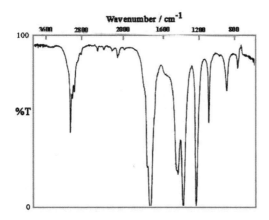

Ethanol contains a hydroxyl (OH) functional group which is assigned the wave number range of 3570 to 3200 cm^{-1} for a OH hydrogen bonded stretch. This is very broad in alcohols.

Functional group identification

The main use of IR spectra, which can be obtained quickly and cheaply, is to identify the presence of functional groups and the carbon backbone type in unknown organic compounds.

Go online

Use the correlation table on page 14 of the data booklet to answer the following questions regarding the IR spectra labelled X, Y and Z.

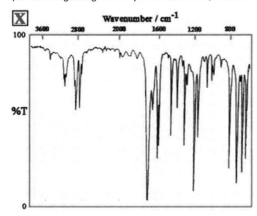

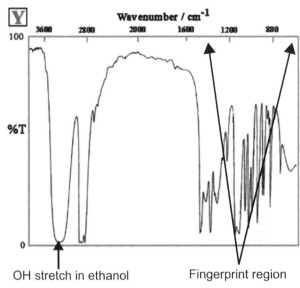

OH stretch in ethanol   Fingerprint region

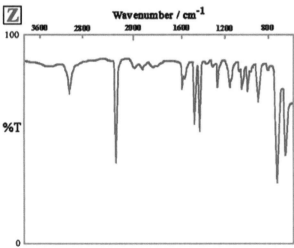

Q13: Which compound contains an alcohol group?

a) X
b) Y
c) Z

. . . .

Q14: Which compound is **not** aromatic?

a) X
b) Y
c) Z

....

Q15: Which class of compound is present in X? Hint: take note of the peaks in the region of $1700cm^{-1}$ and $2800cm^{-1}$.

....

Q16: Which of the following could be Z?

a) Benzyl alcohol ($C_6H_5CH_2OH$)
b) Butan-1-ol (C_4H_9OH)
c) Benzonitrile (C_6H_5CN)
d) Acetonitrile (CH_3CN)

....

5.4   Proton nuclear magnetic resonance (NMR) spectroscopy

Proton nuclear magnetic resonance spectroscopy (proton NMR) can give information about the different environments of hydrogen atoms in an organic molecule, and about how many hydrogen atoms there are in each of these environments.

Hydrogen nuclei (protons) behave like tiny magnets as a result of them spinning in a clockwise direction or anticlockwise direction. When they are placed between the poles of a very powerful magnet some of the protons align with the field of the magnet and some align themselves against the field of the magnet. A superconducting magnet cooled with liquid helium and liquid nitrogen is used.

The protons that are aligned with the direction of the magnetic field have a slightly lower energy than those aligned against and the energy difference between the two corresponds to the radiofrequency region of the electromagnetic spectrum (60 MHz to 1000MHz). When the protons are exposed to radio waves, the absorption of energy promotes those in the lower energy states to the higher energy states. This effectively flips those aligned with the magnetic field to being aligned against it. The protons then fall back to the lower energy state and the same radio frequency that was absorbed is emitted which can be measured with a radio receiver.

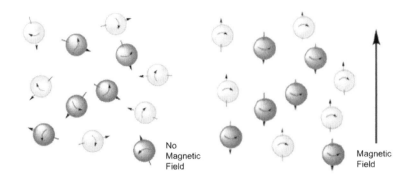

The darker spheres have lower energies than the lighter spheres and are aligned with the direction of the magnetic field in the second diagram.

Diagram of a typical NMR spectrometer

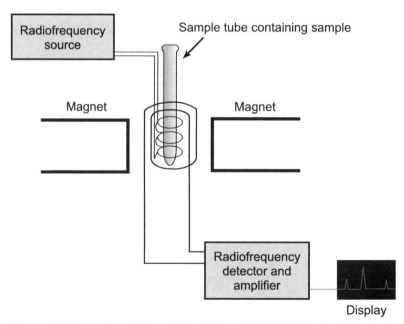

The sample being analysed is dissolved in $CDCl_3$ or CD_3COCD_3, "deuterated" solvents with hydrogens replaced by deuterium (isotope of H which is written as D or ^2H). This is to avoid swamping the NMR spectrum of the sample with a proton signal from the solvent. A referencing substance called tetramethylsilane (TMS) is used, against which all other absorptions due to other proton environments are measured. It has only one proton environment and is assigned a value of 0 ppm (parts per million). The difference between the protons in TMS and other proton environments is called the chemical shift measured in ppm and given the symbol δ. Values for chemical shifts for different proton environments are given on page 16 of the data booklet.

Interpreting NMR spectra

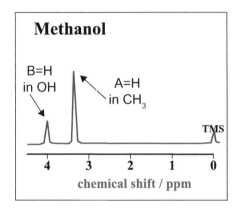

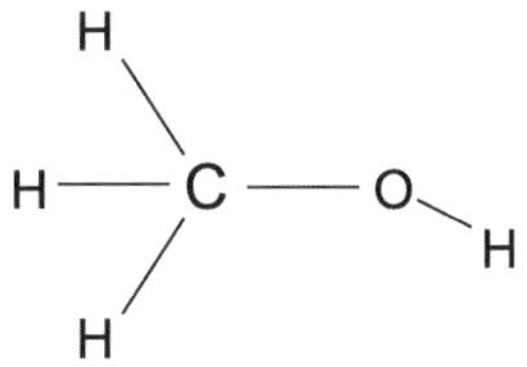

Above is the ^1H NMR spectrum of methanol (CH_3OH). Methanol has 2 different proton environments the H in the CH_3 group and the H in the OH group. The spectrum shows two peaks which correspond to the two different proton environments. The area under each peak is proportional to the number of protons in that particular environment. The area under peaks A and B are in the ratio 3:1 corresponding to the three protons in the CH_3 group and the one proton in the OH group. If you look at the table of chemical shifts in the data book page 16 you will see the chemical shift for R-OH is 1.0-5.0 and the chemical shift for CH_3O is 3.5-3.9 which makes the peak for the protons in CH_3 A and the peak for the proton in OH B.

Determining structure using NMR

The NMR spectrum of a hydrocarbon is shown in Figure 5.3.

Go online

Figure 5.3: NMR spectrum of hydrocarbon W

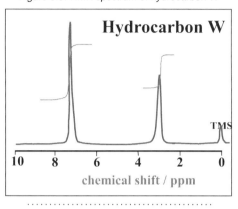

. . . .

Q17: From the spectrum, how many different environments are there for hydrogen atoms?

. . . .

Q18: What is the ratio of areas for these peaks?

. . . .

Q19: From the correlation table (page 16 of the data booklet), can you identify the type of hydrogen in the larger peak?

a) RCH$_3$
b) RCH$_2$R
c) ArCH$_3$
d) ArH

. . . .

Q20: What type of hydrogens are in the smaller peak?

a) RCH$_3$
b) RCH$_2$R
c) ArCH$_3$
d) ArH

. . . .

You might find it useful to summarise the information in a table.

Table 5.1: NMR table

	Peak 1	Peak 2
δ (ppm)	2.3	7.4
Type of H	$ArCH_3$	ArH
Number of H atoms	3	5
Group	CH_3	C_6H_5

. . . .

Q21: What is the hydrocarbon?

. . . .

The next three questions refer to N,N-diethylphenylamine ($C_{10}H_{15}N$), shown in Figure 5.4.

Figure 5.4: N,N-diethylphenylamine

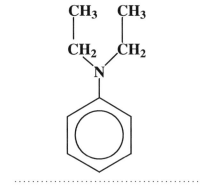

. . . .

Q22: How many different types of H environment are there in the molecule?

a) 1
b) 2
c) 3
d) 5

. . . .

Q23: What is the ratio of H atoms in these groups?

a) 2:1
b) 4:5:6
c) 1:2:3
d) 3:2:5

. . . .

Q24: Can you predict the chemical shift (δ) ranges for this compound?

. . . .

Q25: Sketch the NMR spectrum for N,N-diethylphenylamine.

. . . .

. . . .

High resolution NMR

In a high resolution spectrum, you find that many of what looked like single peaks in the low resolution spectrum are split into clusters of peaks. The amount of splitting tells you about the number of hydrogens attached to the carbon atom or atoms next door to the one you are currently interested in.

The number of sub-peaks in a cluster is one more than the number of hydrogens attached to the next door carbon(s) (n+1 rule where n is the number of hydrogens attached to the neighboring carbon atom).

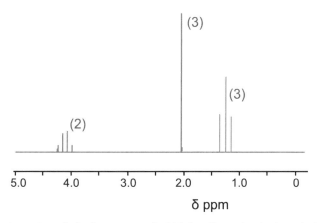

The spectrum above is for the compound which has the molecular formula $C_4H_8O_2$. Treating this as a low resolution spectrum to start with, there are three clusters of peaks and so three different environments for the hydrogens. The hydrogens in those three environments are in the ratio 2:3:3. Since there are 8 hydrogens altogether, this represents a CH_2 group and two CH_3 groups.

The CH_2 group at about 4.1 ppm is a quartet. That tells you that it is next door to a carbon with three hydrogens attached - a CH_3 group.

The CH_3 group at about 1.3 ppm is a triplet. That must be next door to a CH_2 group.

This combination of these two clusters of peaks - one a quartet and the other a triplet - is typical of an ethyl group, CH_3CH_2. It is very common.

Finally, the CH_3 group at about 2.0 ppm is a singlet. That means that the carbon next door doesn't have any hydrogens attached. You would also use chemical shift data to help to identify the environment each group was in, and eventually you would come up with:

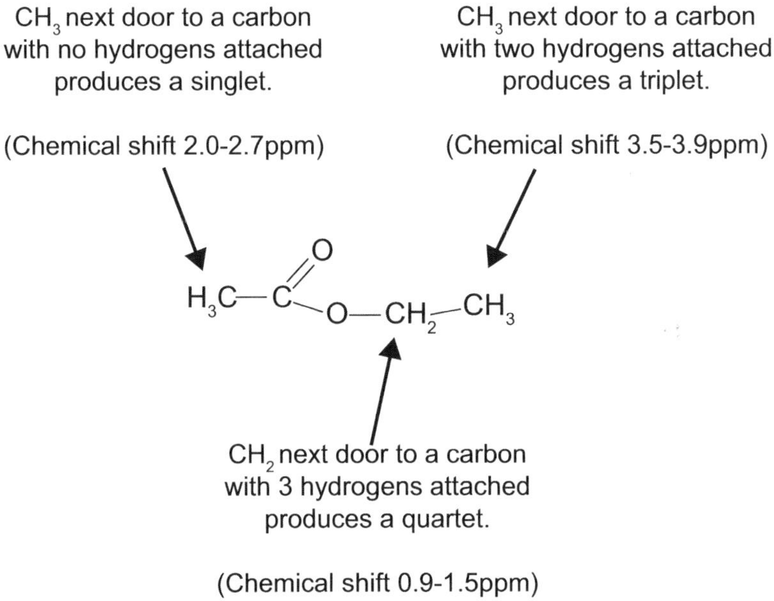

CH$_3$ next door to a carbon with no hydrogens attached produces a singlet.

(Chemical shift 2.0-2.7ppm)

CH$_3$ next door to a carbon with two hydrogens attached produces a triplet.

(Chemical shift 3.5-3.9ppm)

$$H_3C-C\overset{O}{\underset{O}{\diagdown}}O-CH_2-CH_3$$

CH$_2$ next door to a carbon with 3 hydrogens attached produces a quartet.

(Chemical shift 0.9-1.5ppm)

The numbers represent the number of hydrogen atoms within that environment. The splitting gives us information on the number of hydrogens on neighbouring carbon atoms. Chemical shifts have been taken from the data book page 16.

Hydrogen atoms attached to the same carbon atom are said to be equivalent. Equivalent hydrogen atoms have no effect on each other - so that one hydrogen atom in a CH$_2$ group doesn't cause any splitting in the spectrum of the other one. But hydrogen atoms on neighbouring carbon atoms can also be equivalent if they are in exactly the same environment. For example:

Cl-CH$_2$-CH$_2$-Cl.

These four hydrogens are all exactly equivalent. You would get a single peak with no splitting at all.

NMR - Practice

For the high resolution ^{1}H NMR data below, work out the structure of the molecules concerned. You will find a short table of useful chemical shifts at the end of the questions.

Q26: A molecule with the molecular formula C_4H_8O:

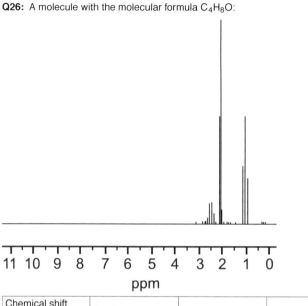

Chemical shift (ppm)	2.449	2.139	1.058
Ratio of area under the peaks	2	3	3
Splitting	quartet	singlet	triplet

. . . .

Q27: A molecule with the molecular formula $C_4H_8O_2$:

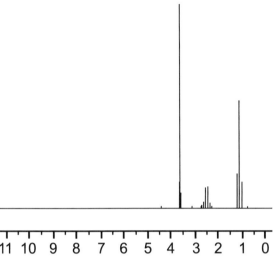

Chemical shift (ppm)	3.674	2.324	1.148
Ratio of area under the peaks	3	2	3
Splitting	singlet	quartet	triplet

. . . .

Q28: Another molecule with the molecular formula $C_4H_8O_2$:

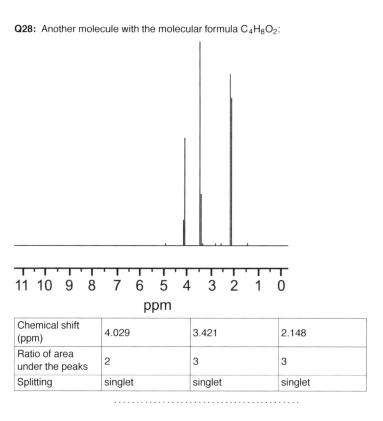

Chemical shift (ppm)	4.029	3.421	2.148
Ratio of area under the peaks	2	3	3
Splitting	singlet	singlet	singlet

. . . .

MRI scanning

Our bodies consist largely of water, which exists in a large number of different environments in each tissue. Just as hydrogen atoms in molecules are shielded to different extents depending on the surrounding atoms, the protons in water in tissues will experience slight differences in a strong external magnetic field, and so will absorb slightly different radiofrequency radiation.

The part of the body being investigated is moved into a strong magnetic field. By using computers to process the absorption data, a series of images of the different water molecule environments, is built up into a picture of the body's tissues. It is assumed that MRI scanning is harmless to health, unlike other imaging processes (such as CAT scans) which involve low doses of ionising X-rays.

MRI scanning is particularly good for brain tissue where there is a large amount of fatty lipids which provides a different environment for water compared with the other tissues.

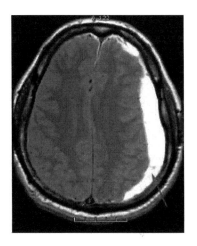

The MRI image opposite shows fluid collection in the region that separates the brain from the skull. This is a blood clot which applies pressure to the brain and is very dangerous, perhaps even fatal.

5.5 Absorption of visible light

Most organic molecules appear colourless because the energy difference between the highest occupied molecular orbital (**HOMO**) and the lowest unoccupied molecular orbital (**LUMO**) is relatively large resulting in the absorption of light in the ultraviolet region of the spectrum. Organic molecules containing only sigma bonds are colourless.

Organic molecules that are coloured contain delocalised electrons spread over a few atoms and they are known as **conjugated** systems. We have looked at conjugation in benzene. For bonds to be conjugated in long carbon chains alternating double and single bonds must be present. An example of this is seen in the structure of Vitamin A.

Conjugated system in retinol (vitamin A)

The conjugated system in Vitamin A is a long chain of σ and π bonds. The molecular orbital contains delocalised electrons which span across the length of the conjugated system. The greater the number of atoms spanned by the delocalised electrons, the smaller the energy gap will be between the delocalised orbital and the next unoccupied orbital. Exciting the delocalised electrons will therefore require less energy. If this energy corresponds to a wavelength that falls within the visible part of the electromagnetic spectrum it will result in the compounds appearing coloured.

Ninhydrin (2,2-Dihydroxyindane-1,3-dione) is a chemical used to detect amino acids. When it reacts with amino acids it produces a highly conjugated product which absorbs light in the visible region and an intense purple colour is seen.

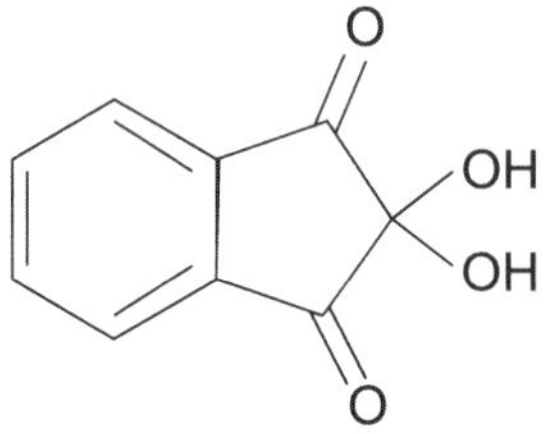

Ninhydrin

Azo dye can be synthesised from aminobenzene (aniline), sodium nitrite and 2-naphthol at low temperatures which then can be used to dye a piece of cotton. Synthetic indigo can also be prepared using a microscale method (see RSC references on methods to carry this out).

5.6 Chromophores

A **chromophore** is a group of atoms within a molecule that is responsible for its colour. Coloured compounds arise because visible light is absorbed by the electrons in the chromophore, which are then promoted to a higher energy molecular orbital.

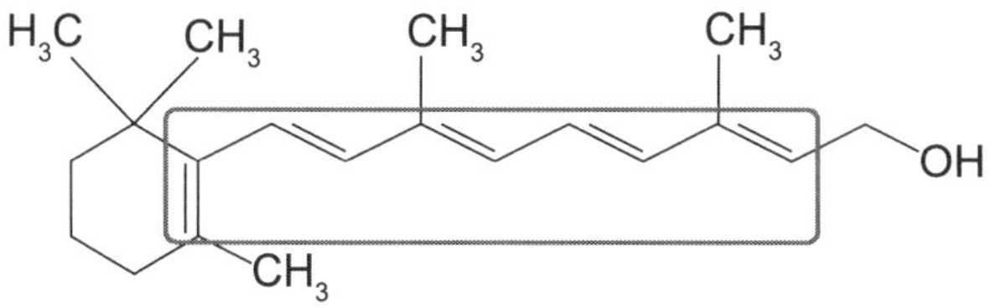

The red box shows the chromophore

Vitamin A (retinol) has a conjugated system that spreads over 5 carbon to carbon double bonds and appears yellow.

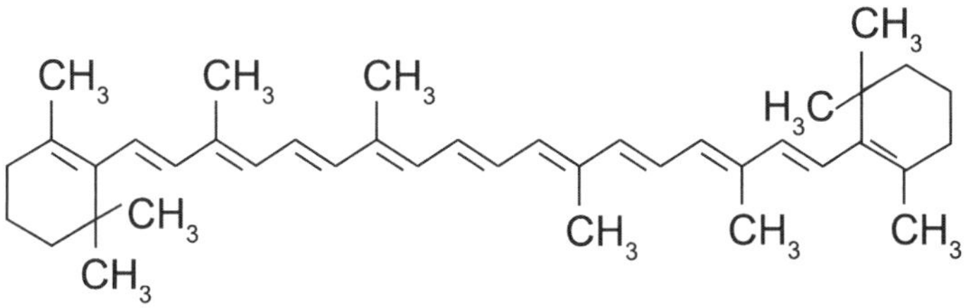

β-carotene found in sweet potatoes, carrots and apricots has a conjugated system that spreads over 11 carbon to carbon bonds and appears orange.

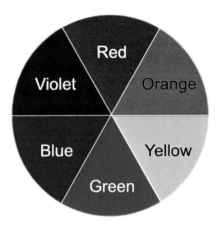

Lycopene is found in watermelon, pink grapefruit and tomatoes and has as conjugated system that spreads over 13 carbon to carbon bonds and appears red.

The colours we observe are not absorbed by the molecule. If the chromophore absorbs light of one colour, then the complementary colour is observed. This means that the Vitamin A will absorb violet light to appear yellow, β-carotene will absorb blue light to appear orange and lycopene will absorb green light to appear red (see the colour wheel below).

The energy associated with a particular colour can be calculated using the equation $E = Lhc/\lambda$ (see Unit 1 topic on 'Electromagnetic radiation and atomic spectra').

5.7   Summary

Summary

You should now be able to:

- explain that elemental microanalysis can be used to determine the masses of C, H, O, S and N in a sample of an organic compound in order to determine its empirical formula;

- work out the empirical formula from data given;

- explain that mass spectrometry can be used to determine the accurate molecular mass and structural features of an organic compound;

- explain infra-red spectroscopy can be used to identify certain functional groups in an organic compound and work out which compound is responsible for a spectra by identifying which functional groups are responsible for peaks;

- proton nuclear magnetic resonance spectroscopy (proton NMR) can give information about the different environments of hydrogen atoms in an organic molecule, and about how many hydrogen atoms there are in each of these environments;

- be able again to identify which compound is responsible for a spectra and be able to draw low resolution NMR spectra;

- explain how absorption of visible light by organic molecules occurs;

- state that the chromophore is the group of atoms within a molecule which is responsible for the absorption of light in the visible region of the spectrum.

5.8   Resources

- Royal Society of Chemistry, Spectroscopy in a Suitcase (http://rsc.li/1VVGlQG)

- LearnChemistry, Introduction to Spectroscopy (http://rsc.li/2aiTp0z)

- LearnChemistry, SpectraSchool (http://rsc.li/2adQ8lo)

- Chemguide, What is Nuclear Magnetic Resonance (NMR)? (http://bit.ly/2acLnnt)

- UV-Visible Absorption Spectra (http://bit.ly/2aMflzk)

- Azo dye synthesis for schools (PDF) (http://bit.ly/2a6w8LH)

- LearnChemistry, Microscale Chemistry - The microscale synthesis of azo dyes (http://rsc.li/2a6vZYl)

- LearnChemistry, Microscale Chemistry - The microscale synthesis of indigo dye (http://rsc.li/2ay2pBa)

- Additive colour model (Flash animation) (http://bit.ly/1IY1zBt)

5.9   End of topic test

End of Topic 5 test

Q29: A mass spectrum contains a line at m/z 28.0312. Given the following accurate atomic masses, what could the formula for the ion be?

Go online

H	1.0078
C	12.0000
N	14.0031
O	15.9949

a) $[NO]^+$
b) $[C_2H_4]^+$
c) $[CO]^+$
d) $[N_2]^+$

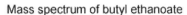

Look at the mass spectrum of butyl ethanoate.

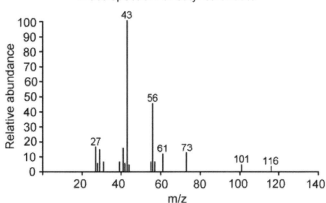

Mass spectrum of butyl ethanoate

Q30: Draw the structural formula for butyl ethanoate.

Q31: The spectrum shows a line at m/z 116. What name is normally given to the species represented by this line?

Q32: The abundance of the m/z 116 ion is small because the parent molecular ion is:

a) unstable and easily fragments to form daughter ions.
b) heavy and easily fragments to form lighter ions.
c) large and easily fragments to form smaller ions.
d) charged and easily fragments to form neutral ions.

....

Q33: Write the formula of an ion which would give the line at **m/z 43**.

....

Q34: Write the formula of an ion which would give the line at **m/z 73**.

....

Q35: From the spectrum, give the m/z value of the ion which you would expect to be deflected most in the magnetic field of the mass spectrometer.

....

Q36: Ethyl butanoate is isomeric with butyl ethanoate but gives a different mass spectrum. By comparing the structures of these two esters, which fragment could **not** be present in both spectra?

a) $[C_4H_9]^+$ at m/z 57
b) M^+ at m/z 116
c) $[C_2H_5]^+$ at m/z 29
d) $[C_3H_7]^+$ at m/z 43

....

Q37: Which **two** ions from the following options could produce a peak with m/z 29 in a mass spectrum?

a) $[CH_3]^+$
b) $[CHO]^+$
c) C_6H_6
d) C_2H_2
e) $[C_2H_5]^+$
f) $[CH_2OH]^+$

....

Q38: Which **two** substances from the following options have an empirical formula CH?

a) $[CH_3]^+$
b) $[CHO]^+$
c) C_6H_6
d) C_2H_2
e) $[C_2H_5]^+$
f) $[CH_2OH]^+$

....

Q39: In mass spectra where a molecular ion loses 15 mass units to produce an $(M-15)^+$ ion, what group is most likely to have been lost?

a) $[CH_3]^+$
b) $[CHO]^+$
c) C_6H_6
d) C_2H_2
e) $[C_2H_5]^+$
f) $[CH_2OH]^+$

....

Q40: Which two fragment ions could be produced when the C - C bond in ethanol breaks?

a) $[CH_3]^+$

b) $[CHO]^+$

c) C_6H_6

d) C_2H_2

e) $[C_2H_5]^+$

f) $[CH_2OH]^+$

....

Q41: In proton nuclear magnetic resonance spectroscopy, which of these would you expect to be used as an internal reference standard?

a) $CDCl_3$
b) Deuterium
c) $Si(CH_3)_4$
d) Hydrogen

....

Q42: Which of the following compounds is most likely to show an infrared absorption at 1725 cm^{-1}?

a) 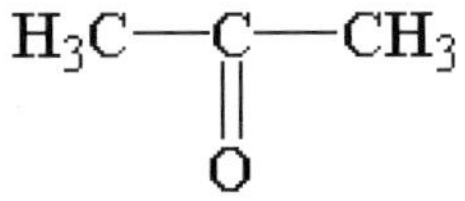 $H_3C-\underset{\underset{O}{\|}}{C}-CH_3$

b) $HOCH_2CH{=}CH_2$

c) 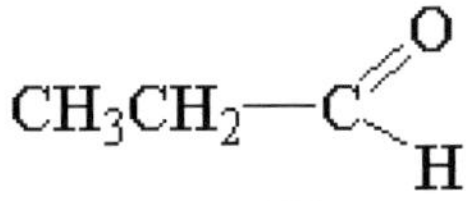 $CH_3CH_2-C\underset{H}{\overset{O}{<}}$

d) 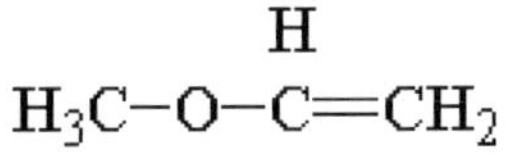 $H_3C-O-\underset{H}{C}{=}CH_2$

....

Q43: How many peaks would you expect in the ^{1}H NMR spectrum of cyclohexane (C_6H_{12})?

a) 1
b) 2
c) 6
d) 12

. . . .

Q44: Consider the ^{1}H NMR spectrum of the ether methoxyethane ($CH_3OCH_2CH_3$). How many peaks would there be?

. . . .

Q45: What is the ratio of the areas of the peaks? If you think there are two peaks in a ratio 1:2, enter your answer in that form.

. . . .

Two aliphatic compounds A and B, which contain carbon, hydrogen and oxygen only, are isomers. They can both be oxidised as follows:

A $\xrightarrow{\text{oxidation}}$ C

B $\xrightarrow{\text{oxidation}}$ D

The table below shows the wavenumbers of the main absorptions in the infrared spectra caused by the functional groups in compounds A to D, between 1500 and 4000 cm^{-1} (absorptions caused by C - H bonds in alkyl components have been omitted):

Compound	Wavenumbers		
A	3300		
B	3350		
C		2750	1730
D			1700

In answering the following questions, you are advised to consult the data on **page 14 of the Data Booklet**.

Q46: Examine the table above and identify the compound present in both A and B.

. . . .

Q47: Which type of compound is compound C?

. . . .

Q48: Which type of compound is compound D?

. . . .

Q49: Compound C will undergo further oxidation to produce compound E. Estimate the wavenumber ranges of the main infrared absorptions caused by the functional groups in compound E (only those between 1500 and 4000 cm^{-1}).

An unpleasant smelling liquid has spilled from a tank which has its label obscured. A sample has an NMR spectrum with only one peak.

Q50: Which **two** substances from the following list could the liquid be?

   a) 1,2-dichloroethane, CH_2ClCH_2Cl

   b) Propylamine, $C_3H_7NH_2$

   c) Ethanenitrile, CH_3CN

   d) Ethanal, CH_3CHO

Q51: The IR spectrum has strong absorbances at 2930, 2245 and 1465 cm^{-1}. The material is most likely to be:

   a)   1,2-dichloroethane, CH_2ClCH_2Cl
   b)   Propylamine, $C_3H_7NH_2$
   c)   Ethanenitrile, CH_3CN
   d)   Ethanal, CH_3CHO

Q52: In what range would you expect the NMR chemical shift of the ^1H signal for this compound to be?

Q53: Anthocyanins are responsible for the colours of many plants and fruits. One such anthocyanin is pelargonidin which produces an orange colour. The structure of this is shown below:

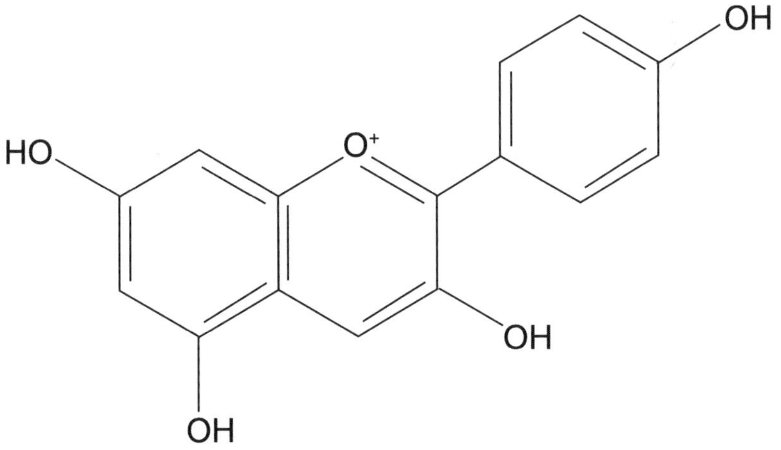

Using your knowledge of chemistry explain why an orange colour is produced. *(3 marks)*

Q54: Calculate the % m/v concentration of sodium ions, if 5 g of sodium chloride is dissolved in 500 cm^3 of water.

. . . .

Q55: Calculate the volume of ethanoic acid in a 250 cm^3 bottle of vinegar, if the % v/v concentration is 3.5%.

. . . .

Q56: A strip of toothpaste weighing 0.75 g was found to contain 1.125 mg of fluoride. Calculate the concentration of fluoride in the toothpaste in ppm.

. . . .

Q57: Calculate the mass of sodium fluoride in mg which would need to be added to 10 g of toothpaste to give this concentration.

. . . .

. . . .

Topic 6

Pharmaceutical chemistry (Unit 2)

Contents

Learning objectives

By the end of this topic, you should be able to state that:

- *drugs are substances which alter the biochemical processes in the body;*

- *many drugs can be classified as agonists or as antagonists at receptors, according to whether they enhance or block the body's natural responses;*

- *drugs bind to receptors in the body so that each drug has a structural fragment which confers the pharmacological activity.*

6.1   Effect of drugs on the body

Drugs are substances which alter the biochemical processes in the body. Drugs which have beneficial effects are used in medicines. A medicine usually contains the drug plus other ingredients.

Paracetamol

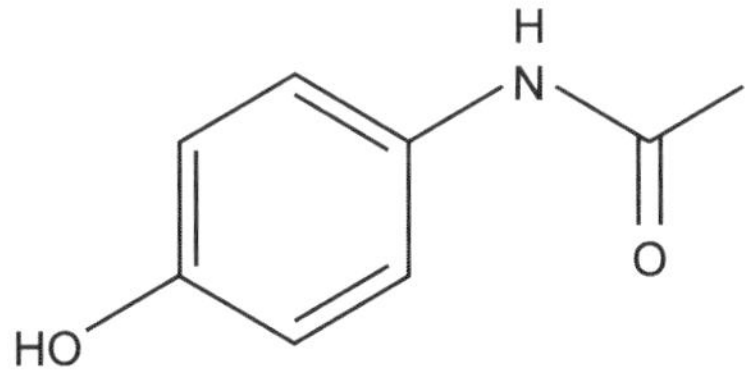

Paracetamol is a mild analgesic (pain reliever) and is used to alleviate pain from headaches and all other minor aches and pains. It is also a major ingredient in many cold and flu remedies. While generally safe for use at recommended doses, even small overdoses can be fatal. Compared to other over-the-counter pain relievers, paracetamol is significantly more toxic in overdose potentially leading to liver failure and death, but may be less toxic when used persistently at the recommended dosage.

Classification of drugs

Most medicines can be classed as **antagonists** or **agonists** according to the response they trigger (enhance or block) when bound to a receptor site. An agonist will mimic the body's naturally active molecule so when bonded to the receptor site it produces the same response as the body's own molecule would do. Antagonists bind strongly to the receptor site, blocking the body's natural molecule from binding and preventing the triggering of the natural response.

An antagonist therefore is a drug which binds to a receptor without stimulating cell activity and prevents any other substances from occupying that receptor. This class of medicine is useful if there is a surplus of natural messengers or where one wants to block a particular message. For example, propranolol is an antagonist which blocks the receptors in the heart that are stimulated by adrenaline. It is called a β-blocker and is used to relieve high blood pressure.

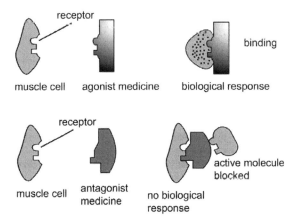

muscle cell   agonist medicine   biological response

muscle cell   antagonist medicine   no biological response   active molecule blocked

An example of an agonist medicine is salbutamol (below) which is used in the treatment of asthma. Asthma attacks are caused when the bronchioles narrow becoming blocked with mucus. The body responds by releasing adrenaline which binds to receptors triggering the dilation of the bronchioles. Unfortunately it also triggers an increase in the heart rate and blood pressure which could lead to a heart attack. Salbutamol binds more strongly to the receptor sites than adrenaline triggering the widening of the bronchioles without an increase in heart rate.

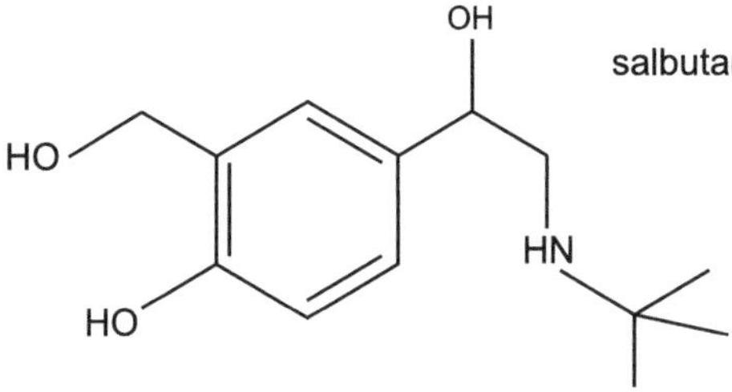

salbutamol

Another example of an antagonist medicine is propranolol (below) which is used to treat high blood pressure.

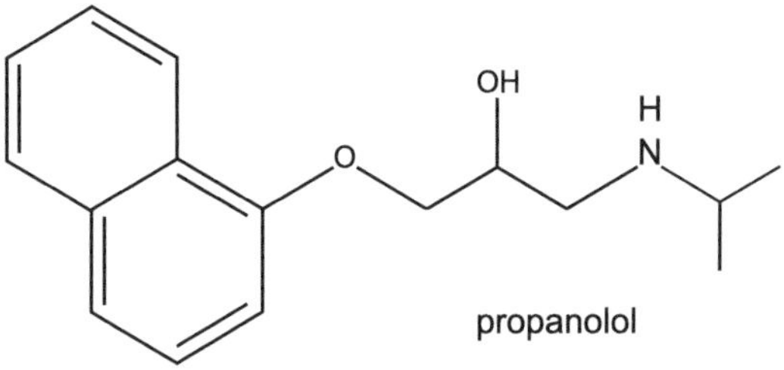

propanolol

Anandamide (bliss molecule) is a recently discovered messenger molecule that plays a role in pain, depression, appetite, memory, and fertility. Anandamide's discovery may lead to the development of an entirely new family of therapeutic drugs. The structure is very similar to that of tetrahydrocannabinol (THC) the active constituent of cannabis and marijuana.

For many years scientists wondered why compounds such as morphine, which were derived from plants, should have a biological effect on humans. They reasoned that there must be a receptor in the brain that morphine could bind to. Scientists suggested that the brain had its own morphine-like molecule, and the receptors were meant for them (morphine just happened to look like the brain's molecule and so had a similar effect). These morphine-like molecules were eventually discovered and called enkephalins, the body's natural painkillers. In 1988, specific receptors were discovered in the brain for THC and scientists started to look for the brain's natural analogue of THC which was isolated in 1992 and called anandamide.

Anandamide has a long hydrocarbon tail which makes it soluble in fat and allows it to easily slip across the blood-brain barrier. Its 3-dimensional shape strongly resembles that of THC. However, THC is a relatively robust molecule, whereas anandamide is fragile and breaks down rapidly in the body. That is why anandamide doesn't produce a continual 'natural high'.

Anandamide is synthesised enzymatically in the areas of the brain that are important in memory, thought processes and control of movement. Research suggests that anandamide plays a part in the making and breaking of short-term connections between nerve cells related to learning and memory. Animal studies suggest that too much anandamide induces forgetfulness suggesting that if substances could be developed that keep anandamide from binding to its receptor, these could be used to treat memory loss or even to enhance existing memory. Anandamide also acts as a messenger between the embryo and the uterus during implantation into the uterine wall.

Anandamide occurs in minute quantities in a number of organisms, from sea urchin roe, pig brains and mice livers. Three compounds that strongly resemble anandamide were found in dark chocolate which is why we may get pleasure from eating chocolate.

Extended information: Lipinski's rule of five

Lipinski's rule of five is a rule of thumb to evaluate drug likeness (qualitative concept used in drug design) or determine if a chemical compound with a certain pharmacological or biological activity has properties that would make it a likely orally active drug in humans. The rule was formulated by Christopher A. Lipinski in 1997, based on the observation that most orally administered drugs are relatively small and moderately lipophilic (tat loving) molecules. Orally active drugs must not break any more than one of the following criteria.

> Molecular mass less than 500 amu.
>
> Not more than 5 hydrogen bond donors.
>
> Not more than 10 hydrogen bond acceptors.
>
> An octanol/water partition coefficient log p not greater than 5.

. . . .

6.2 How drugs work

Most medicines work by binding to receptors within the body which are normally protein molecules. Some of these protein receptor molecules are embedded in the membrane that surrounds cells while others are located within the cell. Those located inside cells are enzymes, globular proteins which act as catalysts in the body. These receptors are referred to as catalytic receptors and their binding sites are often called active sites. Molecules which bind to these active sites are normally called substrate molecules and the catalytic receptors catalyse a reaction on the substrate molecules. It is critical that the shape of the binding molecule complements the shape of the receptor site. Once the molecule fits into the receptor site several forces including hydrogen bonds, London dispersion and ionic bonds hold it in place to prevent it floating away.

Protein receptors that are embedded in membranes have hollows or clefts in their surface into which small biologically active molecules can fit and bind. The binding of the active molecule triggers a response.

6.3 Pharmacological activity

Analgesic medicines (pain relief) including morphine and codeine have a common structural fragment (known as a pharmacophore). These drugs are highly addictive and are strictly controlled legally.

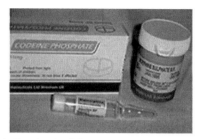

The part of the molecule shown in bold is the structural fragment common to many analgesic medicines. When the shape, dimensions and functional groups present have been identified, this can be cropped, added to and manipulated by chemists to produce compounds which are still analgesic but less addictive. The shape and dimensions of this structural fragment complement the receptor usually through functional groups that are positioned by the shape and size of the structures to allow interaction and binding of the medicine to the receptor.

benzomorphan   etorphine

Etorphine (a synthetic opiate) is almost 100 times as potent as morphine and is used in veterinary medicine to immobilize large animals.

Calculations associated with pharmaceuticals

Percentage solution by mass is the mass of solute made up to 100 cm^3 of solution.
% mass = (mass of solute ÷ mass of solution) × 100

Example 100 g of salt solution has 30 g of salt in it.

% mass = (30 ÷ 100) × 100 = 30%

. . . .

Percentage solution by volume is the number of cm^3 of solute made up to 100 cm^3 of solution.
% volume = (volume of solute ÷ volume of solution) × 100

Examples

1. Wine with 12 ml alcohol per 100 ml of solution.

% volume = $(12 \div 100) \times 100 = 12\%$

. . . .

2. Making 1000 ml of 10% ethylene glycol solution.

$10\% = 0.1 \times 1000 = 100$ ml of ethylene glycol.
The other 900 ml would be made up using water (although this would need to be topped up due to the two liquids not mixing to form exactly 1000 ml).

. . . .

In other calculations, you may come across the unit *parts per million (ppm)* which refers to 1 mg per kg.

6.4 Summary

Summary

You should now be able to state that:

- drugs are substances which alter the biochemical processes in the body;

- many drugs can be classified as agonists or as antagonists at receptors, according to whether they enhance or block the body's natural responses;

- drugs bind to receptors in the body so that each drug has a structural fragment which confers the pharmacological activity.

6.5 Resources

- LearnChemistry, Masterminding molecules: create a medicine (http://rsc.li/2aMhDhY)

- Anandamide: The molecule of extreme pleasure (http://bit.ly/2acWWe9)

- Anandamide (http://bit.ly/1EvLXX0)

- View 3D Molecular Structures (https://www.pymol.org/view)

6.6   End of topic test

Go online

End of Topic 6 test

The end of topic test for *Pharmaceutical chemistry.*

Look at this diagram of four chemical compounds.

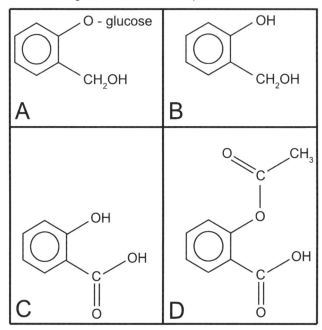

Q1:   Which of the above diagrams represents aspirin?

a) A
b) B
c) C
d) D

. . . .

Q2:   Which reaction type takes place when D is converted to C?

a) Esterification
b) Reduction
c) Hydrolysis
d) Oxidation

. . . .

Q3: Which reaction type takes place when C is converted to B?

a) Esterification
b) Reduction
c) Hydrolysis
d) Oxidation

....

Q4: Which is the molecular formula for the compound in box D?

a) $C_9H_8O_4$
b) $C_7H_4O_4$
c) $C_9H_4O_4$
d) $C_9H_{10}O_4$

....

These structures show isoprenaline and salbutamol. Both can be used in the treatment of asthma.

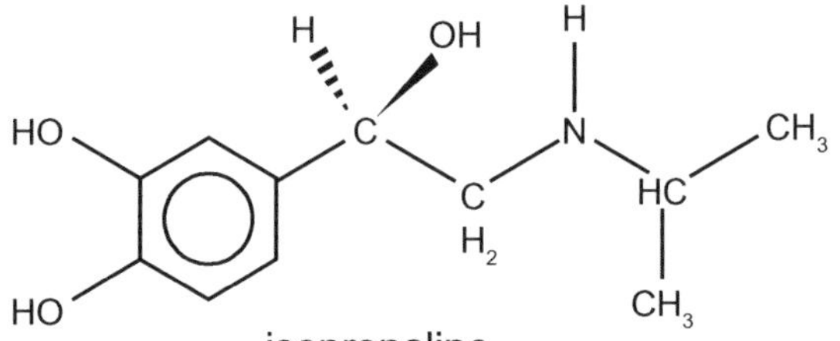

isoprenaline

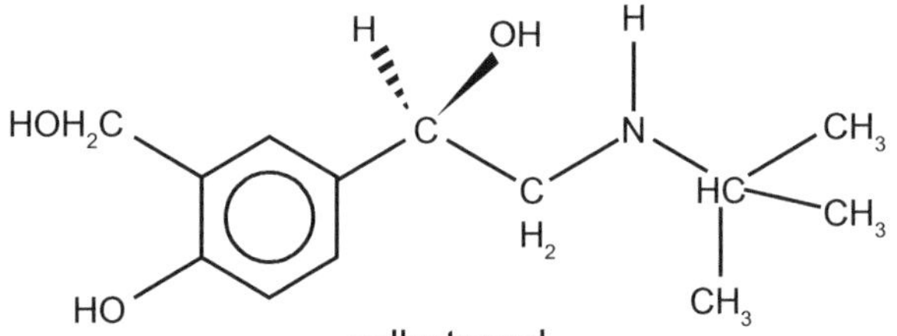

salbutamol

Q5: Which of these functional groups is not present in either molecule?

a) Amine
b) Alkene
c) Phenyl
d) Hydroxyl

....

Q6: Which of these classes of alcohol is present in salbutamol but not in isoprenaline?

a) Primary
b) Secondary
c) Tertiary
d) Aromatic

. . . .

This structure is part of both adrenaline and salbutamol and is the fragment upon which their activity depends.

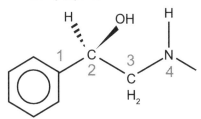

Q7: Salbutamol is a chiral molecule. Which atom indicates the chiral centre?

a) 1
b) 2
c) 3
d) 4

. . . .

Look at these three analgesics. Aspirin and paracetamol act on pain by occupying the enzyme site needed to make the prostaglandins which are produced in response to injury. Phenacetin is metabolised in the body to paracetamol.

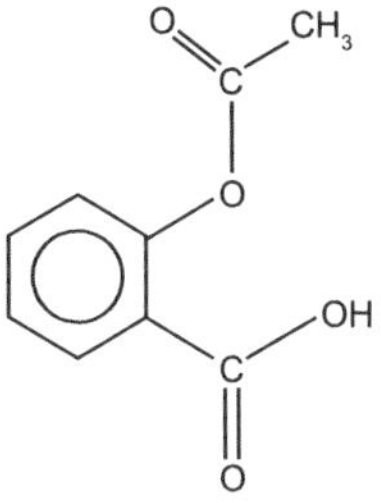

A. aspirin 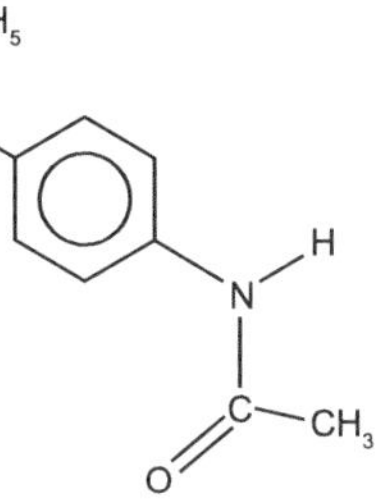   B. phenacetin 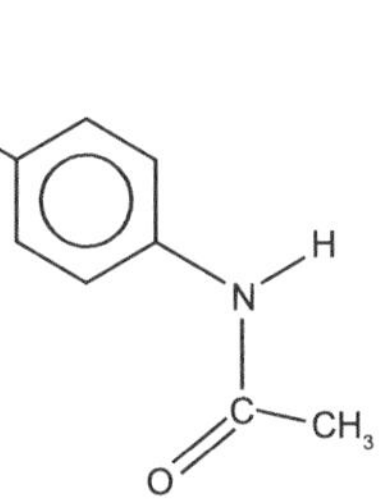   C. paracetamol

Q8: Which of these is an ether?

a) A
b) B
c) C

....

Q9: Which of these is a phenol?

a) A
b) B
c) C

....

Q10: Which of these is a carboxylic acid?

a) A
b) B
c) C

....

Q11: What name is used to describe a medicine which acts to block the body's natural responses?

....

Q12: The metabolism of phenacetin also produces ethanal. Which additional element would have to be added to phenacetin to give paracetamol?

....

Q13: Salbutamol is a **chiral** molecule. One of the **enantiomers** is 68 times more active than the other. Explain fully why one form of salbutamol is much more active than the other. Use the emboldened words in your answer.

....

....

Topic 7

End of Unit 2 test (Unit 2)

Go online

End of Unit 2 test

Q1:   Propene molecules contain:

a) sp^2 hybridised carbon atoms but no sp^3 hybridised carbon atoms.
b)   sigma bonds but no pi bonds.
c) sp^2 hybridised carbon atoms and sp^3 hybridised carbon atoms.
d)   pi bonds but no sigma bonds.

. . . .

Q2:   Which is the correct description of the numbers of sigma and pi bonds in buta-1,3-diene in the table below?

	Number of sigma bonds	Number of pi bonds
A	7	4
B	9	2
C	7	2
D	2	7

a)   A
b)   B
c)   C
d)   D

. . . .

Q3:   The electronic structure of a carbon atom ($1s^2$ $2s^2$ $2p^2$) suggests that carbon has two unpaired electrons, however, in alkanes, all the carbon atoms form four bonds. Explain this observation.

. . . .

Q4:   Which one of these is the skeletal structure of C_4H_{10}?

A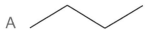

B

C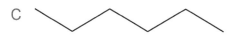

D

a) A
b) B
c) C
d) D

....

Q5: What is the molecular formula of the following skeletal structure? Also name the compound.

....

Q6: Draw the full structure of the following skeletal structure and name the compound.

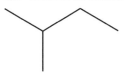

....

Q7: Two optical isomers were made. The isomers could be distinguished using what test?

a) Solubility
b) Effect on plane of polarised light
c) Melting point and boiling points
d) Ease of dehydration

....

Q8: Which of the following isomeric alcohols would exhibit optical isomerism?

a) $CH_3CH_2CH_2CH_2CH_2OH$
b) $CH_3CH_2CH_2CH(OH)CH_3$
c) $CH_3CH_2CH(OH)CH_2CH_3$
d) $(CH_3)_2C(OH)CH_2CH_3$

....

Q9: One of the optical isomers of lactic acid has the following structure:

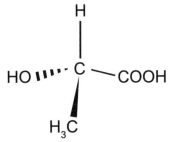

....

Explain why lactic acid can exphibit optical isomerism. Then draw a diagram to show the other optical isomer of lactic acid.

. . . .

Q10: Which of the following pair of compounds react to form the ester methyl ethanoate?

a) CH_3Cl and CH_3COOH
b) CH_3OH and CH_3COCl
c) CH_3CH_2OH and $HCOOH$
d) CH_3COCl and CH_3Cl

. . . .

Q11:

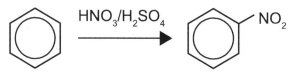

$$HNO_3/H_2SO_4$$

This reaction can be described as:

a) electrophilic addition.
b) electrophilic substitution.
c) nucleophilic addition.
d) nucleophilic substitution.

. . . .

Lactic acid (2-hydroxypropanoic acid) can be prepared by the following 2 step process:

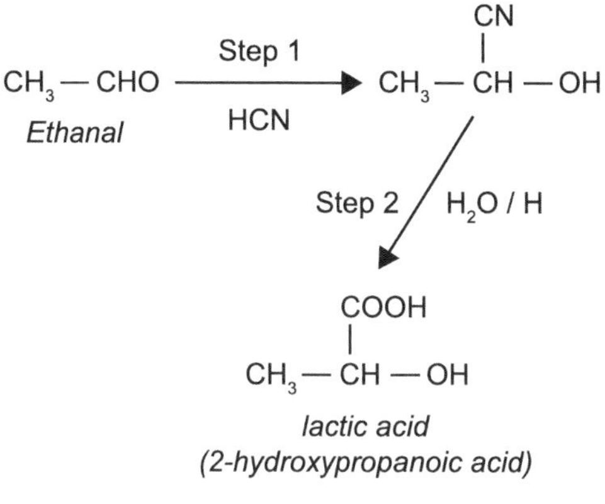

```
                        CN
           Step 1       |
CH₃—CHO  ─────────→  CH₃—CH—OH
Ethanal     HCN                 /
                               /
                  Step 2  /  H₂O / H
                         ↓
                       COOH
                        |
                  CH₃—CH—OH
                    lactic acid
              (2-hydroxypropanoic acid)
```

Q12: Name the type of reaction taking place in Step 2.

. . . .

Q13: Give the systematic name of the final product if butanone had been used instead of ethanal.

. . . .

Butan-2-ol can be prepared from but-2-ene by the following route:

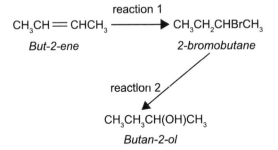

Q14: In reaction 1, but-2-ene undergoes electrophilic addition with hydrogen bromide. Draw diagrams to outline the mechanism for this reaction.

. . . .

Q15: Name a suitable reagent to carry out reaction 2.

. . . .

Q16: The diagram shows a simplified mass spectrum for butanone.

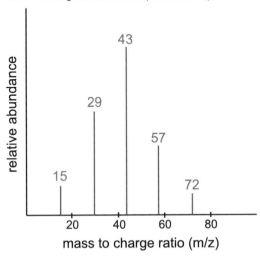

Which fragment could be responsible for the peak at m/z 43?

a) $[CH_3]^+$
b) $[C_2H_5CO]^+$
c) $[C_2H_5]^+$
d) $[CH_3CO]^+$

....

Q17: An organic compound Y has the following NMR spectrum:

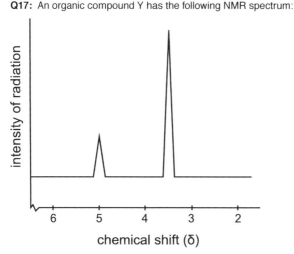

Y could be:

a) CH_3CH_3
b) CH_3NH_2
c) CH_3OH
d) CH_3OCH_3

....

Q18: What technique would be the best way of identifying the presence of a carbonyl group (C=O) in an organic compound?

a) Infra-red spectroscopy
b) Mass spectroscopy
c) Elemental microanalysis
d) Nuclear Magnetic Resonance spectroscopy

....

On subjecting a 0.4440 g sample of an organic compound, **X**, to elemental analysis, it was found to contain 0.2161 g of carbon and 0.0359 g of hydrogen. The only other element present in **X** was oxygen. The infra-red spectrum of **X** was also recorded.

Q19: Calculate the empirical formula of compound **X**.

....

Q20: What effect does the absorption of infra-red radiation have on the bonds in a molecule of compound **X**?

....

Q21: The drug *ranitidine* has proved effective in healing stomach ulcers. It binds to cells in the same way as histamine and so prevents histamine triggering the release of excess hydrochloric acid which is the cause of the ulcers. In healing ulcers ranitidine is acting as:

a) agonist.
b) antagonist.
c) receptor.
d) base.

....

Morphine, metazocene and pethidine are analgesic drugs.

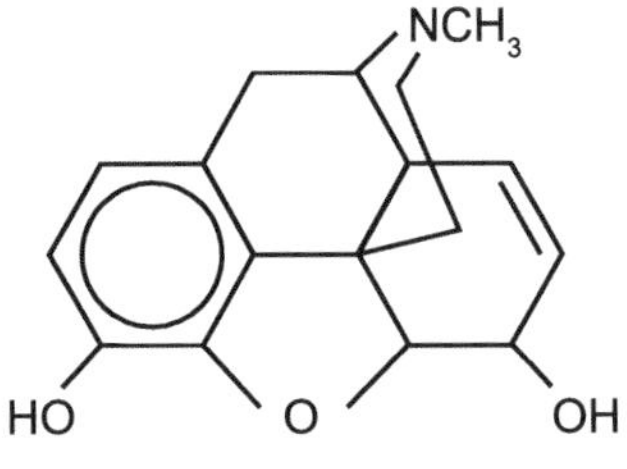

morphine

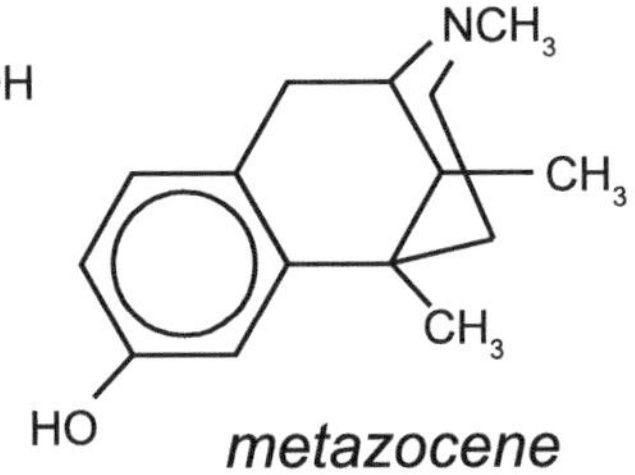

metazocene

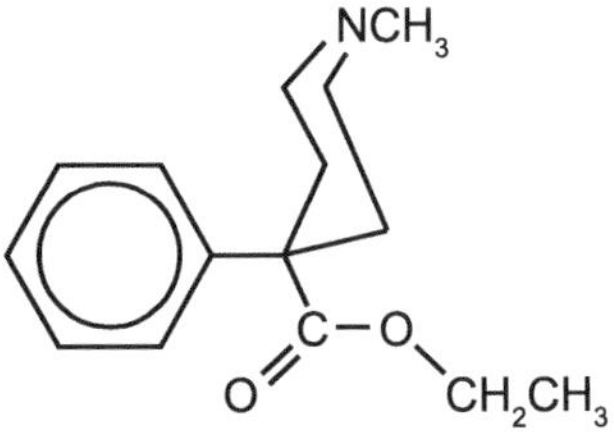

pethidine

Q22: Identify the pharmacophore in these drugs and draw its structural formula.

....

Q23: Morphine, metazocene and pethidine are agonists. What is meant by the term agonist?

. . . .

. . . .

Topic 8

Gravimetric analysis (Unit 3)

Contents

Prerequisite knowledge

Before you begin this topic, you should be able to work out the empirical formula from data given. (Advanced Higher Chemistry Unit 2)

Learning objectives

By the end of this topic, you should be able to:

- *use an accurate electronic balance;*
- *explain what is meant by "weighing by difference";*
- *explain what is meant by "heating to constant mass".*

8.1   Gravimetric analysis

Measurements for gravimetric analysis are masses in grams and are typically determined using a digital balance that is accurate to two, three or even four decimal places. Gravimetric analysis is a quantitative determination of an analyte based on the mass of a solid.

The mass of the analyte present in a substance is determined by changing that chemical substance into solid by precipitation (with an appropriate reagent) of known chemical composition and formula. The precipitate needs to be readily isolated, purified and weighed. Often it is easier to remove the analyte by evaporation.

The final product has to be dried completely which is done by "heating to constant mass". This involves heating the substance, allowing it to cool in a desiccator (dry atmosphere) and then reweighing it all in a crucible. During heating the crucible lid should be left partially off to allow the water to escape. A blue flame should be used for heating to avoid a build up of soot on the outside of the crucible which could affect the mass. Heating should be started off gently and then more strongly.

The mass of the crucible is measured before adding the substance and the final mass of the substance is determined by subtracting the mass of the crucible from the mass of the crucible and dried substance (weighing by difference). This process of heating, drying and weighing is repeated until a constant mass is obtained. This shows that all the water has been driven off.

8.2   Gravimetric analysis experiments

Experiment 1: Gravimetric analysis of water in hydrated barium chloride

This experiment determines the value of n in the formula $BaCl_2.nH_2O$ using gravimetric analysis through accurate weighing. The hydrated barium chloride is heated until all the water has been removed as in the equation:

$$BaCl_2.nH_2O \rightarrow BaCl_2 + nH_2O$$

From the masses measured in the experiment it is possible to calculate the relative number of moles of barium chloride and water and hence calculate the value of n (must be a whole number).

The experiment is carried out in a crucible which is heated first to remove any residual water, cooled in a desiccator and then weighed accurately. Approximately 2.5 g of hydrated barium chloride is then added to the crucible and again weighed accurately (crucible plus contents). The hydrated barium chloride is heated in the crucible using a blue Bunsen flame after which the crucible and contents are placed in a desiccator to cool (desiccator contains a drying agent which reduces the chances of any moisture from the air being absorbed by the sample). After cooling the crucible and contents are reweighed accurately.

The heating, cooling and reweighing is repeated until a constant mass is achieved and the assumption that all the water has been removed.

Example of results obtained

Mass of empty crucible = 32.67 g
Mass of crucible + hydrated barium chloride = 35.03 g
Mass of hydrated barium chloride = 35.03 - 32.67 = 2.36 g

Mass of crucible and anhydrous barium chloride = 34.69 g (heated to constant mass)
Mass of anhydrous barium chloride = 34.69 - 32.67 = 2.02 g

Mass of water removed = 2.36 - 2.02 = 0.34 g
Moles of water removed = 0.34 / 18 = 0.0189

Number of moles of barium chloride $BaCl_2$ = 2.02 / 208.3 = 0.00970

Ratio of moles = 0.00970:0.0189 = $BaCl_2:H_2O$ = 1:2

Formula of $BaCl_2.nH_2O$ is therefore $BaCl_2.2H_2O$ where n=2.

Experiment 2: Gravimetric determination of water in hydrated magnesium sulfate

As for the gravimetric determination of water in barium chloride the water in magnesium sulfate ($MgSO_4.nH_2O$) can also be determined using this technique. The same experimental procedure is used substituting hydrated barium chloride for hydrated magnesium sulfate.

Experiment 3: Determination of nickel using butanedioxime (dimethylglyoxime)

Dimethylgloxime is used as a chelating agent in the gravimetric analysis of nickel. It has the formula $CH_3C(NOH)C(NOH)CH_3$.

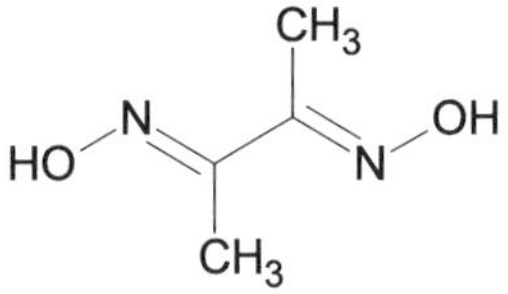

Dimethylglyoxime

$$2 \quad \text{[structure]} \quad + Ni^{2+} \quad \rightarrow \quad \text{[complex structure]} \quad 2H^+$$

Dimethylglyoxime   *Insoluble complex*

The nickel is precipitated as red nickel dimethylglyoxime by adding an alcoholic solution of dimethylglyoxime and then adding a slight excess of aqueous ammonia solution. When the pH is buffered in the range of 5 to 9, the formation of the red chelate occurs quantitatively in a solution. The chelation reaction occurs due to donation of the electron pairs on the four nitrogen atoms, not by electrons on the oxygen atoms. The reaction is performed in a solution buffered by either an ammonia or citrate buffer to prevent the pH of the solution from falling below 5.

The mass of the nickel is determined from the mass of the precipitate which is filtered, washed and dried to constant mass as in the other examples.

> Mass of crucible + lid = w1 (w=weight)
> Mass of crucible + lid + dried precipitate = w2
> Mass of precipitate = w2 - w1 = w3
>
> GFM of precipitate $Ni(C_4H_7O_2N_2)_2$ = 288.7 g
> GFM of nickel (Ni) = 58.7 g
>
> Mass of nickel = (w3 × 58.7) / 288.7

Go online

Gravimetric analysis

Q1: Describe the process of heating to constant mass.

. . . .

Q2: During the process of heating to constant mass a desiccator is used. Explain why a desiccator is used.

. . . .

Q3: Explain why gravimetric analysis is a suitable technique for determining nickel with dimethylglyoxime.

. . . .

Q4: When 0·968 g of an impure sample of nickel(II) sulfate, $NiSO_4.7H_2O$ was dissolved in water and reacted with dimethylglyoxime, 0·942 g of the red precipitate was formed.

Calculate the percentage of nickel in the impure sample.

. . . .

8.3 Summary

Summary

You should now be able to:

- use an accurate electronic balance;
- explain what is meant by "weighing by difference";
- explain what is meant by "heating to constant mass".

Topic 9

Volumetric analysis (Unit 3)

Contents

Prerequisite knowledge

Before you begin this topic, you should know:

- *about the key area of chemical analysis; (National 5 Chemistry Unit 3);*

- *how and when to use pipettes, burettes and standard/volumetric flasks (National 5 and Higher Chemistry Unit 3);*

- *about ligands/complexes and indicator/choice of indicator (Advanced Higher Chemistry Unit 1).*

Learning objectives

By the end of this topic you should:

- *be able to prepare a standard solution;*

- *understand that standard solutions can be prepared from primary standards;*

- *be able to state that a primary standard must have the following characteristics:*

    - *a high state of purity;*
    - *stability in air and water;*
    - *solubility;*
    - *reasonably high formula mass;*

- *understand the role of a control in experiments to validate a technique;*

- *understand the use of complexometric titrations in quantitative analysis of solutions containing a metal ion;*

- *understand the use of back titrations;*

- *understand how to use calculations associated with back titrations.*

9.1 Volumetric analysis

Volumetric analysis involves using a solution of known concentration (standard solution) in a quantitative reaction to determine the concentration of the other reactant. The procedure used to carry out volumetric analysis is titration whether in the form of standard, complexometric or back titrations.

Titrations involve measuring one solution quantitatively into a conical flask using a pipette. The other solution is added from a burette until a permanent colour change of an indicator is seen in the conical flask.

A "rough" titration is carried out first followed by more accurate titrations until concordant titre values are achieved (titre volumes added from burette should be ± 0.1 cm^3 of each other). The mean or average value of the concordant titres is used in calculations.

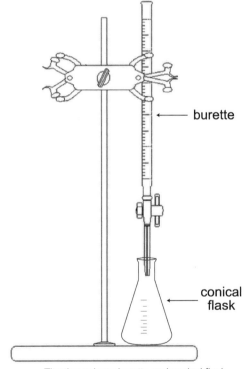

— burette

conical flask

Titration using a burette and conical flask

9.2   Preparation of a standard solution

A standard solution is one of which the concentration is known accurately and can be prepared directly from a primary standard.

A primary standard must have, at least, the following characteristics:

- a high state of purity;
- stability in air and water;
- solubility;
- reasonably high formula mass;

The standard solution is prepared as follows:

- calculate the mass of the primary standard required to make the concentration of solution required in the appropriate volume of solution;
- weigh out the primary standard as accurately as possible;
- dissolve the primary standard in a small volume of deionised water in a beaker;
- transfer the solution and all the rinsings into a standard flask;
- make the solution up to the mark with more deionised water;
- invert the stoppered standard flask several times to ensure thorough mixing.

A standard (volumetric) flask

Substances that are used as primary standards include:

- Oxalic acid ($H_2C_2O_4.2H_2O$);
- Sodium carbonate (Na_2CO_3);
- Potassium hydrogen phthalate ($KH(C_8H_4O_4)$);
- Potassium iodate (KIO_3);
- Potassium dichromate ($K_2Cr_2O_7$).

It is not possible to use substances such as sodium hydroxide (NaOH) as primary standards as they readily absorb water and carbon dioxide from the atmosphere.

> **Experiment: Preparation of 0.1 mol l^{-1} oxalic acid**
>
> Oxalic acid $H_2C_2O_4.2H_2O$ formula mass = 126 g
>
> In 1000 cm^3 (1 litre) you would need to weigh out **12.6 g** (0.1 × 126) which would make a solution with a concentration of 0.1 mol l^{-1}.
>
> In 250 cm^3 you would need to weigh out $n = c \times v$ **moles** of oxalic acid to make a 0.1 mol l^{-1} solution.
>
> n = 0.1 × 0.25 moles = 0.025 moles
>
> 0.025 moles of oxalic acid = **0.025 × 126 = 3.15 g**

Other possible experiments include:

a) Standardisation of approximately 0.1 mol l^{-1} NaOH

b) Determination of ethanoic acid content of vinegar

c) Preparation of a standard solution of 0.1 mol l^{-1} Na_2CO_3

d) Standardisation of approximately 0.1 mol l^{-1} HCl

e) Determination of purity of marble by back titration (see section 2.5)

Standardisation of approximately 0.1 mol l^{-1} NaOH/0.1 mol l^{-1} HCl

Standardisation is the process of determining the exact concentration of a solution. Titration is one type of analytical procedure often used in standardisation.

The point at which the reaction is complete in a titration is referred to as the endpoint. A chemical substance known as an indicator is used to indicate the endpoint. The indicator used in this experiment is phenolphthalein. Phenolphthalein, an organic compound, is colourless in acidic solution and pink in basic solution.

This experiment involves two separate acid-base standardisation procedures. In the first standardisation the concentration of a sodium hydroxide solution (NaOH) will be determined by titrating a sample of potassium acid phthalate (KHP; $HKC_8H_4O_4$) with the NaOH. This is the primary standard.

Experiment: first procedure

A 0.128 g sample of potassium acid phthalate (KHP, $HKC_8H_4O_4$) required 28.5 ml of NaOH solution to reach a phenolphthalein endpoint. Calculate the concentration of the NaOH.

$$HKC_8H_4O_4 \rightarrow NaKC_8H_4O_4 + H_2O$$

Moles of KHP (mass / gfm) = 0.128 / 204.1 = 6.271×10^{-4} mol

Moles of NaOH (KHP:NaOH = 1:1) = 6.271×10^{-4} mol

Concentration of NaOH (moles / volume in litres)
= 6.271×10^{-4} / 0.0285 = ***0.0220 mol l^{-1} NaOH***

In the second procedure the standardised NaOH will be used to determine the concentration of a hydrochloric solution (HCl).

Experiment: second procedure

A 20.00 ml sample of HCl was titrated with the NaOH solution from experiment 1. To reach the endpoint required 23.7 ml of the NaOH. Calculate the concentration of the HCl.

$$HCl + NaOH \rightarrow NaCl + H_2O$$

Moles of NaOH (concentration × volume) = $0.0237 \times 0.0220 = 5.214 \times 10^{-4}$ mol

Moles of HCl (HCl:NaOH = 1:1) = 5.214×10^{-4} mol

Concentration of HCl (moles / volumes in litres)
= 5.214×10^{-4} / 0.02000 = ***0.0261 mol l^{-1}***

9.3   Use of controls in chemical reactions

The use of a control in chemical reactions validates a technique and may consist of carrying out a determination on a solution of known concentration.

In the determination of the percentage of acetyl salicylic acid in commercial aspirin tablets a sample of pure aspirin (100% aspirin) would also be analysed to validate the techniques being used. In the determination of the vitamin C content in fruit juice a sample of ascorbic acid (pure vitamin C) would also be analysed again to validate the techniques being used in the determination and to give a referencing point on which to base all other results from impure samples.

9.4 Complexometric titrations

Complexometric titration is a form of volumetric analysis in which the formation of a coloured complex is used to indicate the end point of a titration. Complexometric titrations are particularly useful for the determination of a mixture of different metal ions in solution.

EDTA (ethylenediaminetetraacetic acid) is a hexadentate ligand and an important complexometric reagent used to determine the concentration of metal ions in solution forming an octahedral complex with the metal 2+ ion in a 1:1 ratio. In particular it can be used to determine the concentration of nickel ions in a nickel salt. To carry out metal ion titrations using EDTA, it is almost always necessary to use a complexometric indicator to determine when the end point has been reached.

Murexide is used and, compared to its colour when it is attached to the Ni^{2+} ions, is a different colour when free. Murexide is a suitable indicator as it binds less strongly to the Ni^{2+} ions than the EDTA does and so is no longer attached to the Ni^{2+} ions at the end point of the titration where the colour is changed from yellow to blue-purple. Titrations would be carried out until concordant results were obtained.

In this experiment ammonium chloride and ammonia solutions are used as a buffer to keep the pH constant as murexide is a pH dependent indicator.

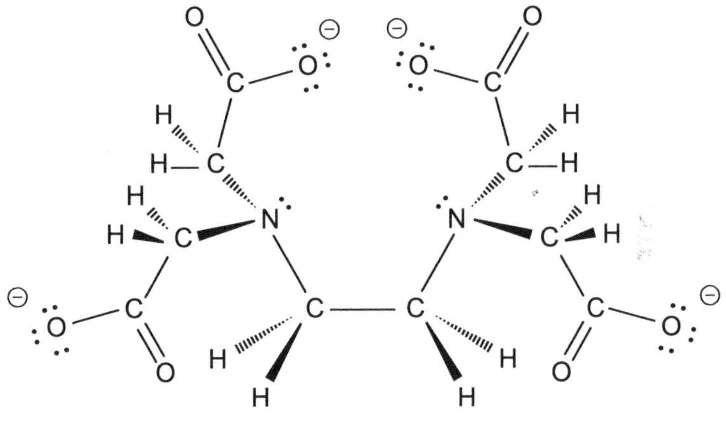

EDTA

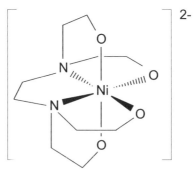

EDTA complexed with nickel

Worked example for Ni-EDTA titration

Approximately 2.6 g of hydrated nickel sulfate ($NiSO_4.6H_2O$) was weighed accurately and dissolved in a small volume of deionised water, transferred to a 100 cm^3 standard flask and made up to the mark with deionised water. 20cm3 samples of this solution were titrated with 0.112 mol l^{-1} EDTA solution using murexide as an indicator. The following results were obtained.

Mass of weighing boat + $NiSO_4.6H_2O$ = 4.076 g

Mass of weighing boat after transferring $NiSO_4.6H_2O$ = 1.472 g

Titre	Initial burette reading (cm³)	Initial burette reading (cm³)	Volume of EDTA added(cm³)
1	1.4	20.5	19.1
2	20.5	38.6	18.1
3	15.3	33.5	18.2

Calculate the % of nickel in the hydrated nickel sulfate.

Average titre volume = 18.15 cm^3

Number of moles of EDTA used = C × V = 0.112 × 0.01815 = 0.00203 mol

Since EDTA complexes with nickel ions in a 1:1 ratio, the number of moles of Ni^{2+} in 20 cm^3 = 0.00203 mol

Therefore, in 100 cm^3, there are 0.00203 × 5 = 0.0102 mol

Mass of nickel present = n × GFM = 0.0102 × 58.7 = 0.599 g

Mass of $NiSO_4.6H_2O$ = 4.076 - 1.472 = 2.604 g

% nickel = (0.599/2.604) × 100 = 23.0 %

9.5   Back titrations and associated calculations

A back titration is used to find the number of moles of a substance by reacting it with an excess volume of reactant of known concentration. The resulting mixture is then titrated to work out the number of moles of the reactant in excess. A back titration is useful when trying to work out the quantity of substance in an insoluble solid.

From the initial number of moles of that reactant the number of moles used in the reaction can be determined, making it possible to work back to calculate the initial number of moles of the substance under test.

An experiment which uses a back titration is in the determination of aspirin due to it being insoluble in water. A sample of aspirin of accurately known mass is treated with an excess of sodium hydroxide (the actual volume is known). The sodium hydroxide catalyses the hydrolysis of aspirin to ethanoic acid and salicylic acid and then neutralises these two acids.

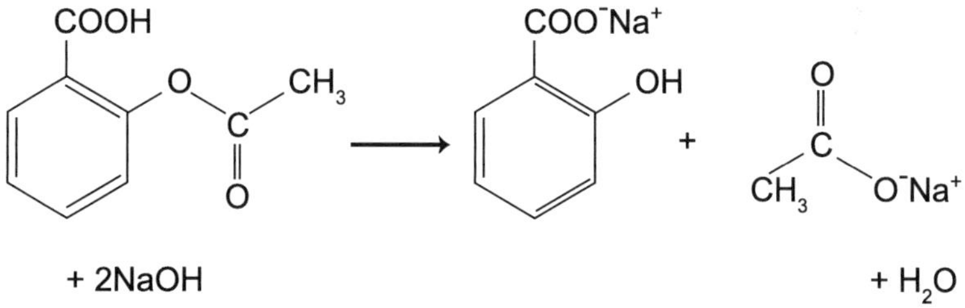

+ 2NaOH

Determination of aspirin

As an excess of sodium hydroxide is used the volume remaining is determined by titrating it against a standard solution of sulfuric acid. The difference between the initial and excess volumes of sodium hydroxide allows the mass of aspirin in the tablet to be determined.

The reaction taking place is:

$2NaOH + H_2SO_4 \rightarrow Na_2SO_4 + 2H_2O$

Determination of aspirin

Three aspirin tablets of approximately 1.5 g were added to a conical flask. Sodium hydroxide (25.0 cm^3 1.00 mol l^{-1}) was pipetted into the flask along with 25.0 cm^3 water. The resulting mixture was simmered gently on a hot plate for approximately 30 minutes and after cooling transferred along with rinsings to a 250 cm^3 standard flask. The solution was made up to the graduation mark with water, stoppered and inverted several times to ensure through mixing of the contents. 25.0 cm^3 was pipetted into a conical flask; phenolphthalein indicator added and titrated against sulfuric acid (0.05 mol l^{-1}). The end point of the titration was indicated by the colour change pink to colourless and they were repeated until concordant results were obtained (± 0.1 cm^3).

Titre	Starting volume cm^3	End volume cm^3	Titre volume cm^3
Rough	0.0	15.6	15.6
1	15.6	30.8	15.2
2	0.0	15.1	15.1

Titration results

Average titre volume = (15.1 + 15.2) / 2 = 15.15 cm^3

Number of moles of sulfuric acid used = 0.01515 × 0.05 = 7.575 x 10^{-4} mol

Number of moles of NaOH left in the 25.0 cm^3 hydrolysed solution = 2 × 7.575 × 10^{-4} = 1.515 × 10^{-3} mol

Number of moles of NaOH left in 250.0 cm^3 of the hydrolysed solution = 10 × 1.515 × 10^{-3} = 1.515 × 10^{-2} mol

Number of moles of NaOH added to aspirin initially = 0.025 × 1 = 2.5 x10^{-2} mol

Number of moles NaOH reacted with aspirin = 2.5 × 10^{-2} - 1.515 × 10^{-2} = 9.85 × 10^{-3} mol

2 moles of NaOH reacts with 1 mole aspirin therefore number of moles of aspirin in 3 tablets = 9.85 × 10^{-3} / 2 = 4.925 × 10^{-3} mol

Number of moles of aspirin in 1 tablet = 4.925 × 10^{-3} / 3 = 1.642 × 10^{-3} mol

Mass of aspirin in each tablet = n × gfm = 1.642 × 10^{-3} x 180 = 0.296 g = 296 mg

9.6 Summary

Summary

You should now:

- be able to prepare a standard solution;

- understand that standard solutions can be prepared from primary standards;

- be able to state that a primary standard must have the following characteristics:

  ○ a high state of purity;

  ○ stability in air and water;

  ○ solubility;

  ○ reasonably high formula mass;

- understand the role of a control in experiments to validate a technique;

- understand the use of complexometric titrations in quantitative analysis of solutions containing a metal ion;

- understand the use of back titrations;

- understand how to use calculations associated with back titrations.

9.7 Resources

- Determination of the molarity of an acid or base solution (http://bit.ly/N5tutz)

- Titration experiment (http://bit.ly/1ucRlHo)

9.8   End of topic test

Go online

End of Topic 2 test

Q1: The equation for the reaction between benzoic acid solution and sodium hydroxide solution is:

$$C_6H_5COOH(aq) + NaOH(aq) \rightarrow C_6H_5COONa(aq) + H_2O(l)$$

A student used a standard solution of 0.0563 mol l^{-1} benzoic acid to standardise 20.0 cm^3 of approximately 0.05 mol l^{-1} sodium hydroxide solution.

The results for the titration are given in the table.

	1st attempt	2nd attempt	3rd attempt
Final burette reading/cm^3	17.2	33.8	16.6
Initial burette reading/cm^3	0.0	17.2	0.1
Titre/cm^3	17.2	16.6	16.5

Calculate the accurate concentration of the sodium hydroxide solution (to four decimal places). *(3 marks)*

. . . .

Q2: Four aspirin tablets of approximately 2.0 g were added to a conical flask. Sodium hydroxide (25.0 cm^3 1.00 mol l^{-1}) was pipetted into the flask along with 25.0 cm^3 water. The resulting mixture was simmered gently on a hot plate for approximately 30 minutes and after cooling transferred along with rinsings to a 250 cm^3 standard flask. The solution was made up to the graduation mark with water, stoppered and inverted several times to ensure through mixing of the contents. 25.0 cm^3 was pipetted into a conical flask; phenolphthalein indicator added and titrated against sulfuric acid (0.02 mol l^{-1}). The end point of the titration was indicated by the colour change pink to colourless and they were repeated until concordant results were obtained ($\pm$0.1 cm^3).

Titre	Starting volume cm^3	End volume cm^3	Titre volume cm^3
Rough	0	14.9	14.9
1	14.9	30.8	15.9
2	0	15.8	15.8

Titration results

Calculate the mass of aspirin in each tablet in grams. *(3 marks)*

. . . .

Topic 10

Practical skills and techniques (Unit 3)

Contents

Learning objectives

By the end of this topic, you should be able to:

- *prepare standard solutions using accurate dilution technique;*

- *form and use calibration curves to determine an unknown concentration using solutions of appropriate concentration;*

- *gain knowledge of the appropriate use of distillation, reflux, vacuum filtration, recrystallisation and use of a separating funnel in the preparation and purification of an experimental product;*

- *gain knowledge of the appropriate uses of thin-layer chromatography, melting point and mixed melting point determination in evaluating the purity of experimental products;*

- *calculate Rf values from relevant data and their use in following the course of a reaction.*

10.1   Colorimetry and accurate dilution

Colorimetry uses the relationship between colour intensity of a solution and the concentration of the coloured species present to determine the concentration.   A colorimeter consists essentially of a light source, a coloured filter, a light detector and a recorder. The filter colour is chosen as the complementary colour to that of the solution resulting in maximum absorbance. The light passes through the filter and then through the coloured solution and the difference in absorbance between the coloured solution and water is detected and noted as an absorbance value.

Colorimetry is a useful method of analysis for coloured compounds since at low concentrations, coloured solutions obey the Beer-Lambert law. The Beer-Lambert law is usually quoted as $A = \varepsilon cl$, where

- ε = molar absorptivity (this is a constant dependent on the solution at a particular wavelength)

- c = the concentration of solution

- l = path length (the distance the light travels through the sample solution).

The absorbance of the solution is calculated using $A = \log (I_0/I)$ where I_0 is the initial intensity of light before passing through the solution and I is the intensity of light after passing through the solution.

A calibration curve must be prepared using solutions of known concentrations (standard solutions). The unknown concentration of the solution is determined from its absorbance and by referring to the calibration curve. The straight line section of the calibration graph should cover the dilution range likely to be used in the determination.

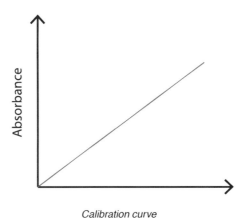

Calibration curve

Making of standard solutions:

a) Dissolve the accurately weighed substance in a small volume of water in a beaker.

b) Pour the solution into a standard flask.

c) Rinse the beaker with distilled water and add the rinsings to the standard flask.

d) Add distilled water to the standard flask making the volume up to the mark.

e) Stopper the flask and invert several times to ensure thorough mixing.

Experiments including the determination of manganese in steel (see Advanced Higher Chemistry. A Practical Guide Support Materials (http://bit.ly/2aOKlJP)) and determination of nickel could both be used to gain skills in this practical technique.

Document

- Determination of manganese in steel by permanganate colorimetry (http://bit.ly/1l 6fPTY)

10.2 Distillation

In the modern organic chemistry laboratory, distillation is a powerful tool, both for the identification and the purification of organic compounds. The boiling point of a compound which can be determined by distillation is well-defined and thus is one of the physical properties of a compound by which it is identified.

Distillation is used to purify a compound by separating it from a non-volatile or less-volatile material.

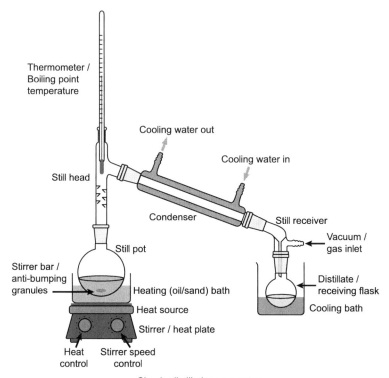

Simple distillation apparatus

Steam distillation is a special type of distillation for temperature sensitive materials like natural aromatic compounds.

Fractional distillation is the separation of a mixture into its component parts, or fractions, such as in separating chemical compounds by their boiling point by heating them to a temperature at which one or more fractions of the compound will vaporise. Used in the oil industry in the separation of crude oil into fractions that are used mostly as fuels.

Possible experiments for distillation include:

- Preparation of benzoic acid by hydrolysis of ethyl benzoate;

- Preparation of ethyl ethanoate;

- Preparation of cyclohexene from cyclohexanol.

All of these experiments are shown in the practical guide for revised Advanced Higher Chemistry (see Advanced Higher Chemistry. A Practical Guide Support Materials (http://bit.ly/2aOKIJP) .

Videos

- Steam distillation (http://bit.ly/1neY5aK)
- Fractional distillation (http://bit.ly/1ZY6rl6)
- Preparation of ethyl ethanoate (http://bit.ly/1PbLXO5)

10.3 Refluxing

Refluxing is a technique used to apply heat energy to a chemical reaction mixture over an extended period of time. The liquid reaction mixture is placed in a round-bottomed flask along with anti-bumping granules with a condenser connected at the top. The flask is heated vigorously over the course of the chemical reaction; any vapours given off are immediately returned to the reaction vessel as liquids when they reach the condenser.

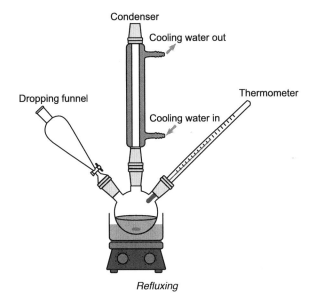

Refluxing

Possible experiments to use this technique include:

- Preparation of benzoic acid by hydrolysis of ethyl benzoate;
- Preparation of ethyl ethanoate.

Both of these experiments are shown in the practical guide for revised Advanced Higher Chemistry (see Advanced Higher Chemistry. A Practical Guide Support Materials (http://bit.ly/2aOKIJP) .

10.4   Vacuum filtration

Vacuum filtration can be carried out using a Buchner, Hirsch or sintered glass funnel. These methods are carried out under reduced pressure and provide a faster means of separating the precipitate from the filtrate. The choice of filtering medium depends on the quantity and nature of the precipitate.

Possible experiments include:

- Preparation of potassium trioxolatoferrate(III);

- Preparation of aspirin;

- Preparation of benzoic acid by hydrolysis of ethyl benzoate;

- Identification by derivative formation.

The first three of these experiments are shown in the practical guide for revised Advanced Higher Chemistry (see Advanced Higher Chemistry. A Practical Guide Support Materials (http://bit.ly/2aOKlJP) .

Hirsch funnel

Buchner flask

Sinter funnel

Buchner funnel

Hirsch funnels are generally used for smaller quantities of material. On top of the funnel part of both the Hirsch and Buchner funnels there is a cylinder with a fritted glass disc/perforated plate separating it from the funnel. A funnel with a fritted glass disc can be used immediately. For a funnel with a perforated plate, filtration material in the form of filter paper is placed on the plate, and the filter paper is moistened with a solvent to prevent initial leakage. The liquid to be filtered is poured into the cylinder and drawn through the perforated plate/fritted glass disc by vacuum suction.

Hot filtration method is mainly used to separate solids from a hot solution. This is done in order to prevent crystal formation in the filter funnel and other apparatuses that comes in contact with the solution. As a result, the apparatus and the solution used are heated in order to prevent the rapid decrease in temperature which in turn, would lead to the crystallization of the solids in the funnel and hinder the filtration process.

Videos

- How to use a Buchner funnel - http://bit.ly/1TV62wb

- Vacuum filtration - http://bit.ly/22ZuoLm

10.5   Recrystallisation

Recrystallisation is a laboratory technique used to purify solids, based upon solubility. The solvent for recrystallisation must be carefully selected such that the impure compound is insoluble at lower temperatures, yet completely soluble at higher temperatures. The impure compound is dissolved gently in the minimum volume of hot solvent then filtered to remove insoluble impurities. The filtrate is allowed to cool slowly to force crystallisation. The more soluble impurities are left behind in the solvent.

Recrystallisation can also be achieved where the pure compound is soluble in the hot solvent but not the cold solvent and the impurities are soluble in the hot and cold solvent. The impure compound is dissolved in the hot solvent. As it cools down the impurities stay in the solvent and can be filtered off.

Selection of a suitable solvent is crucial to achieve a satisfactory recrystallisation. Neither the compound nor the impurities should react with the solvent. Other factors to consider will be the solubility of the compound and the impurities in the solvent, and the boiling point of the solvent (if the boiling point is too low, it will not be possible to heat the solvent to dissolve the impurities as it will evaporate off too easily).

Possible experiments include:

- Preparation of benzoic acid by hydrolysis of ethyl benzoate;

- Preparation of potassium trioxalatoferrate(III);

- Preparation of acetylsalicylic acid.

All of these experiments are shown in the practical guide for revised Advanced Higher Chemistry (see Advanced Higher Chemistry. A Practical Guide Support Materials (http://bit.ly/2aOKIJP) .

Video

- Recrystallization - http://bit.ly/1Or2vCi

10.6   Use of a separating funnel

Solvent extraction can be an application of the partition of a solute between two liquids. It is based on the relative solubility of a compound in two different immiscible liquids, usually water and an organic solvent. The partition coefficient is expressed as the concentration of a solute in the organic layer over that in the aqueous layer. The two solvents form two separate layers in the separating funnel and the lower layer is run off into one container and the upper layer is poured out into another container. The quantity of solute extracted depends on the partition coefficient and on the number of times that the process is repeated.

Again, selection of a suitable solvent is crucial to achieve a high concentration of the solute after the extraction. The solute should be more soluble in the solvent than in the aqueous solution and it should not react with the solvent.

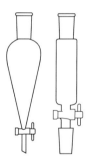

Separating funnels

See also the section in Unit 1 of Advanced Higher Chemistry on 'Phase equilibria'.

Possible experiments include:

- Preparation of ethyl ethanoate;
- Extraction of caffeine from tea.

The largest risk when using a separating funnel is that of pressure build-up. Pressure accumulates during mixing if gas evolving reactions occur. This problem can be easily handled by simply opening the stopper at the top of the funnel routinely while mixing. This should be done with the top of the funnel pointed away from the body. When shaking, hold the stopper in place or it can become dislodged causing the liquids will spill. To account for this, simply hold the stopper in place with one hand.

Videos

- Liquid-liquid extraction using a separating funnel - http://bit.ly/1OPnUZo
- Industrial extraction of caffeine - http://bit.ly/22ZuvXg

10.7 Thin-layer chromatography (TLC)

Thin-layer chromatography can be used to assess product purity. Instead of chromatography paper, thin-layer chromatography (TLC) uses a fine film of silica or aluminium oxide spread over glass or plastic.

When setting up a TLC plate, a pencil line is drawn, usually 1 cm up from the bottom of the plate. The sample solution is then spotted on several times to get a concentrated spot on the plate. The plate is then placed in a suitable solvent (which acts as the mobile phase) making sure that the solvent is below the level of the spot. The solvent then travels up the plate and the components of the sample separate out according to their relative attractions to the stationary phase on the plate and the mobile phase (see the section on Chromatography in the 'Chemical equilibrium' topic in Unit 1).

Retardation factor (R_f) values (distance travelled by compound/distance travelled by solvent) can be calculated and under similar conditions a compound will always have the same R_f value within experimental error. R_f values can be used to follow the course of a reaction by spotting a TLC plate with the authentic product, the authentic reactant and the reaction mixture at that point. Comparison of the R_f values of all three spots will allow the progress of the reaction to be determined.

Since a pure substance will show up as only one spot on the developed chromatogram, TLC can be used to assess the purity of a product prepared in the lab.

Possible experiments include:

- Preparation of aspirin;
- Hydrolysis of ethyl benzoate.

Chromatogram example

An organic chemist is trying to synthesis a fragrance compound by reacting compound X and compound Y together. After an hour of reaction, a sample is removed and compared with pure samples of X and Y using TLC. If the experiment produces pure product, then only one spot will be seen for the product sample. If impurities are present they will show up as additional spots on the chromatogram.

In the following chromatogram you can see that there are two spots produced from the product sample which have travelled the same distance as the spots for both compounds X and Y. This means the product sample only contains both compounds X and Y suggesting the chemical reaction has not taken place between X and Y.

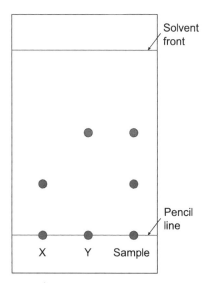

In the following chromatogram you can again see two spots in the product sample. This time, one of the spots has travelled further and is not in the same position as the spots for either compound X and compound Y. This suggests this spot is for the fragrance compound. The product sample still contains a spot at the same distance for compound X suggesting this is still present as an impurity.

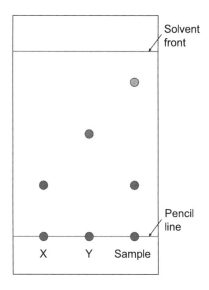

In the chromatogram below only one spot is seen for the product sample which has travelled a different distance from the spots for both compound X and compound Y. This suggests the fragrance compound is pure.

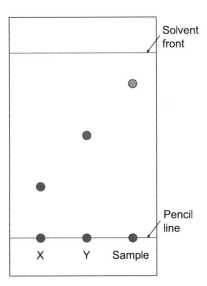

To check if a sample is pure, a co-spot can also be used. A 50:50 mixture of the sample and a known pure sample can be spotted on. If the sample is pure, then only one spot will be present on the chromatogram. If the sample is impure then more than one spot will be observed.

Measuring R_f values

R_f is measured by calculating the distance travelled by the substance divided by the distance travelled by the solvent front, as illustrated in the diagram below.

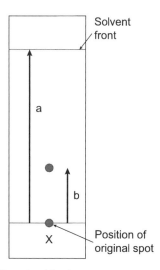

a = distance travelled by the solvent front

b = distance travelled by the substance

$R_f = b/a$

Videos

- Thin-layer chromatography (TLC) - http://bit.ly/1niBs5q

- Thin-layer chromatography (TLC) theory - http://bit.ly/1mW7x2n

10.8   Melting point and mixed melting point

The melting point of an organic compound is one of several physical properties by which it can be identified. A crystalline substance has a sharp melting point falling within a very small temperature range.

Determination of the melting point can also give an indication of the purity of an organic compound, as the presence of impurities lowers the melting point and extends its melting temperature range. Since impurities lower the melting point, the technique of mixed melting point determination can be used as a means of identifying the product of a reaction.

In this case the product can be mixed with a pure sample of the substance and a melting point taken of the mixture. If the melting point range is lowered and widened, it means that the two are different compounds. If the melting point stays the same it means that the two compounds are likely identical.

The melting point of an organic solid can be determined by introducing a tiny amount into a small capillary tube and placing inside the melting point apparatus. A window in the apparatus allows you to determine when the sample melts in the capillary tube and the temperature can be determined from the thermometer. Pure samples usually have sharp melting points, for example 149.5 - 150 °C or 189 - 190 °C; impure samples of the same compounds melt at lower temperatures and over a wider range, for example 145 - 148 °C or 186 - 189 °C.

Possible experiments include:

- Preparation of benzoic acid by hydrolysis of ethyl benzoate;

- Identification by derivative formation (see *PPA 2 Unit 3 Traditional Advanced Higher Chemistry*);

- Preparation of aspirin.

'Preparation of benzoic acid' and 'Preparation of aspirin' are shown in the practical guide for revised Advanced Higher Chemistry (see Advanced Higher Chemistry. A Practical Guide Support Materials (http://bit.ly/2aOKIJP) .

10.9   Summary

Summary

You should now be able to:

- prepare standard solutions using accurate dilution technique;

- form and use calibration curves to determine an unknown concentration using solutions of appropriate concentration;

- gain knowledge of the appropriate use of distillation, reflux, vacuum filtration, recrystallisation and use of a separating funnel in the preparation and purification of an experimental product;

- gain knowledge of the appropriate uses of thin-layer chromatography, melting point and mixed melting point determination in evaluating the purity of experimental products;

- calculate Rf values from relevant data and their use in following the course of a reaction.

10.10   Resources

- Determination of manganese in steel by permanganate colorimetry (http://bit.ly/1l 6fPTY)

- Fractional distillation (http://bit.ly/1ZY6rl6)

- Steam distillation (http://bit.ly/1neY5aK)

- Advanced Higher Chemistry - A Practical guide - Support Materials (http://www.e ducationscotland.gov.uk/resources/nq/c/nqresource_tcm4723691.asp)

- Preparation of ethyl ethanoate (http://bit.ly/1PbLXO5)

- LearnChemistry, Aspirin screen experiment (http://rsc.li/1EZAhew)

- How to use a Buchner funnel (http://bit.ly/1TV62wb)

- Vacuum filtration (http://bit.ly/22ZuoLm)

- Recrystallization (http://bit.ly/1Or2vCi)

- Using a separating funnel (http://bit.ly/1OPnUZo)

- Extraction of caffeine from coffee (http://bit.ly/1Kdgiun)

- Thin-layer chromatography (TLC) (http://bit.ly/1niBs5q)

- Thin-layer chromatography (TLC) theory (http://bit.ly/1mW7x2n)

10.11   End of topic test

End of Topic 3 test

Q1: Small amounts of manganese are added to aluminium in the making of drink cans to prevent corrosion. Using colorimetry the concentration of manganese in the alloy can be determined through the conversion of manganese to permanganate ions.

Go online

Describe how this is done. *(3 marks)*

....

Q2: A student was measuring the percentage of calcium carbonate in different types of egg shell. The egg shells were ground and approximately 0.5 g were weighed accurately of each.

What is meant by weighing accurately approximately 0.5 g? *(1 mark)*

....

Q3: The above egg shells were placed in a beaker and 30.0 cm^3 of 0.1 mol l^{-1} hydrochloric acid was added. Once the reaction was complete, the solution was made up to 250 cm^3 in a standard flask.

Describe the steps to prepare the 250 cm^3 solution. *(2 marks)*

....

10.0 cm^3 aliquots of the solution were titrated against 0.1 mol l^{-1} standardised sodium hydroxide solution using phenolphthalein as an indicator until concordant results were obtained.

Q4: What do we mean by concordant results? *(1 mark)*

....

Q5: Why did the sodium hydroxide solution have to be standardised? *(1 mark)*

....

Benzocaine (see structure diagram below) is used to relieve pain and itching caused by conditions such as sunburn, insect bites or stings.

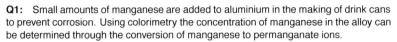

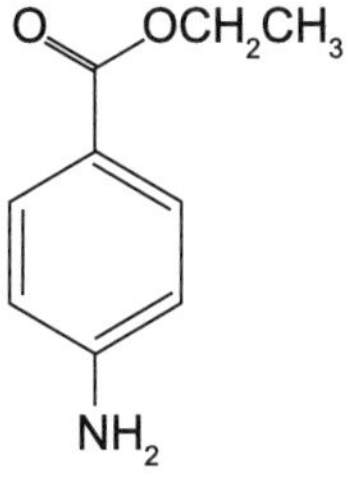

A student was carrying out a project to synthesise benzocaine. Part of the procedure to isolate the synthesised benzocaine is given below:

1. Add 20 cm^3 of diethyl ether to the reaction mixture and pour into a separating funnel.

2. Add 20 cm^3 of distilled water to the separating funnel.

3. Stopper the funnel, invert and gently shake.

4. Allow the aqueous layer to settle to the bottom.

Q6: Name the technique used above. *(1 mark)*
....

Q7: Outline the next steps required to obtain a maximum yield of benzocaine. *(3 marks))*
....

Q8: State two properties the solvent must have to be appropriate. *(2 marks)*
....

Q9: Suggest a second technique that could be used to purify a solid sample of benzocaine. *(1 mark)*
....

Thin layer chromatography (TLC) was used to help confirm the identity of the product. A small volume of solvent was used to dissolve a sample of product which was then spotted onto a TLC plate. The plate was allowed to develop (see diagram below).

Q10: Name the substance spotted at A on the TLC plate. *(1 mark)*

....

Q11: How pure is the student's product? *(1 mark)*

....

....

Topic 11

Stoichiometric calculations (Unit 3)

Contents

Learning objectives

By the end of this topic, you should be able to:

- *carry out calculations from balanced equations (including multi-step reactions, reactant excess, and empirical formulae from given data);*

- *calculate the theoretical and actual yield from experimental data and provide explanations comparing the two.*

11.1   Stoichiometry

Stoichiometry is the study of quantitative relationships involved in chemical reactions. The ability to balance and interpret equations is required to enable calculations to be carried out using any of the techniques covered in Topic 3 of this unit.

11.2   Stoichiometric calculation: preparation of benzoic acid by hydrolysis of ethylbenzoate

This practical preparation of benzoic acid can be used to demonstrate the types of stoichiometric calculations that are required at Advanced Higher Chemistry. From the measurements obtained during the practical and from the balanced equation the percentage yield of benzoic acid can be calculated and compared with the theoretical yield which is calculated from the balanced equation.

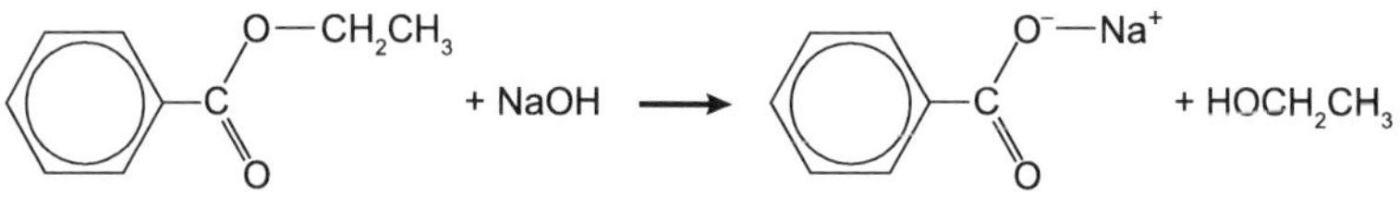

Ethyl benzoate   **Sodium benzoate**

The first step involves the alkaline hydrolysis of the ester ethyl benzoate to form sodium benzoate.

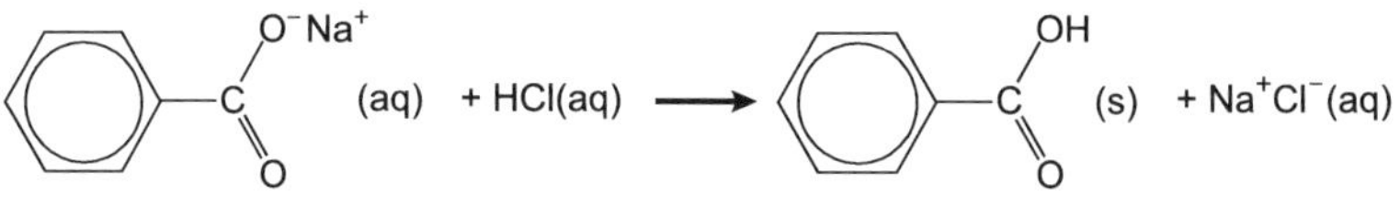

Sodium benzoate   **Benzoic acid**

The second step involves adding hydrochloric acid (strong acid) to precipitate out benzoic acid (weak acid). The crude benzoic acid is separated by filtration and recrystallised from water. The pure sample of benzoic acid is weighed and the percentage yield calculated. The melting point of the pure benzoic acid is determined.

The percentage yield of a product can be lowered due to many reasons including mass transfer of reactants/products, mechanical losses, purification of the product, side reactions, position of the equilibrium and the purity of the reactants.

Examples of results obtained from an experiment:

Mass	Weight
Mass of round bottomed flask	40.25 g
Mass of round bottomed flask + ethyl benzoate	45.61 g
Mass of ethyl benzoate	5.36 g
Mass of clock glass	9.62 g
Mass of clock glass + benzoic acid	12.86 g
Mass of benzoic acid	3.24 g

From the two-step balanced equations for this reaction it can be noted that 1 mole of ethyl benzoate is required to produce 1 mole of benzoic acid. From this we can work out the theoretical yield of benzoic acid.

$$1 \text{ mole of ethyl benzoate} \rightarrow 1 \text{ mole of benzoic acid}$$
$$150.0 \text{ g} \rightarrow 122.0 \text{ g}$$
$$5.36 \text{ g (quantity used in experiment)} \rightarrow (5.36 \times 122.0) / 150.0 = \textbf{4.36 g}$$

4.36 g is the theoretical yield of benzoic acid that will be produced from 5.36 g of ethyl benzoate according to the molar ratios of the balanced equations.

As reactions are rarely 100% effective due to the reasons given above we can calculate the actual percentage yield of benzoic acid by comparing the theoretical yield to the mass actually produced in the experiment.

Theoretical yield = 4.36 g
Actual yield = 3.24 g

Percentage (%) yield
= Actual yield/Theoretical yield × 100
= 3.24/4.36 × 100
= **74.3%**

Other experiments including the preparation of aspirin and potassium trioxalatoferrate (III) can be used in the same way to calculate the percentage yield of product by comparing the theoretical yield calculated from the balanced equations to the actual yield which is weighed at the end of the experiment.

11.3   Reactant in excess

In the reaction between calcium carbonate and hydrochloric acid they react according to the following equation:

$$CaCO_3 + 2HCl \rightarrow CaCl_2 + CO_2 + H_2O$$

The calcium carbonate and hydrochloric acid react in a molar ratio of 1:2 where 1 mole of calcium carbonate ($CaCO_3$) reacts with 2 moles of hydrochloric acid (HCl). In this case both reactants would be used up completely in the reaction. If there was a shortage of calcium carbonate for example then the reaction would stop when it runs out and there would be hydrochloric acid left over. In other words the hydrochloric acid would be in excess.

Excess calculation example

Calculate which reactant is in excess, when 10 g of calcium carbonate reacts with 50 cm^3 of 2 mol l^{-1} hydrochloric acid. This calculation involves the reactants only.

Firstly work out the number of moles of each reactant involved.

Calcium carbonate - number of moles
= mass/gfm
= 10/100
= **0.1 moles**

Hydrochloric acid - number of moles
= concentration $\times$ volume (litres)
= 2 $\times$ (50/1000)
= 2 $\times$ 0.05
= **0.1 moles**

From the balanced equation 1 mole of calcium carbonate reacts with 2 moles of hydrochloric acid and therefore 0.1 moles of calcium carbonate would react with 0.2 moles of hydrochloric acid.

From the above calculation of the number of moles of hydrochloric acid you can see that there is only 0.1 moles available which is insufficient and therefore the calcium carbonate is in excess and some will be left over at the end of the reaction which will terminate once the hydrochloric acid has been used up.

Further examples of excess calculations can be found below.

Excess calculations

Go online

Q1:   Calculate which reactant is in excess when 0.654 g of zinc reacts with 20 cm^3 of hydrochloric acid, concentration 0.5 mol l^{-1}, using the following equation:

$$Zn + 2HCl \rightarrow ZnCl_2 + H_2$$

a) Zinc
b) Hydrochloric acid

. . . .

Q2: Calculate which reactant is in excess when 25 cm^3 of 1 mol l^{-1} sulfuric acid is mixed with 25 cm^3 of 1 mol l^{-1} sodium hydroxide, using the following equation:

$$H_2SO_4 + 2NaOH \rightarrow Na_2SO_4 + 2H_2O$$

a) Sulfuric acid
b) Sodium hydroxide

. . . .

11.4   Stoichiometric calculation: nickel(II) ions and dimethylglyoxime

Nickel(II) ions and dimethylglyoxime

Nickel(II) ions react quantitatively with dimethylglyoxime ($C_4H_8O_2N_2$) forming a complex which precipitates out as a red solid. The gram formula mass of the complex is 288.7. The equation for the reaction and the structure of the complex are shown below.

Go online

$$Ni^{2+} + 2C_4H_8O_2N_2 \rightarrow Ni(C_4H_7O_2N_2)_2 + 2H^+$$

Equation for nickel(II) ions and dimethylglyoxime reaction

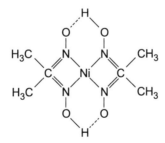

Structure of nickel(II) dimethylglyoxime complex

Q3: What is the coordination number of nickel in the complex?

. . . .

11.5 Summary

Summary

You should now be able to:

- carry out calculations from balanced equations, including multi-step reactions, reactant excess, and empirical formulae from given data (empirical formulae are covered in Unit 2 Topic 5);

- calculate the theoretical and actual yield from experimental data and provide explanations comparing the two.

11.6 End of topic test

End of Topic 4 test

Go online

Aspirin tablet back titration

The acetylsalicylic acid ($C_9H_8O_4$) content of an aspirin tablet was determined using a back titration.

Five aspirin tablets were crushed and added to 25.0 cm^3 of 1.00 mol l^{-1} sodium hydroxide solution. The mixture was heated and allowed to simmer for 30 minutes.

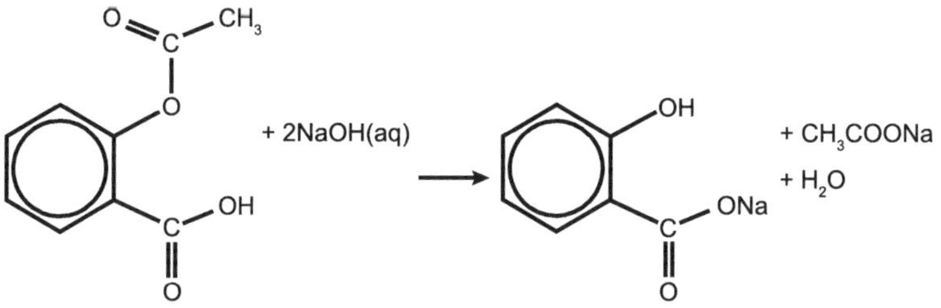

The resulting mixture (see diagram above) was allowed to cool before being transferred to a 250 cm^3 standard flask and made up to the mark with deionised water. 25.0 cm^3 samples of this solution were titrated with 0.050 mol l^{-1} sulfuric acid, using the following equation:

$$H_2SO_4 + 2NaOH \rightarrow Na_2SO_4 + 2H_2O$$

Results of the titrations are included in the table below.

	Rough titration	First titration	Second titration
Initial burette reading/cm^3	0.0	9.0	17.7
Final burette reading/cm^3	9.0	17.7	26.3
Volume used/cm^3	9.0	8.7	8.6

Q4: Which of the following indicators would be the best one to use in the back titration? *(1 mark)*

a) screened methyl orange
b) bromocresol green
c) thymolphthalein
d) phenolphthalein

. . . .

Q5:   Calculate the number of moles of sulfuric acid in the average titre. *(2 marks)*

....

Q6:   Calculate the number of moles of sodium hydroxide in the standard flask. *(2 marks)*

....

Q7:   Calculate the number of moles of sodium hydroxide which reacted with the acetylsalicylic acid. *(2 marks)*

....

Q8:   The mass of one mole of acetylsalicylic acid is 180 g. Use this along with the answer to the previous question to calculate the mass of acetylsalicylic acid in one aspirin tablet. *(3 marks)*

....

Q9:   A compound contains only carbon, hydrogen and sulfur. Complete combustion of the compound gives 3.52 g of carbon dioxide, 2.16 g of water and 2.56 g of sulfur dioxide.

Show by calculation that the empirical formula is C_2H_6S. *(4 marks)*

....

Benzoic acid is prepared from ethyl benzoate by refluxing with sodium hydroxide solution (the diagram below shows a shortened version of the equation provided to carry out the calculation).

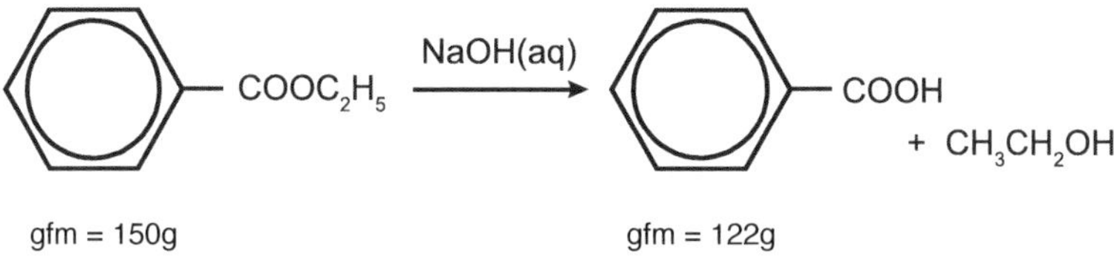

gfm = 150g   gfm = 122g

A yield of 73.2% of benzoic acid was obtained from 5.64 g of ethyl benzoate.

Q10:   Calculate the mass of benzoic acid produced. *(2 marks)*

....

....

Topic 12

End of Unit 3 test (Unit 3)

Go online

End of Unit 3 test

Q1: 2.58 g of hydrated barium chloride $BaCl_2.nH_2O$, was heated until constant mass was achieved, leaving 2.22 g of anhydrous barium chloride $BaCl_2$.

Calculate the value of **n** in the formula $BaCl_2.nH_2O$. Give your answer to the nearest whole number. *(2 marks)*

....

Q2: Oxalic acid, a dicarboxylic acid has the formula $H_2C_2O_4$ and can be made by reacting calcium oxalate with sulfuric acid.

$$H_2SO_4(aq) + CaC_2O_4(s) + xH_2O \rightarrow CaSO_4.xH_2O(s) + H_2C_2O_4(aq)$$

4.94 g of $CaSO_4.xH_2O$ was dehydrated to form 3.89 g of $CaSO_4$.

Calculate the value of **x** in the above formula. Give your answer to the nearest whole number. *(2 marks)*

....

Q3: 10.0 cm^3 of a liquid drain cleaner containing sodium hydroxide was diluted to 250 cm^3 in a standard flask. 25.0 cm^3 samples of this diluted solution were pipetted into a conical flask and titrated against 0.220 mol l^{-1} sulfuric acid solution. The average of the concordant titres was 17.8 cm^3.

Calculate the mass of sodium hydroxide in 1 litre of drain cleaner. *(3 marks)*

....

Sulfa drugs are compounds with antibiotic properties. Sulfa drugs can be prepared from a solid compound called sulfanilamide which is prepared in a six stage synthesis. The equation for the final step in the synthesis is shown below.

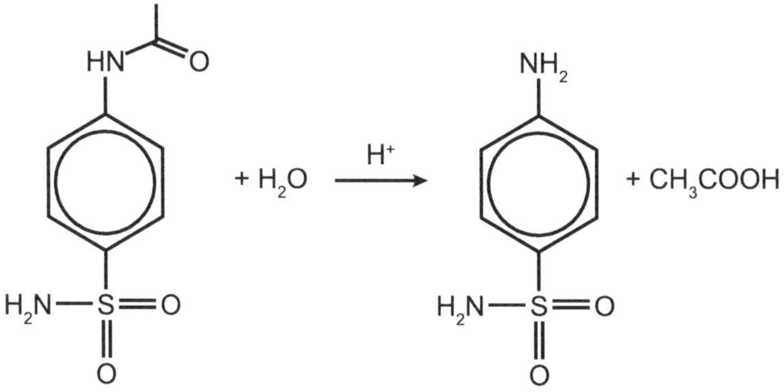

Q4: What type of reaction is this? *(1 mark)*

....

Q5: The culfanilamide is separated from the reaction mixture and recrystallised from boiling water. Why is the recrystallisation necessary? *(1 mark)*

. . . .

Q6: Calculate the percentage yield of sulfanilamide if 4.282 g of 4-acetamidobenzenesulfonamide produced 2.237 g of sulfanilamide. *(2 marks)*

. . . .

Q7: Describe how a mixed melting point experiment would be carried out and the result used to confirm the product was pure. *(1 mark)*

. . . .

Q8: Suggest another analytical technique which could be used to indicate whether the final sample is pure. *(1 mark)*

. . . .

The formula of potassium hydrogen oxalate can be written as $K_xH_y(C_2O_4)_z$.

In an experiment to determine x, y and z, 4.49 g of this compound was dissolved in water and the solution made up to one litre. 20.0 cm^3 of the solution was pipetted into a conical flask and then titrated with 0.0200 mol l^{-1} acidified potassium permanganate at 60 °C. Average titre volume was 16.5 cm^3.

Equation for the reaction taking place is:

$$5C_2O_4^{2-} + 16H^+ + 2MnO_4^- \rightarrow 2Mn^{2+} + 10CO_2 + 8H_2O$$

Q9: What colour change would indicate the end of the titration? *(1 mark)*

. . . .

Q10: Calculate the number of moles of oxalate ions, $C_2O_4^{2-}$, in 20.0 cm^3 of the solution. *(2 marks)*

. . . .

Q11: Calculate the mass of oxalate ions in 1 litre of the solution. *(1 mark)*

. . . .

Q12: Using another analytical procedure, 4.49 g of the potassium hydrogen oxalate was found to contain 0.060 g of hydrogen. Use this information from this and part c to calculate the mass of potassium in this sample. *(1 mark)*

. . . .

Q13: Calculate the value of x, y, and z. *(2 marks)*

. . . .

. . . .

Glossary

Agonist

a drug which enhances the body's natural response or mimics the natural response of the body

Antagonist

a drug which blocks the natural response of the body

Aromatic

aromatic compounds contain a benzene ring in their structure

Carbocation

an ion with a positively charged carbon atom

Chiral

A chiral molecule is one which has a non-superimposable mirror image, i.e. optical isomers exist. All chiral molecules have a chiral carbon atom, i.e. a carbon atom with four different atoms or groups bonded to it

Chromophore

the group of atoms within a molecule which is responsible for the absorption of light in the visible spectrum. Molecules containing a chromophore are coloured

Cis

cis molecules have two of the same atom or group on the same side of a carbon-to-carbon double bond.

Conjugated

conjugated systems contain delocalised electrons spread over a few atoms. This is often through alternating single and double bonds

Electrophile

a species which is attracted to an electron rich site; electrophiles are electron deficient and have a positive charge

Electrophilic addition

addition across a carbon-to-carbon double bond

Electrophilic substitution

substitution of a hydrogen atom on a benzene ring for an electrophile

Enantiomers

optical isomers (non-superimposable mirror images) are known as enantiomers.

HOMO

highest occupied molecular orbital

LUMO

lowest unoccupied molecular orbital

Nucleophile

a species which is attracted to a positive charge; nucleophiles are electron rich species, i.e. they have a negative charge or lone pairs of electrons

Nucleophilic substitution

a reaction in which one atom or group of atoms is substituted by a nucleophile

Parent ion

a molecular ion produced during electron bombardment in mass spectrometry. In a mass spectrum the peak with the largest m/z value represents the parent ion and therefore the molecular mass of the molecule

Racemic mixture

a racemic mixture contains equal concentrations of both optical isomers. Racemic mixtures have no effect on plane polarised light

S_N1

nucleophilic substitution, 1st order. Tertiary halogenoalkanes are likely to take part in S_N1 reactions, although the kinetics of these reactions can only be determined through experimental results and not by the structure alone

S_N2

nucleophilic substitution, 2nd order. Primary halogenoalkanes are likely to take part in S_N2 reactions, although the kinetics of these reactions can only be determined through experimental results and not by the structure alone

Trans

trans molecules have two of the same atom or group on opposite sides of a carbon-to-carbon double bond

Answers to questions and activities

1 Molecular orbitals (Unit 2)

Bonding in hydrocarbons (page 7)

Q1: c) sp

Q2: c) There are two π bonds and three σ bonds.

Q3: linear

End of Topic 1 test (page 10)

Q4: d) 8 sigma bonds, 1 pi bond

Q5: The mixing of one s and three p orbitals.

Q6: d) 8

Q7: a) sp^2 hybridised carbon atoms but no sp^3 hybridised carbon atoms.

Q8: Electronegativity difference between Ti and Cl is 1.5 (value of 3.0 for chlorine and 1.5 for titanium), therefore the bonding is probably polar covalent and the bonding orbitals will lie around the chlorine atom due to it having the higher electronegativity.

3 Stereochemistry (Unit ?)

Optical isomers of alanine (page 24)

Q1: a) Structure 1

Q2: b) non-superimposable

Q3: b) asymmetric

Q4: b) chiral

End of Topic 3 test (page 28)

Q5: d) C_2F_4

Q6: d)

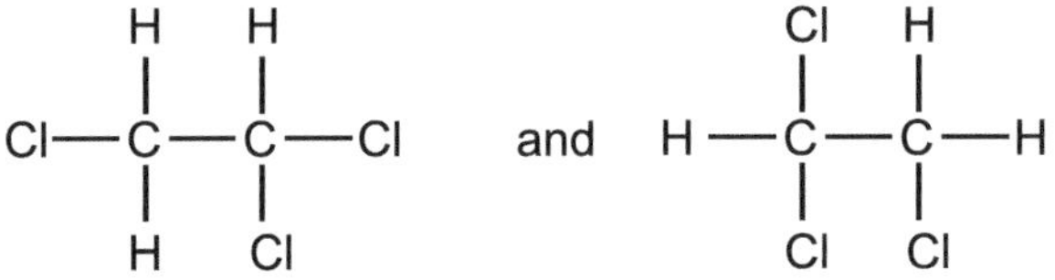

$$Cl-\underset{\underset{H}{|}}{\overset{\overset{H}{|}}{C}}-\underset{\underset{Cl}{|}}{\overset{\overset{H}{|}}{C}}-Cl \quad \text{and} \quad H-\underset{\underset{Cl}{|}}{\overset{\overset{Cl}{|}}{C}}-\underset{\underset{Cl}{|}}{\overset{\overset{H}{|}}{C}}-H$$

Q7: c) Butan-2-ol

Q8: b) 1-chloroethanol

Q9: c) C

Q10: c) A and C

Q11: a) A

Q12: d) D

Q13: d) D

Q14: a) A

Q15: c) sample tube of chiral molecules.

Q16: c) Y

Q17: c) Y

Q18: b) Two

Q19: Optical isomers are also called enantiomers.

Q20: b) Geometric isomers

Q21: a) Structural isomers

Q22: c) A and B

Q23: c)

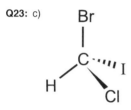

Q24: A and D

4 Synthesis (Unit 2)

The reaction of chlorine with benzene (page 63)

Q1: c) Electrophilic substitution

Q2: b) Chlorobenzene

Q3: d) Heterolytic

Q4: carbocation

Di- and trinitrobenzene structures (page 65)

Q5: d) Electrophile

Q6: b) Carbocation

Q7: The intermediate is stabilised by delocalisation of the positive charge around the benzene ring.

Q8: c) 1,3-dinitrobenzene

Benzene (page 66)

Q9: electrophile

Q10: substitution

Methylbenzene (page 67)

Q11: catalyst

Q12: c) heterolytic

Q13: chloroethane

Extra practice questions (page 68)

Q14:

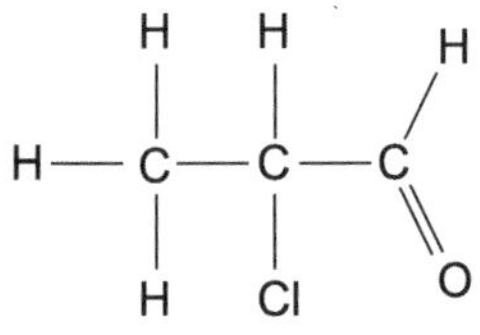

Q15: Lithium aluminium hybride/LiAlH$_4$
Sodium borohybride/sodium
tetrahydroborate/NaBH$_4$

Q16: Electrophilic substitution or alkylation

Q17: Light/UV radiation

Q18:

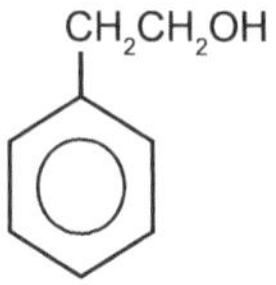

CH_2CH_2OH

Q19: (Base-induced) elimination

Q20:

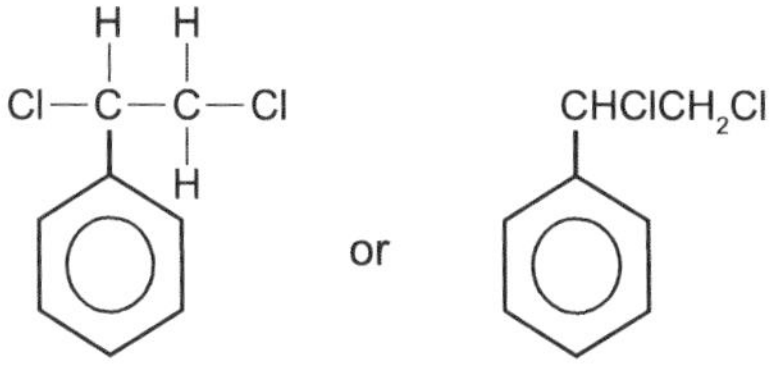

or

Q21: 2-hydroxypropanoic acid

Q22: Carbon atom 2 because 4 different groups are attached.

Q23: KCN or NaCN or HCN

Q24: Hydrolysis, acid hydrolysis

Q25: But-2-ene has two different groups attached to each of the carbon atoms of the double bond.

Q26:

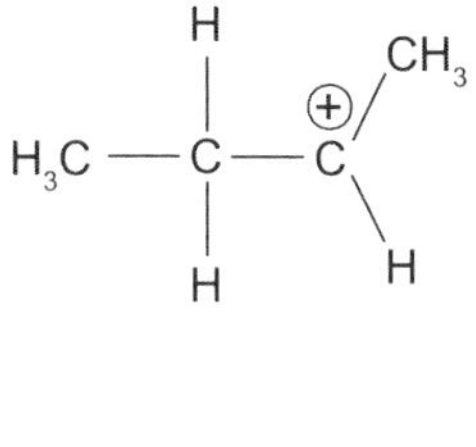

Q27: Potassium (or sodium) hydroxide solution

Q28: Aluminium chloride

Q29:

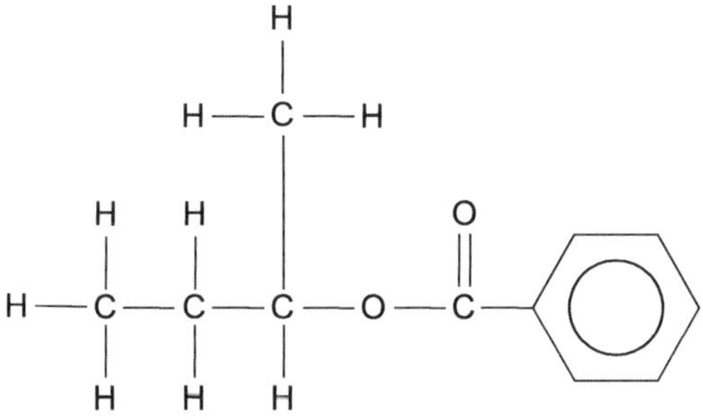

End of Topic 4 test (page 73)

Q30: a) ethene acting as an nucleophile and Br⁻ acting as an nucleophile.

Q31: Propane

Q32: 1,2-dibromopropane

Q33: b) Secondary

Q34: 1-ethoxy-2-methylpropane

Q35: Secondary

Q36: b) 2-methylbut-1-ene

Q37: An acid chloride

Q38: The reaction is too slow.

Q39: c) 3

Q40: c) The ethanoate ion is more stable than the ethoxide ion due to electron delocalisation.

Q41: a) $MgCO_3$

Q42: They are both salts

Q43: An amide

Q44: Propan-1-ol

Q45: Propyl propanoate

Q46: d) resists addition reactions.

Q47: b)

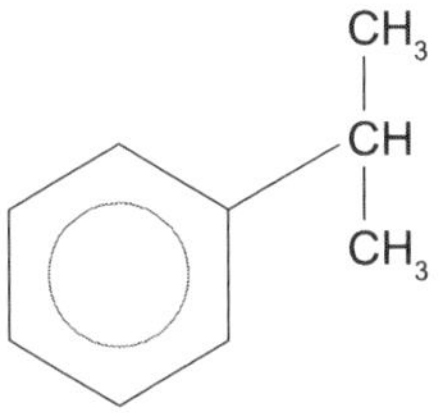

Q48: 1-bromo-3-methylbenzene

Q49: 3-chloro-4-methylphenol

Q50: 1,2-dimethylbenzene, 1,4-dimethylbenzene, ethylbenzene are all possible aromatic isomers. There are some elaborate isomers which are not aromatic and you may like to while away a few minutes finding some!

Q51: Alkylation and sulfonation

Q52: d) Nitration

Q53: b) 2

Q54: a) Primary

Q55: c) 2-aminobutane

Q56: Ethylamine and nitric acid (hydrogen nitrate) in the gas state give ethylammonium nitrate as small white crystals.

Q57: Ethylamine behaves as a proton acceptor (a base) and is neutralised by the acidic nitric acid.The reaction is a neutralisation.

Q58: The hydrogen of the nitric acid is attracted to the ethylamine which is basic. The hydrogen is therefore acting as an electrophile.

Q59: The lone pair on the ethylamine is attracted to the hydrogen of the nitric acid and the hydrogen of the nitric acid is therefore acting as an electrophile, the nitrogen a nucleophile.

Q60: The longer the alkyl chain in an amine, the lower the solubility in water. There is only one primary amine with a shorter alkyl chain than ethylamine, and that is methylamine.

5 Experimental determination of structure (Unit 2)

Calculate the empirical formula (page 84)

Q1: C_2H_6O

Questions about fragmentation (page 89)

Q2:

butan-1-ol	$CH_3CH_2CH_2CH_2OH$

butan-2-ol

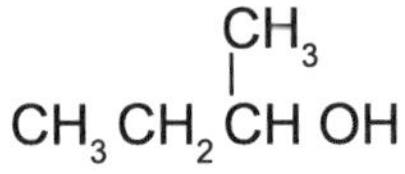

$$CH_3CH_2\overset{\overset{\displaystyle CH_3}{\mid}}{CH}OH$$

2-methylpropan-2-ol

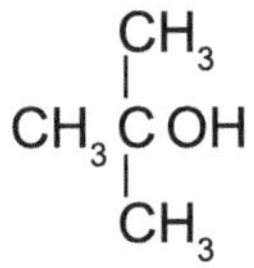

$$CH_3\overset{\overset{\displaystyle CH_3}{\mid}}{\underset{\underset{\displaystyle CH_3}{\mid}}{C}}OH$$

2-methylpropan-1-ol

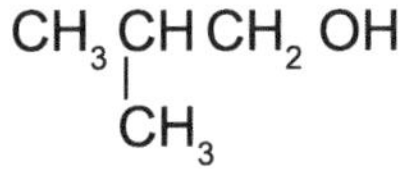

$$CH_3\overset{}{\underset{\underset{\displaystyle CH_3}{\mid}}{CH}}CH_2OH$$

methoxopropane

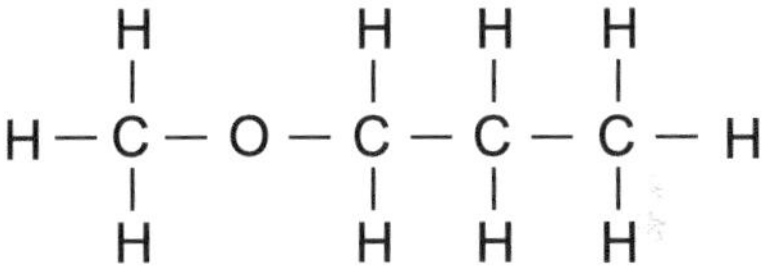

$$H-\overset{\overset{\displaystyle H}{\mid}}{\underset{\underset{\displaystyle H}{\mid}}{C}}-O-\overset{\overset{\displaystyle H}{\mid}}{\underset{\underset{\displaystyle H}{\mid}}{C}}-\overset{\overset{\displaystyle H}{\mid}}{\underset{\underset{\displaystyle H}{\mid}}{C}}-\overset{\overset{\displaystyle H}{\mid}}{\underset{\underset{\displaystyle H}{\mid}}{C}}-H$$

ethoxyethane

$$H-\overset{\overset{\displaystyle H}{\mid}}{\underset{\underset{\displaystyle H}{\mid}}{C}}-\overset{\overset{\displaystyle H}{\mid}}{\underset{\underset{\displaystyle H}{\mid}}{C}}-O-\overset{\overset{\displaystyle H}{\mid}}{\underset{\underset{\displaystyle H}{\mid}}{C}}-\overset{\overset{\displaystyle H}{\mid}}{\underset{\underset{\displaystyle H}{\mid}}{C}}-H$$

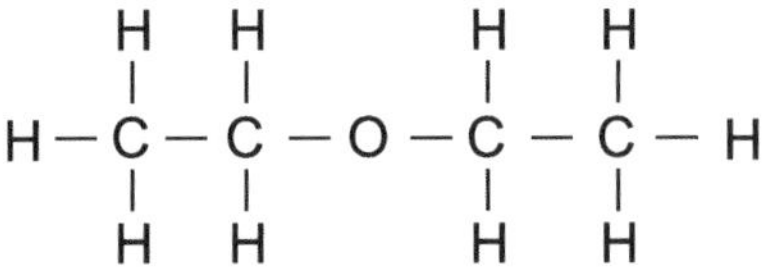

Q3: 74

Q4: CH_3

Q5: 2-methylpropan-2-ol has three methyl groups attatched to the carbon with the OH group attatched, and is likely most easily to lose one of them. Butan-2-ol also has a single CH_3 attatched to the carbon with the OH on, so would be expected to have a reduced m/z 59 ion. (Possibly spectrum Y)

Q6: Loss of the CH_3CH_2 group from the carbon with the OH, in a similar manner to loss of the CH_3 above, will leave a fragment of this m/z value, with structure $[HC(CH_3)OH]^+$. This suggests further that spectrum Y is butan-2-ol.

Q7: $[CH_3CH_2CH_2]^+$ with m/z 43 and $[CH_2OH]^+$ with m/z 31.

Q8: m/z 31 is the base peak in spectrum Z, and m/z 43 is also a major peak. In other spectra they are less prominent. Since these two fragments are easily produced from the structure of butan-1-ol, spectrum Z applies to this.

Q9: Mass spectrum X is for 2-methylpropan-2-ol (tertiary alcohol). The peak at m/z 59 is due to the $[C(CH_3)_2OH]^+$ fragment. A $[CH_3]^+$ fragment adjacent to the C-OH has been lost from the molecule to give this fragment.
Mass spectrum Y is for butan-2-ol (secondary alcohol). The peak at m/z 45 is due to the $[CH_3CHOH]^+$ fragment and the peak at m/z 59 is due to the $[C_2H_5CHOH]^+$ fragment.
Mass spectrum Z is for butan-1-ol. A $[CH_3CH_2CH_2]^+$ fragment has been lost to give the peak at m/z 31 which is due to the $[CH_2OH]^+$ fragment.

Questions on bond vibration (page 92)

Q10: a) HCl

Q11: c) HI

Q12: a) HCl

Functional group identification (page 95)

Q13: b) Y

Q14: b) Y

Q15: aldehyde

Q16: c) Benzonitrile (C_6H_5CN)

Determining structure using NMR (page 99)

Q17: 2

Q18: 5:3

Q19: d) Ar**H**

Q20: c) ArC**H**$_3$

Q21: Toluene, $C_6H_5CH_3$, with 5 aromatic Hs and 3 methyl Hs.

Q22: c) 3

Q23: b) 4:5:6

Q24: For the 6 CH_3, δ range 0.8 - 1.3 (actually found at 1.2);
for the 4 CH_2 - N, δ range 2.5 - 3.0 (actually found at 3.5);
for the 5 aromatic H, δ range 6.5 - 8.3 (actually found at 7.2).

Q25:

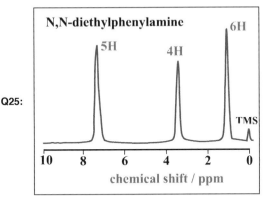

NMR - Practice (page 103)

Q26: The triplet at 1.058 is due to a CH_3 group adjacent to a CH_2 group. The singlet at 2.139 is due to a CH_3 group which is not adjacent to any other atoms with hydrogen atoms attached. The quartet at 2.449 is due to a CH_2CO group adjacent to a CH_3 group. This is also consistent with the ratio of the peak areas of 3:2:3. Therefore the molecule must contain two different CH_3 groups and a CH_2CO. This leads to the following structure...

$$CH_3CH_2CCH_3$$
$$\underset{O}{\overset{\|}{}}$$

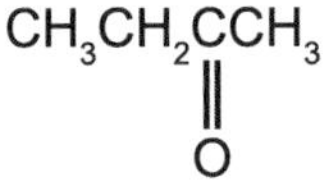

Q27: The triplet at 1.148 is due to a CH_3 group adjacent to a CH_2 group. The quartet at 2.324 is due to a CH_2CO group adjacent to a CH_3 group. The singlet at 3.674 is due to a CH_3-O group. This is also consistent with the ratio of the peak areas of 3:2:3. Therefore the molecule must contain a CH_3 group, a CH_2CO group and a CH_3-O group. This leads to the following structure...

$$CH_3CH_2C\underset{OCH_3}{\overset{O}{\diagdown}}$$

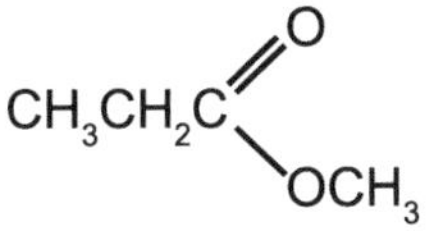

Q28: All three peaks here are singlets which means that none of the groups are adjacent to any atoms with hydrogen atoms attached. The ratio of the peak areas is consistent with two different CH_3 groups and a CH_2 group. The peak at 2.148 is due to a CH_3CO group. The peak at 3.421 is due to a CH_2-O group and the peak at 4.029 is due to a CH_3-O group. This leads to the following structure...

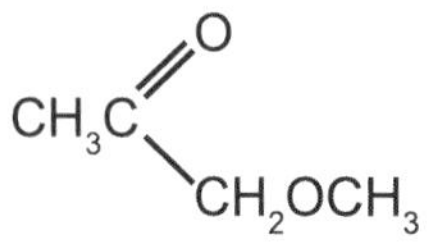

End of Topic 5 test (page 111)

Q29: b) $[C_2H_4]^+$

Q30: $CH_3COOCH_2CH_2CH_2CH_3$

Q31: The molecular or parent ion, $[C_6H_{12}O_2]^+$

Q32: a) unstable and easily fragments to form daughter ions.

Q33: $[C_3H_7]^+$ or $[CH_3CO]^+$

Q34: $[CH_2OCOCH_3]^+$ or $[CH_3COOCH_2]^+$ or $[C_4H_9O]^+$; all three have m/z = 73.

Q35: 27

Q36: a) $[C_4H_9]^+$ at m/z 57

Q37: b) $[CHO]^+$ and e) $[C_2H_5]^+$

Q38: c) C_6H_6 and d) C_2H_2

Q39: a) $[CH_3]^+$

Q40: a) $[CH_3]^+$ and f) $[CH_2OH]^+$

Q41: c) $Si(CH_3)_4$

Q42: c)

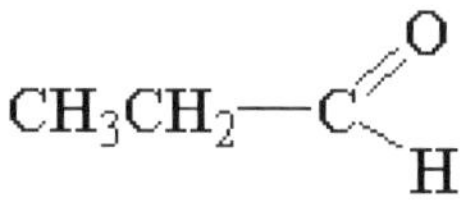

Q43: a) 1

Q44: There are three different hydrogen environments, therefore three peaks.

Q45: The peak areas are in the ratio 3:2:3.

Q46: Both are alcohols.

Q47: C is an aldehyde.

Q48: D is a ketone.

Q49: E is a carboxylic acid, with a broad band in the 3500 to 2500 cm^{-1} range and one in the 1725 to 1700 cm^{-1} range.

Q50: a) 1,2-dichloroethane, CH_2ClCH_2Cl and c) Ethanenitrile, CH_3CN

Q51: c) Ethanenitrile, CH_3CN

Q52: The NMR chemical shift for RCH_2CN is 2.5 - 2.0 ppm.

Q53: This is an open-ended question and as such it does not have only one correct answer. A variety of answers are acceptable.

In your answer to this question, you may wish to discuss the following - Chromophores, conjugation, colour related to wavelength of light, $E = Lhc/\lambda$, absorption of light, complementary colour, HOMO to LUMO transitions.

Open-ended questions are always marked out of a maximum of three marks according to the following criteria:

- 3 marks: The maximum available mark would be awarded to a student who has demonstrated, at an appropriate level, a good understanding, of the chemistry involved. The student shows a good comprehension of the chemistry of the situation and has provided a logically correct answer to the question posed. This type of response might include a statement of the principles involved, a relationship or an equation, and the application of these to respond to the problem. This does not mean the answer has to be what might be termed an "excellent" answer or a "complete" one.

- 2 marks: The student has demonstrated, at an appropriate level, a reasonable understanding of the chemistry involved. The student makes some statement(s) which is/are relevant to the situation, showing that the problem is understood.

- 1 mark: The student has demonstrated, at an appropriate level, a limited understanding of the chemistry involved. The student has made some statement(s) which is/are relevant to the situation, showing that at least a little of the chemistry within the problem is understood.

- Zero marks should be awarded if: The student has demonstrated no understanding of the chemistry involved at an appropriate level. There is no evidence that the student has recognised the area of chemistry involved or has given any statement of a relevant chemistry principle. This mark would also be given when the student merely restates the chemistry given in the question.

N.B. It is not necessary to discuss everything to gain a maximum three marks. Remember that the answer does not have to be "excellent" or "complete".

Q54: 58.5 g of NaCl contains 23 g of sodium.
Therefore 5 g of NaCl contains $(23/58.5) \times 5 = 1.97$ g of Na^+ ions.
% m/v = $(1.97/500) \times 100 = 0.39\%$.

Q55: 3.5% of 250 $cm^3 = (3.5/100) \times 250 = 8.75$ cm^3

Q56: 0.75 g toothpaste contains 1.125 mg F^- ions ppm = mg per kg, so 1 kg toothpaste (1000 g) contains $(1.125/0.75) \times 1000 = 1500$ ppm

Q57: 1000 g of toothpaste contains 1500 mg (1.5 g) of F^- ions.
10 g of toothpaste contains 15 mg (0.015 g) of F^- ions.

42 g of NaF contains 19 g of F^- ions.
Therefore, 0.015 g of F^- ions is contained in $(42/19) \times 0.015 = 0.033$ g.
33 mg of sodium fluoride is required.

6 Pharmaceutical chemistry (Unit 2)

End of Topic 6 test (page 124)

Q1: d) D

Q2: c) Hydrolysis

Q3: b) Reduction

Q4: a) $C_9H_8O_4$

Q5: b) Alkene

Q6: a) Primary

Q7: b) 2

Q8: b) B

Q9: c) C

Q10: a) A

Q11: Antagonist or inhibitor

Q12: Ethanal has the formula C_2H_4O and the difference between phenacetin and paracetamol is C_2H_4. This means that there is an oxgen atom required for the change.

Q13:

- A tetrahedral carbon atom with four different groups attached to it gives rise to two possible arrangements, each a mirror image of the other. Such isomers are known as enantiomers. They have no centre of symmetry and are described as chiral.
- Since salbutamol functions as an agonist it has to be sufficiently similar to the natural molecule which initiates a biological response to mimic this molecule. The shape of one of the isomers does this well, the other much less well. (one mark for each of the two parts)

7 End of Unit 2 test (Unit 2)

End of Unit 2 test (page 130)

Q1: c) sp^2 hybridised carbon atoms and sp^3 hybridised carbon atoms.

Q2: b) B

Q3: In an isolated carbon atom there are only two unpaired electrons. However, when carbon forms bonds with other atoms the 2s and the 2p orbitals mix to form four degenerate orbitals known as sp^3 hybrid orbitals. These four orbitals each contain one unpaired electron therefore carbon is able to form four bonds.

Q4: a) A

Q5: C_2H_5OH, ethanol or C_2H_6O

Q6:

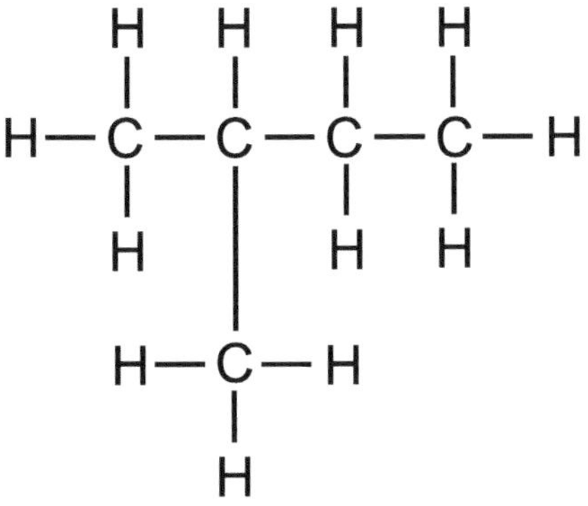

$$
\begin{array}{c}
\quad\ H \quad\ H \quad\ H \quad\ H \\
\quad\ | \qquad | \qquad | \qquad | \\
H-C-C-C-C-H \\
\quad\ | \qquad | \qquad | \qquad | \\
\quad\ H \qquad | \qquad H \quad\ H \\
\qquad\quad H-C-H \\
\qquad\qquad\ | \\
\qquad\qquad\ H
\end{array}
$$

2-methylbutane

Q7: b) Effect on plane of polarised light

Q8: b) $CH_3CH_2CH_2CH(OH)CH_3$

Q9: Four different groups around a carbon atom or mirror images are non-superimposable.

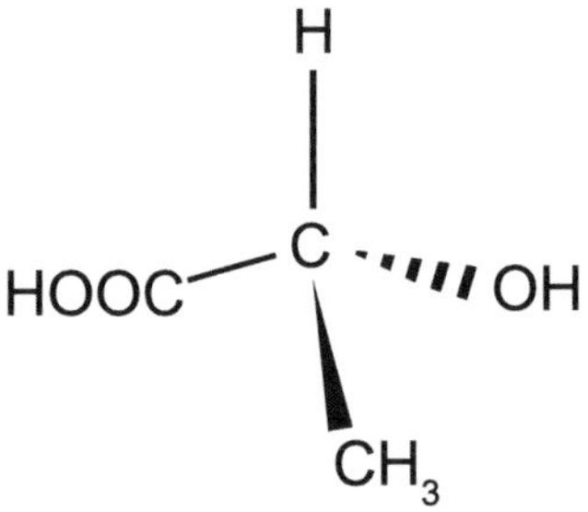

$$
\begin{array}{c}
H \\
| \\
HOOC-C\cdots OH \\
| \\
CH_3
\end{array}
$$

Q10: b) CH_3OH and CH_3COCl

Q11: b) electrophilic substitution.

Q12: Hydrolysis

Q13: 2-hydroxy-2-methylbutanoic acid

Q14:

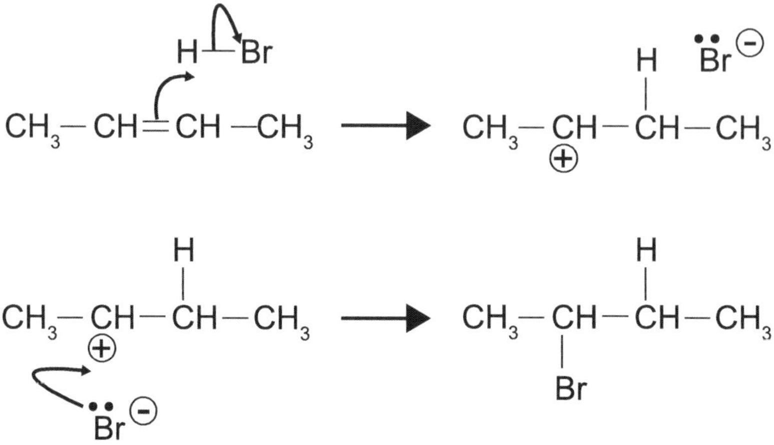

Q15: Aqueous KOH/NaOH or aqueous alkali

Q16: d) $[CH_3CO]^+$

Q17: c) CH_3OH

Q18: a) Infra-red spectroscopy

Q19: Mass of oxygen = 0·1920g

Element	C	H	O
moles	0.0180	0.0359	0.0120
mole ratio	3	6	2
or	1.5	3	1

Empirical formula $C_3H_6O_2$

Q20: It causes them to vibrate.

Q21: b) antagonist.

Q22:

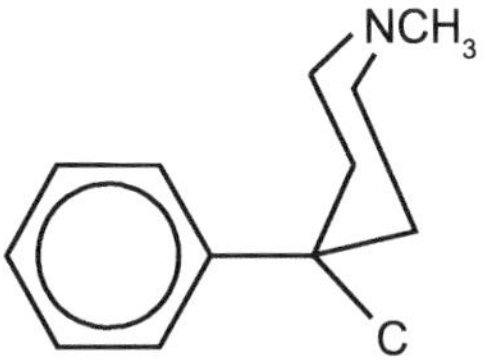

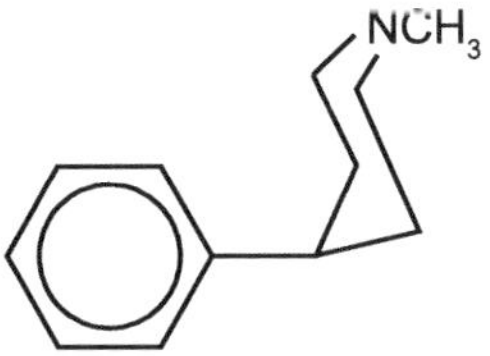

Q23: Produces a response like the body's natural active compound/enhances body's natural compound.

8 Gravimetric analysis (Unit 3)

Gravimetric analysis (page 140)

Q1: The compound is heated repeatedly until the mass does not change.

Q2: Prevent the compound from absorbing moisture (keep it dry).

Q3: Insoluble product is formed (precipitate formed).

Q4: Number of moles of nickel dimethylglyoxime (number of moles of nickel) = 0.942 / 288.7

Mass of nickel = $n \times GFM = (0.942 / 288.7) \times 58.7 = 0.192$ g

Percentage of nickel = $0.192 / 0.968 \times 100 = $ **19·8%**

9 Volumetric analysis (Unit 3)

End of Topic 2 test (page 154)

Q1: n for benzoic acid = $16.55 \times 10^{-3} \times 0.0563 = 9.32 \times 10^{-4}$

20 cm^3 of NaOH was used. The number of moles of benzoic acid is equal to the number of moles of sodium hydroxide (a 1:1 ratio).

c for NaOH = $(9.32 \times 10^{-4}) / 0.020 =$ ***0.0466 mol l^{-1}***

Q2: Average titre volume = $(15.9 + 15.8) / 2 = 15.85$ cm^3

Number of moles of sulfuric acid used = $0.001585 \times 0.02 = 3.17 \times 10^{-4}$ mol

Number of moles of NaOH left in the 25.0 cm^3 hydrolysed solution = $2 \times 3.17 \times 10^{-4} = 6.34 \times 10^{-4}$ mol

Number of moles of NaOH left in 250.0 cm^3 of the hydrolysed solution = $10 \times 6.34 \times 10^{-4} = 6.34 \times 10^{-3}$ mol

Number of moles of NaOH added to aspirin initially = $0.025 \times 1 = 2.5 \times 10^{-2}$ mol

Number of moles NaOH reacted with aspirin = $2.5 \times 10^{-2} - 6.34 \times 10^{-3} = 1.866 \times 10^{-2}$ mol

2 moles of NaOH reacts with 1 mole aspirin, therefore number of moles of aspirin in 4 tablets = $1.866 \times 10^{-2} / 2 = 9.33 \times 10^{-3}$ mol

Number of moles of aspirin in 1 tablet = $9.33 \times 10^{-3} / 4 = 2.333 \times 10^{-3}$ mol

Mass of aspirin in each tablet = n $\times$ gfm = $2.333 \times 10^{-3} \times 180 =$ ***0.420 g***

10 Practical skills and techniques (Unit 3)

End of Topic 3 test (page 169)

Q1: Make up several solutions accurately of various concentrations of permanganate and measure the absorbance of each one. *(1 mark)*

Draw a calibration curve of concentration versus absorbance. *(1 mark)*

The absorbance of the unknown permanganate concentration is measured and using the calibration curve the concentration of the solution is determined. *(1 mark)*

Q2: The exact mass must be in the region of 0.5 g and the exact mass must be known.

Q3: Pour mixture into standard flask, rinse beaker with water and add rinsings to the flask. *(1 mark)*

Make up to the mark with distilled water, stopper and invert flask. *(1 mark)*

Q4: Results that are ± 0.1 cm^3 in range from each other.

Q5: Low gram formula mass (GFM) or unstable in air or absorbs moisture or not a primary standard.

Q6: Solvent extraction

Q7:

1. Drain layers into separate beakers. *(1 mark)*
2. Return lower/aqueous layer to the separating funnel add more diethyl ether and repeat. *(1 mark)*
3. Evaporate/distill all diethyl ether layers. *(1 mark)*

Q8:

1. Immiscible in water. *(1 mark)*
2. Benzocaine is soluble within it. *(1 mark)*

Q9: Recrystallisation or chromatography

Q10: Pure benzocaine

Q11: It has a small impurity so is not 100% pure.

11 Stoichiometric calculations (Unit 3)

Excess calculations (page 176)

Q1: a) Zinc

Q2: a) Sulfuric acid

Nickel(II) ions and dimethylglyoxime (page 177)

Q3: 4

End of Topic 4 test (page 179)

Q4: d) phenolphthalein

Q5: Average titre in litres = ((first titration + second titration) / 2) / 1000 = ((8.7 + 8.6) / 2) / 1000 = 0.00865

Number of moles of sulfuric acid in the average titre = 0.00865 $\times$ 0.050 (concentration) = **4.325 $\times$ 10^{-4} moles**

Q6: Moles of sodium hydroxide in 25 cm^3 sample = 4.325 $\times$ 10^{-4} $\times$ 2 = 8.65 $\times$ 10^{-4} (reacts in a 2:1 ratio with the sulfuric acid)

Moles of sodium hydroxide in 250 cm^3 sample = 8.65 $\times$ 10^{-4} $\times$ 10 = **8.65 $\times$ 10^{-3} moles**

Q7: Starting moles of sodium hydroxide = 0.0025 $\times$ 1 = 0.025

Moles reacting = 0.025 - 8.65 $\times$ 10^{-3} = **0.01635** or **1.635 $\times$ 10^{-2} moles**

Q8: Moles of acetylsalicylic acid = 0.01635 / 2 = 0.008175 (molar ratio with sodium hydroxide from equation 1:2)

Mass of acetylsalicylic acid = 0.008175 $\times$ 180 = 1.4715 g (5 tablets)

Mass in one tablet = 1.4715 / 5 = **0.2943 g**

Q9: Calculate the mass of each element in the compounds formed from combustion of the compound as follows:

- Carbon = 3.52/44 $\times$ 12 = 0.96 g
- Hydrogen = 2.16/18 $\times$ 2 = 0.24 g
- Sulfur = 2.56/64.1 $\times$ 32.1 = 1.28 g

Molar ratio:

C = 0.96 /12 = 0.08 = **2**   H = 0.24 / 1 = 0.24 = **6**   S = 1.28 / 32.1 = 0.04 = **1**

Q10: 5.64 g of ethyl benzoate produces (5.64 $\times$ 122) / 150 = 4.59 g which is the theoretical yield.

Mass of benzoic acid produced = 73.2% of theoretical yield = 73.2% of 4.59 g = **3.36 g**

12 End of Unit 3 test (Unit 3)

End of Unit 3 test (page 182)

Q1: Mass of water driven off = mass of hydrated barium chloride - mass of anhydrous barium chloride = 2.58 - 2.22 = 0.36 g

Number of moles of H_2O = 0.36 / 18 = 0.020 mol
Number of moles of $BaCl_2$ = 2.22 / 208.3 = 0.011 mol

Ratio of moles $BaCl_2:H_2O$ = 0.011 / 0.02 = 1:1.8 which is 1:2 to the nearest whole number.
Therefore **n=2** and the formula is $BaCl_2.2H_2O$

Q2: Number of moles of $CaSO_4$ = 3.89 / 136.2 = 0.0286

Number of moles of H_2O = (4.94 - 3.89) / 18 = 1.05 / 18 = 0.0583

Ratio of moles = 0.0286:0.0583 = 1:2 therefore **x=2**.

Q3: Number of moles of sulfuric acid = $v \times c$ = 0.0178 (litres) $\times$ 0.22 = 0.003916 moles

Balanced stoichiometric equation shows that 2 moles of NaOH reacts with 1 mole of H_2SO_4

2NaOH +	H_2SO_4	$\rightarrow$	$Na_2SO_4 + 2H_2O$
2 mol	1 mol		
2 $\times$ 0.003916	0.003916		

In the 25.0 cm^3 sample of diluted drain cleaner there was 2 $\times$ 0.003916 = 0.007832 moles of NaOH.
In the 250.0 cm^3 standard flask there would have been 0.007832 $\times$ 10 = 0.07832 moles of NaOH which is the number of moles in the 10.0 cm^3 undiluted drain cleaner.

In 1 litre there would be 0.07832 $\times$ 100 = 7.832 moles of NaOH

Mass of NaOH = $n \times gfm$ = 7.832 $\times$ 40.0 = **313 g in 1 litre**.

Q4: Acid hydrolysis

Q5: To purify the sulfanilamide

Q6:

	1 mole of 4-acetamidobenzenesulfonamide	$\rightarrow$	1 mole of sulfanilamide
	214.1 g	$\rightarrow$	172.1 g
	4.282 g		(4.282 $\times$ 172.1)/214.1 = 3.442 g

Percentage yield = (actual yield / theoretical yield) $\times$ 100 = (2.237 / 3.442) $\times$ 100 = 0.65 $\times$ 100 = **65%**

Q7: Sample is mixed with pure sulfanilamide and if pure the melting point of the mixture will be the same as the pure sample.

Q8: Any one of:

- thin-layer chromatography (TLC)
- nuclear magnetic resonance (NMR) spectroscopy
- infrared (IR) spectroscopy

Q9: Colourless to pink/purple

Q10: Number of moles of acidified potassium dichromate $n = c \times v = 0.02 \times 0.0165 = 0.00033$

Oxalate reacts in a 5:2 ratio so the number of moles of oxalate $= 0.00033 \times 2.5 = $ ***0.000825*** or ***8.25×10^{-4}***

Q11: Mass in 1 litre = number of moles of oxalate $\times$ original conc in 20 cm^3 $\times$ gfm = $0.000825 \times 50 \times 88 = $ ***3.63 g***

Q12: Mass of potassium = total mass - mass of oxalate - mass of hydrogen = 4.49 - 3.63 - 0.06 = ***0.8 g***

Q13: $K = 0.8 / 39 = 0.02$
$H = 0.06 / 1 = 0.06$
$C_2O_4 = 3.63 / 88 = 0.04$

Ratio of x, y, z = 0.02:0.06:0.04 = 1:3:2
x=1, y=3 and z=2

The Global Smartphone

The Global Smartphone

Beyond a youth technology

Daniel Miller, Laila Abed Rabho, Patrick Awondo,
Maya de Vries, Marília Duque, Pauline Garvey,
Laura Haapio-Kirk, Charlotte Hawkins, Alfonso
Otaegui, Shireen Walton and Xinyuan Wang

First published in 2021 by
UCL Press
University College London
Gower Street
London WC1E 6BT

Available to download free: www.uclpress.co.uk

ISBN: 978-1-78735-963-5 (Hbk.)
ISBN: 978-1-78735-962-8 (Pbk.)
ISBN: 978-1-78735-961-1 (PDF)
ISBN: 978-1-78735-964-2 (epub)
ISBN: 978-1-78735-965-9 (mobi)
DOI: https://doi.org/10.14324/111.9781787359611

Contents

Chapter summaries

Chapter 1: Introduction

The ASSA project is presented as a study of 'smart from below', intended to learn from the creativity and practices of smartphone users all around the world.

The term smartphone is misleading. Firstly, it should no longer be regarded as primarily a type of phone, since traditional phone calls now represent only a small part of usage.

Secondly, the smartphone, as encountered in this project, is not a good example of 'smart', in the sense of being a device that can learn from its employment. Such autonomous learning is far less important in creating the smartphone we actually encounter than is the way smartphones are transformed by users.

Smartphones are now employed across all age groups. It is just as reasonable to consider them mainly from the perspective of older people as from that of youth.

The project involved 11 researchers working in 10 fieldsites. Each spent approximately 16 months carrying out an ethnography focused upon ageing, smartphone use and the potential of smartphones for health.

A brief history of the smartphone is followed by a short survey of previous approaches, first by anthropologists and then by other disciplines.

This book concentrates on that which is apparent from our ethnographic method. We acknowledge that we lack evidence about significant externalities such as environmental consequences, exploited labour and the study of relevant corporations.

Chapter 2: What people say about smartphones

What people say about smartphones is often full of contradictions – an ambivalence that reflects the way in which smartphones mostly create simultaneous benefits and problems.

These discourses about the smartphone are distinct from what people actually *do* with smartphones, being mainly dictated by moral and political debates.

Instead it is best to regard these discourses as independent properties of the smartphone, whose consequences need to be examined in their own right.

The state, the media and commerce all add to such contradictions. For example, states condemn overuse of smartphones, but then make it difficult for citizens to deal with the state without recourse to digital processes.

Older people in the Chinese fieldsite tend to identify with the smartphone as being part of their duty as citizens to assist their country's technological progress. They stand in contrast to the more general conservatism of older people elsewhere.

Certain topics dominate discussion, such as fake news, addiction and surveillance. By contrast, there is limited public discussion of more general usage and consequences of smartphones.

Academic evidence for the most common claims made about the consequences of smartphones is equally contradictory.

Chapter 3: The smartphone in context

Smartphones are material objects that can be used as fashion accessories or as markers of status or religion. They can also be stolen.

There remains a global divide. Studies of smartphones may exclude people who cannot afford them or, as in Japan, still focus on established feature phones.

For people with low incomes, the costs of handsets, plans, Wi-Fi or data may be of considerable concern. They are often resourceful in finding ways to gain access.

The term 'Screen Ecology' refers to how smartphones work in tandem with other screens, such as tablets, laptops and smart televisions. The usage of any one of these only makes sense relative to the others.

The term 'Social Ecology' is used to consider how smartphones can reflect the form of social relations in a particular society. For example, some families in Kampala share their smartphones.

Smartphones may facilitate the rise of networks based around an individual. Equally, however, they may reinforce traditional social groups, such as the family or community.

Smartphones are just starting to have an impact as a remote-control hub on the 'Internet of Things'.

Chapter 4: From apps to everyday life

In general, users of smartphones are focused upon tasks rather than individual apps. Often they simply combine bits of different apps to achieve their goals.

Taking the example of health, we can see that bespoke apps for health are usually less important to users than combining generic apps such as WhatsApp with googling.

The term 'Scalable Solutionism' describes the spectrum of what people actually do with their apps. This ranges from apps with a single function ('there is an app for that'), or used as though they only had one function, through to apps such as WeChat that aim to be useful for all tasks.

To know a smartphone and its user properly involves going through every single app on that smartphone and finding out whether it is used and how it is used.

Understanding apps also involves an exploration of the way in which companies develop apps and respond to unexpected ways in which those apps are then deployed.

A consideration of apps includes an investigation of the different ways in which people organise the screens of their smartphones.

Chapter 5: Perpetual Opportunism

The term 'Perpetual Opportunism' refers to the smartphone being always available and the ways in which this changes people's relationship to the world around them.

Smartphone photography, for example, has become almost the exact opposite of analogue photography. Photography was traditionally concerned with representation and creating a permanent record. Smartphone photography is more about being alert to the moment and engaging in transient sharing.

Older people have varied responses to being photographed. The real person might be considered as 1) the person they feel they are inside, 2) their external appearance or 3) the crafted image that they can produce using filters and apps.

Perpetual Opportunism changes our relationship to location and transport systems, making it easier to travel on a whim. Map apps also facilitate holidays and leisure.

Thanks to Perpetual Opportunism, news flows in real time and may become a constant preoccupation. News and information take on new roles in relation to community.

Smartphones make entertainment available at any time of potential boredom, such as waiting in queues or travelling. Music, for example, may be accessed in many different ways.

Chapter 6: Crafting

The smartphone is unprecedented in its malleability and intimacy. It can be moulded into a close correspondence with the character or interests of its user.

The algorithms and artificial intelligence (AI) developed for this purpose remain less important than the ability of the individual to select apps, change settings and create or curate content.

The individual's creation of their smartphone can be considered as an artisanal craft.

Smartphones are also crafted to fit relationships, rather than individuals. Examples may include those between partners, parent and child or between an employee and employer.

Individuals generally manifest the cultural norms and values of society, which are then the foundation for what smartphones become. However, individuals may be typical or eccentric in respect to such norms.

Smartphones may then mainly conform to consensual norms, as in Japan or among a religious community.

Smartphones may also be important in facilitating change in those cultural values – for example, in creating the values of the Cameroonian middle class.

Chapter 7: Age and smartphones

Smartphones serve to reflect but also transform social parameters such as gender and class – or, in this chapter, age.

Smartphones may facilitate transformations, for instance helping 'second-generation' youth in Italy who are exploring aspects of their identity or people crafting a new everyday life on retirement.

For older people, smartphones can represent a loss of respect for knowledge accumulated over decades that may now be considered redundant.

Younger people often incorrectly claim that smartphones are intuitive when teaching usage to older people.

Older people may struggle with smartphone use where tasks require digital dexterity or terms used in unfamiliar ways. There are also hurdles in learning deployment and appropriate usage.

Although initially older people may feel excluded, those who master their smartphones may feel more aligned with youth as a result.

Companies may sometimes devise apps specifically for use by older people, for example the Meipian app in China.

Chapter 8: The heart of the smartphone: LINE, WeChat and WhatsApp

Apps such as LINE, WeChat and WhatsApp may become so dominant that users view smartphones essentially as devices for gaining access to these platforms.

Visual media such as stickers have joined speech and text as an integral part of conversation. They provide new ways to facilitate care and affection at a distance.

These apps may also be elements in the transformation of family relationships, for example partly reversing the historical shift from extended to nuclear families.

These apps have also gained an important presence in the functioning and organisation of community.

Smartphones thereby extend 'scalable sociality', matching usage to different size groups and different degrees of privacy.

In turn, corporations may learn from the social incorporation of these apps and adapt technology accordingly. One example is the development of a kinship app as part of WeChat.

Chapter 9: General and theoretical reflections

We refer to the smartphone as the 'Transportal Home', since we may understand it better by thinking of it as a place within which we live, rather than as a device that we use. There are many ways in which people treat the smartphone as a domestic space.

The 'Death of Distance' has been followed by the 'Death of Proximity'.

The smartphone has moved 'Beyond Anthropomorphism' because intimacy is achieved not by trying to look like people, but by complementing human capacities such as cognitive functions. As a result, a smartphone has come to feel like an integral part of a person.

Smartphones can equally assume every unpleasant characteristic of our inhumanity, with traits ranging from bullying to addiction.

The rise of the Covid-19 pandemic clarified a key contradiction. Smartphones considerably extend the possibilities of surveillance, but are simultaneously a means for developing 'Care Transcending Distance'.

This project shows why in response to Covid-19 we should appreciate people's relevant experience as an asset in making decisions about future smartphone usage. We call this perspective 'smart from below'.

List of figures

List of abbreviations

4G   The fourth generation of broadband cellular network technology.

5G   The fifth generation of broadband cellular network technology.

AI   Artificial intelligence.

ARPANET   The Advanced Research Projects Agency Network.

ASSA   The Anthropology of Smartphones and Smart Ageing.

BBC   British Broadcasting Corporation.

Covid-19 Coronavirus Disease (2019).

DRC   The Democratic Republic of the Congo.

ESPM   Escola Superior de Propaganda e Marketing.

GPS   Global Positioning System.

HDR   High-dynamic-range imaging (a technique used in photography).

IBM   International Business Machines (corporation).

ICT4D   Information and Communications Technology for Development.

LATAM   LATAM Airlines Group S.A. is an airline with headquarters in Santiago, Chile. The name is a result of the merger between the operations of Chile's Línea Aérea Nacional (LAN) and Transportes Aéreos Meridionais (TAM).

NoLo   North of Loreto (or Nord Loreto in Italian), a district in Milan located in the northeast of the city, from Piazzale Loreto including the areas around via Padova, Pasteur and Parco Trotter.

OTT   Over The Top – a tax on the use of social media platforms introduced in Uganda in 2018.

S.M.A.R.T. Self-Monitoring, Analysis and Reporting Technology.

SUS   Sistema Único de Saúde (Brazil's publicly funded healthcare system).

UGX   Ugandan shilling – the currency of Uganda.

List of contributors

Laila Abed Rabho is a researcher at the Harry S. Truman Institute for the Advancement of Peace. Laila received her PhD from the Hebrew University of Jerusalem, Department of Islam and Middle East Studies. She is also a litigator in the Shari'a court in al-Quds.

Patrick Awondo is Research Fellow at UCL Anthropology and a lecturer at the University of Yaoundé 1. He is the author of *Le Sexe et ses Doubles* (2019). Before focusing on digital anthropology, he worked on gender and migrants with a specific interest in members of the LGBTI community fleeing homophobia in sub-Saharan Africa and trying to take refuge in France. Patrick has published articles in both French and English journals, including *Politique Africaine, Diasporas, Société contemporaine, African Studies Review, Review of African Political Economy* and *Archives of Sexual Behavior.*

Maya de Vries is Postdoctoral Researcher at UCL and the Hebrew University of Jerusalem. She received her PhD in Communication from the Hebrew University of Jerusalem in 2019. Her areas of research are digital ethnography, new media, activism and ethno-political conflicts in Israel/Palestine.

Marília Duque is a researcher at ESPM (Escola Superior de Propaganda e Marketing) São Paulo and author of the book *Learning from WhatsApp: Best practices for health*. She has also worked as a Research Assistant at UCL Anthropology. Her research interests focus on ethics, technology consumption, ageing and health in Brazil.

Pauline Garvey is Associate Professor in the Department of Anthropology, Maynooth University, National University of Ireland, Maynooth, Co. Kildare, Ireland. She is the author of *Unpacking Ikea: Swedish design for the purchasing masses* (2018) and editor of *Home Cultures: The journal of architecture, design and domestic space.*

Laura Haapio-Kirk is a PhD student at UCL Anthropology and RAI/Leach Fellow in Public Anthropology. Her research interests include ageing and the life course, wellbeing and digital technologies. She has a Masters in Visual Anthropology from the University of Oxford and integrates illustration into her research.

Charlotte Hawkins is a PhD student at UCL Anthropology. Her research interests include the determinants of health, intersubjectivity and storytelling, age and intergenerational care, collaborative ethnography, media and morality.

Daniel Miller is Professor of Anthropology at UCL. He is director of the ASSA project and was director of the *Why We Post* project. He is author/editor/co-author of 42 books including *How the World Changed Social Media* (with eight others), *Social Media in an English Village*, *Tales from Facebook*, *Digital Anthropology* (ed. with H. Horst), *The Comfort of Things*, *Stuff*, *A Theory of Shopping* and *Material Culture and Mass Consumption*.

Alfonso Otaegui is Lecturer at the Center for Intercultural and Indigenous Research (Pontifical Catholic University of Chile). He received his PhD in Anthropology at the EHESS in 2014. His research addresses verbal art among the peoples of Gran Chaco (South America), digital literacy of older adults and the religious and communicative practices of Latin American migrants.

Shireen Walton is Lecturer in Anthropology at Goldsmiths, University of London. She received her DPhil in Anthropology from the University of Oxford before joining UCL Anthropology as Teaching Fellow and then Postdoctoral Researcher as part of the ASSA project. Her work engages with media and social change, mobilities and migration, and digital-visual anthropology. Shireen has also carried out ethnographic research in Iran, the UK, Italy and online.

Xinyuan Wang is Postdoctoral Researcher at UCL. She is the author *of Social Media in Industrial China* and co-author of *How the World Changed Social Media*. Xinyuan is also the 2020 Daphne Oram Award Lecture winner for her contributions to UK Science.

Series Foreword

This book series is based on a project called 'The Anthropology of Smartphones and Smart Ageing', or ASSA. This project focused on the experiences of ageing among a demographic who generally do not regard themselves as either young or elderly. We were particularly interested in the use and consequence of smartphones for this age group, as these devices are today a global and increasingly ubiquitous technology that had previously been associated with youth. We also wanted to consider how the smartphone has impacted upon the health of people in this age group and to see whether we could contribute to this field by reporting on the ways in which people have adopted smartphones as a means of improving their welfare.

The project consists of 11 researchers working in 10 fieldsites across 9 countries as follows: Alfonso Otaegui (Santiago, Chile); Charlotte Hawkins (Kampala, Uganda); Daniel Miller (Cuan, Ireland); Laila Abed Rabho and Maya de Vries (al-Quds [East Jerusalem]); Laura Haapio-Kirk (Kōchi and Kyoto, Japan); Marília Duque (Bento, São Paulo, Brazil); Patrick Awondo (Yaoundé, Cameroon); Pauline Garvey (Dublin, Ireland); Shireen Walton (NoLo, Milan, Italy) and Xinyuan Wang (Shanghai, China). Several of the names used for these fieldsites are pseudonyms.

Most of the researchers were based at the Department of Anthropology, University College London. The exceptions are Alfonso Otaegui at the Pontificia Universidad Católica de Chile, Pauline Garvey at Maynooth University, the National University of Ireland, Maynooth, Marília Duque at Escola Superior de Propaganda e Marketing (ESPM) in São Paulo, Laila Abed Rabho, an independent scholar, and Maya de Vries, based at the Hebrew University of Jerusalem. The ethnographic research was conducted simultaneously, other than that of al-Quds which started and ended later.

This series comprises a comparative book about the use and consequences of smartphones called *The Global Smartphone*. In addition we intend to publish an edited collection presenting our work in the area of mHealth. There will also be nine monographs representing our ethnographic research, with the two Irish fieldsites combined within a single volume. These ethnographic monographs will all have the same chapter headings with the exception of chapter 7 – a repetition that will enable readers to consider our work comparatively.

The project has been highly collaborative and comparative from the beginning. We have been blogging since its inception at https://blogs.ucl.ac.uk/assa/. Our main project website can be found at https://www.ucl.ac.uk/anthropology/assa/, where further information about the project may be found. The core of this website is translated into the languages of our fieldsites. The comparative book and several of the monographs will also appear in translation. As far as possible, all our work is available without cost, under a creative commons licence. The narrative is intended to be accessible to a wide audience, with detailed information on academic discussion and references being supplied in the endnotes.

We have included films within the digital version of these books; almost all are less than three minutes long. We hope they will help convey more of our fieldsites and allow you to hear directly from some of our research participants. If you are reading this in eBook format, simply click on each film to watch them on our website. If you are reading a hard copy of this book, the URLs for each film are provided in each caption so you can view them when you have internet access.

Acknowledgements

Our primary acknowledgements are to the thousands of participants who took part in this research and gave of their time and experience. A commitment to anonymity means that they cannot be thanked individually, but we thank each and every one of them deeply for their collaboration in this research. We are particularly grateful to Georgiana Murariu, our research assistant on this project, who has organised us, edited the manuscripts, created many of the infographics and helped in countless other ways. We are grateful to Sasaki Lise and Alum Milly who provided research assistance in two fieldsites. Thanks to all those who read earlier versions of this manuscript, including the anonymous readers from UCL Press, Rik Adriaans, Wendy Alexander, Rickie Burman, Andrew Cropper, Justin Davis, Marcus Fedder, Heather Horst, Victoria Irisarri, Suzana Jovicic, Katrien Pype, Simin Walton and Christopher Welbourn. Also specific thanks to academic colleagues Kimura Yumi, Marjorie Murray, David Prendergast, Elizabeth Schroeder-Butterfill and Jay Sokolovsky. In addition we would like to thank Ben Collier, who produced many of our short films, and other associated film-makers, including Daniel Balteanu. We are also grateful for the support of UCL Press and to Catherine Bradley for her thoughtful copy-editing.

Most of the research and researchers were funded by the European Research Council (ERC) under the European Union's Horizon 2020 research and innovation programme (grant agreement No. 740472). In addition Alfonso Otaegui was funded by the Center for Intercultural and Indigenous Research in Santiago, Chile, Grant CIIR, ANID – FONDAP15110006. Laila Abed Rabho and Maya de Vries received funding from the Humanitarian Trust Committee and the Swiss Center for Conflict Research, Management and Resolution and The Smart Family Institute of Communications, The Hebrew University of Jerusalem. Laura Haapio-Kirk received additional funding from the Osaka-UCL Partnership fund for her mHealth project with Kimura Yumi and Sasaki Lise. Marília Duque received additional funding from CAPES Brazil (grant agreement N° 88881.362032/2019-01).

1
Introduction

The smart and the phone

Sato san from Japan is a 90-year-old master of flower arrangement (*ikebana*). She is still practising and also teaches her traditional craft from her Kyoto home. In the three years since she obtained a smartphone, it has become central to her work and life. Sato san arranges her students' lessons via the messaging application LINE, on which she has over 100 contacts. She likes the fact that LINE tells her if a message has been read, and she sometimes follows up emails with a LINE message informing a student that she has emailed them. The calendar on her smartphone tells her when she needs to replace flower displays in various shops around Kyoto. She also writes a blog about flower arranging and her exhibitions, through which many students find her.

Outside of her work, Sato san's smartphone makes everyday tasks, such as checking the weather or bus times, easier to do. She orders groceries such as lunch boxes (*bento*), pickles and tofu from her local Seven Eleven store via LINE. They check her order by sending a picture of the products back to her. Describing herself as passionately curious about the world, Sato san uses her smartphone to maintain her mental health by doing brain training every day through dedicated apps; she also learns one new English word per day on a language app. Physical wellbeing is also important: Sato san checks her step-counter daily to see how many calories she has burned up. She sometimes researches why her legs have swollen up or looks up a healthy recipe that she has been told about. She has replaced her previous custom of phoning her niece to ask her about things she hears on the television with asking Google. Sato san admits to getting frustrated that most friends around her age, and even those who are younger, still have the more limited feature phones (*garakei*). They are not as curious about new technology as she is, even though she tries to encourage them. Her adoption of the smartphone reflects a lifelong attitude of being ahead of her peers in embracing the new.

Sato san has a student called Midori san, a woman in her mid-sixties who is a professional musician. In the short film featured below (Fig. 1.1), Midori san explains why she finally decided to get a smartphone after hesitating for so long.

Mary, aged 80 and living in Ireland, makes extensive use of Pinterest for assisting with her hobby of drawing flowers; she both looks for examples of the way people draw and also checks the spelling of botanical plant names. She videocalls a friend she is planning to visit in the Netherlands via WhatsApp and uses an app called Measure to help a grandchild with maths homework. YouTube music helps Mary practise for the choir she sings in, and she uses RTÉ player to catch up on programmes she missed on the radio. She has only recently stopped using her step-counter.

On Instagram she follows not only her daughter, who posted a drawing to her account every day for six months, but also a whole set of local Irish artists, as well as googling background information on exhibitions at the national gallery. Mary uses the camera on her smartphone to take pictures of quirky scenes, for example a hen that appears to be waiting at a bus stop. She looks at a range of newspapers on her phone, keeps up with Facebook and makes use of apps for the bus and

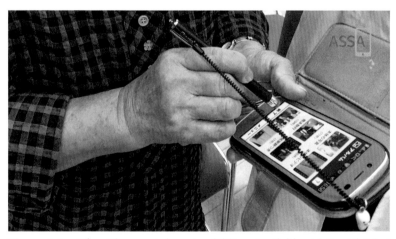

Figure 1.1   Film: *The smartphone is a lifeline*. Available at http://bit.ly/smartphoneisalifeline.

rail timetable, as well as the Realtime app for other transport information. She has clear views on how time spent on digital devices should be used. For example, she thinks that her friend wastes far too much time on Facebook, although she also concedes that 'everyone has to have a hobby'. Mary dislikes Facetime, mainly because it means she has not got time to get her face made up when someone wants to speak with her. She has had her iPhone for five years and prefers to work things out for herself, rather than ask for help or go on a course. With regard to smartphones more generally, she argues that 'you shouldn't look a gift horse in the mouth'.

As a final example, the short film featured below (Fig. 1.2) gives a more general impression of smartphone use by older people in China.

The title and subtitle of this book are 'The Global Smartphone: Beyond a youth technology'. It has only recently become possible to write a general survey of the use and consequence of the smartphone as an everyday device that is taken for granted. Previously usage was more restricted. The smartphone has now found a place among the blossoms for these 80- and 90-year-olds, shorn of any previous assumption that it belonged naturally to youth. During the project that preceded this one, a comparative study of social media called *Why We Post*,[1] people in many regions were adamant that smartphones and social media would never be used by those over 40. That barrier has now been demolished. As a result, it feels imperative that we try to understand a device so central to the lives of 3.5 billion users;[2] in the UK, for example, 84 per cent of the adult population own a smartphone.[3]

Figure 1.2   Film: *How can I live my life without you?* Available at http://bit.ly/lifewithoutyou.

The smartphone is arguably becoming the ubiquitous appendage to humanity. We wrote this volume during the social isolation that followed the Covid-19 pandemic – a time when, for many people around the world, the smartphone became the sum total of their social communication. While most people were thus made acutely aware of the shortcomings of smartphone communication compared to face-to-face encounters, they were simultaneously aghast at the very thought that they might have had to experience a lockdown without a smartphone. The range of uses for which they deployed smartphones is astonishing, yet has been almost immediately taken for granted.

The primary aim of this book is to understand the consequences of smartphones for people around the world. The secondary aim is to gain a better understanding of what a smartphone really is. The final chapter provides a series of conclusions that combine these two aims. The terms introduced there and also in the conclusions of prior chapters, such as the 'Transportal Home', 'Beyond Anthropomorphism' and 'Perpetual Opportunism', are claims made about the way in which smartphones are used and how they have transformed, for example, our sense of where we live.

Our method of ethnographic research primarily consisted of each of us living inside a community for 16 months, participating in the daily lives of people living there. Ethnographic research may also be the only means by which we can discover what the smartphone is. This is because the device may be unprecedented in the degree to which it is open to transformation. We have to be able to examine what each user has turned their smartphone into and how they subsequently deploy it, mostly in private. Ethnography is a method for establishing the long-term trust required for such research. Apart from chapter 2, based largely on what people say about smartphones, this book is mainly based on direct observation of how people use smartphones in everyday life, in addition to our discussions with users.

If most of the book is a description of what a smartphone is and its consequences for users, it nevertheless begins with a brief discussion of what the smartphone is not. This is necessary because the term smartphone itself is misleading. To understand this device we need to reconsider both the terms 'smart' and 'phone'. When 'smart' is used within terms such as smartphone, smart cities or smart homes, it comes from the acronym S.M.A.R.T., standing for 'Self-Monitoring, Analysis and Reporting Technology'. The term was introduced in 1995 when the multinational technology company IBM (together with Compaq) started producing discs that could provide the user with warnings of failure well in advance of this occurring.[4] Today, with the rise of artificial intelligence (AI), 'smart' is conceived as a form of intelligence based on autonomous learning. It reflects the idea that machines and devices can adapt to their user through autonomous monitoring and processing.

By contrast, this volume is based on an entirely different concept of smart, for which we will borrow the phrase 'smart from below'.[5] The evidence presented throughout this book suggests that the ability of smartphones to learn from the way they are being used through self-monitoring turns out to be quite a minor feature. Vastly more important in determining what a smartphone ends up being is the way in which the device is transformed by its user, a form of craftsmanship that only begins after purchase. It is such craftsmanship that turns the smartphone into an extraordinarily intimate and personal tool.

A smartphone developer, for example, may have had in mind a teenager in Korea. They may not have been particularly concerned about the fact that Peruvian migrants in Chile may end up using their smartphones in markedly different ways to middle-class retirees in Cameroon or Ireland, let alone a 90-year-old flower arrangement master from Japan. What we document here is not a smartphone smart enough to learn about flower arranging. Rather, we studied a 90-year-old smart enough to reconfigure her smartphone to become an asset in her flower arranging. This is the creativity signified by the term 'smart from below'.

We propose an equally radical departure from the word 'phone'. The semantics imply that the smartphone is essentially the latest iteration of the device we know as a phone. But to what degree is a smartphone in fact a phone? The telephone was a device for making phone calls; a smartphone has developed so many functions that voice calling is quite a small segment of its usage, especially among younger people.[6] Even the use of voice has become dispersed into other functions such as dictation.

Thinking about the smartphone as merely an extension of the phone may therefore be misleading; indeed, calling it any kind of phone may be considered unhelpful. A history and literature review of the telephone[7] would exclude most of what a smartphone now represents. The smartphone is an aggregate of dozens of prior practices, each of which presumes a huge historical literature. For example, a significant portion of contemporary day-to-day photo taking, sharing and viewing across the globe is carried out via the smartphone. Is the device better described as a smartcamera that can also make calls? To understand it, should we focus on the history of (digital) photography rather than that of the phone?[8] Is that perhaps the more dramatic shift?

Photography was initially more concerned with representation and archiving. The smart-camera-phone is more used for sharing visual materials through social media. Yet photography itself is only one aspect of the visual elements of the smartphone – consider Google Maps – so the smartphone invokes a more general history of the use of visual media.[9]

If photographs are now shared on social media, it implicates the smartphone as the device through which we now share information about our lives. The historian Lee Humphreys has recently published a fascinating history of such sharing.[10] For example, the diary in Victorian times was often intended to be for the eyes of others; it was the way in which a wife might inform her parents about her new life in her husband's home, while children's diaries might be read aloud in the evening. Humphreys relates this practice to the subsequent development of blogging and then vlogging, as well as the contemporary use of Facebook. It is a neat corrective to assumptions that telling people what you had for dinner is necessarily an example of contemporary narcissism.[11] Similarly, Humphreys notes precedents for social media such as Pinterest in the Victorian scrapbook, recognising the long history behind telling others about our holidays abroad or creating commemoration through images.

As we consider ever more uses, it feels pretty hard even to scratch the Gorilla Glass surface of the contemporary smartphone. This is in part because even thinking of the device as a smart-camera-phone-diary is still very incomplete. Why are those uses more important than considering the way the smartphone has changed our sense of location?[12] Are all places now much the same since our smartphone is with us wherever we are, as one commentator has suggested?[13] The smartphone is also the place in which we store, listen to and share music. It has become a device for translation,[14] a key link to transport and tourism and a place for gaming. It has become a primary source of knowledge, the place where people seek information on every conceivable topic, from celebrity gossip to news, fake or otherwise, to discoveries in physics.[15]

All of this means that this introductory chapter cannot follow convention. It cannot simply include a history of every single thing that the smartphone is, from the history of how we obtain information to the history of entertainment. Nor is there any possibility of a literature review that would cover everything from locational technology through to photography. About the only plausible introduction might have been a history and literature review of the development of our capacity to do many things from one device. A commonplace analogy has become the Swiss Army penknife, but it is a very limited one. Another is that of 'the computer in my pocket' – but the computer was never the main device for making telephone calls or taking pictures, let alone an object so intimate that it feels like an extension of one's body. Furthermore, this research included several fieldsites where either only a few people or only young people had used computers; for many users the smartphone has been their first access to the internet. There really is no plausible precedent to the smartphone.

Yet these references to past devices bring us closer to what will become a key to unlocking our understanding of the smartphone and its capacity to blur the prior boundaries between the activities for which it is employed. Photography has been mentioned, as has the use of the smartphone for recording information. But what is really new is the way in which these have become linked through the smartphone, to the extent that the camera can facilitate the storing of information. People may photograph an advert for a concert on a shop window, or a page in a magazine, or a shopping list. Similarly, because the smartphone's calendar function is so easily linked to group communication through WhatsApp it has become much easier to change the time and place for a meeting on the hoof. The smartphone has become as important for linking capabilities as for aggregating them.

Confronting the magnitude of possibilities contained within the smartphone is a further justification for our ethnographic approach. Unlike more technologically orientated studies, we will not attempt to examine the smartphone simply as a device with capacities. Many of these capacities remain unused. For us the smartphone consists only of the observable uses found among the particular populations with whom we worked. In our 'smart from below' perspective, it is the ingenuity of *people* rather than the device that will be to the fore, irrespective of whether their deployments were envisaged by the smartphones' creators. The volume is replete with amazing invention, design and application by people integrating smartphones within their everyday lives.

The Global Smartphone is also a work of comparative anthropology that constantly acknowledges cultural differences. Sometimes it is about the global smartphone, where generalisations across the different fieldsites seem warranted; at other times it is very much a local smartphone. No population represents the natural or normal users of smartphones, from which others are then variations or deviations. Within each population we may discuss typical usage. Yet that implies there will also be individuals who are not typical. Generalisation may be made in relation to gender or class, but every individual is more than a type or token of a category. Consequently much of this book consists of short portraits of individuals, sometimes to present them as an example, but also to acknowledge them as unique.

An outline of our project

This project consisted of 11 researchers who carried out 10 ethnographies across 9 countries.[16] Each ethnography lasted at least 16 months. Apart from al-Quds, they took place from the start of February 2018 to the end of June 2019, with the researchers living inside or next to the communities they studied. An outline of the methodology, funding, ethics and wider context is provided in an appendix to this volume. The fieldsites were not chosen as representative or as a sample. There is no logic hidden in their selection. They were simply chosen according to the backgrounds and interests of the research team. What mattered was only that they included sufficient diversity to exemplify the range of behaviours and values found in our contemporary world. Nor should these fieldsites be viewed as a sample of their country. A middle-class community within Yaoundé cannot stand for Cameroon, nor a Peruvian migrant in Santiago for Chile.

The research included three primary topics: ageing, health and smartphones. We are also planning to publish nine monographs that focus on the experience of ageing for those who consider themselves neither young nor elderly. Most of the research focused upon that demographic. This varied considerably, a result of the much higher proportion of older people in, for example, Japan compared to in Uganda. For this reason the term used throughout this book is 'older people'. We did not study the more elderly, frail population that is the concern of gerontology. There is a far more extensive discussion of these issues of ageing within our individual monographs. Many of the researchers are themselves around 30, and they naturally also came to know people of their own age.

At first an emphasis upon older people may appear strange because we have become so used to concentrating upon youth, once thought the natural users of smartphones. Just as those earlier studies would have included topics specific to youth, so in this volume there are sections more specific to older people, for example chapter 7. More generally, however, a focus upon older people has helped to extract the study of smartphones from any specific demographic niche so that they may be considered as the possession of humanity as a whole. If anything, what will come to the fore in this volume is the importance of the smartphone within intergenerational relations.

In addition to studying ageing and smartphones, we connect these together through research on mHealth[17]– the use of smartphones for health purposes. The focus upon mHealth is a more applied component, as the intention was to contribute directly to people's welfare. Our starting point for this last component was the huge rise of bespoke apps currently being developed for a range of activities: helping patients obtain information,[18] the transformation of care,[19] delivering public health interventions[20] or online health communities.[21] Health-related apps are being introduced to the market for managing almost any condition, ranging from monitoring fitness to period tracking or diabetes.

As the research developed, however, our 'smart from below' approach resulted in a major change in direction. We found relatively limited use of such bespoke health apps. Instead the majority of research participants (subsequently referred to as participants) regularly used *other* apps for health purposes, such as setting up WhatsApp groups to coordinate the care of an elderly relative. The focus then shifted to the use of general apps such as Google and WeChat for health. We have focused upon uses of the smartphone for health benefits that are free, as opposed to the commercial development of mHealth. The projects that have developed from this part of our project will be published elsewhere,[22] but they account for the frequent references to issues of care and health within this volume. An additional component has been a result of Covid-19. All the team have naturally remained in contact with their friends and participants within the fieldsites during this period. We have been thereby researching by default the way people utilised their smartphones in response to the pandemic and the condition of social isolation.

The methodology used in this research has been described in an Appendix. This is because in most respects it was the standard ethnographic method as employed in anthropology. But if this is not familiar to you, or you want to start by establishing your sense of where the evidence presented in this volume comes from, then of course feel free to read the Appendix prior to the other chapters. The research protocol stipulated that each fieldsite should include at least 25 interviews concerned with each of our three main research topics of smartphones, ageing and health. But our primary method consisted of participant observation, with researchers joining in many local activities. Several members of this team also taught smartphone use to older people, in some cases for over a year.

The fieldsites

One can collect anecdotes in a visit of two weeks. The reason for participating in people's lives for a full 16 months is firstly that this allows the ethnographer to determine whether what is being observed is common and typical or unusual and specific to an individual or particular group. This time frame is also essential for building the necessary trust and friendship to participate in online conversations and activities, as these are mainly private and often dominated by communication within families. One important method was going through each smartphone examining every single app to establish whether and how it was used. Trust was based on the assurance of anonymity and explanations of why it is impossible to teach about the use and consequence of smartphones unless we establish a form of scholarship based on the direct observation of those uses and consequences. In some of the larger fieldsites our promise to retain anonymity was compatible with using the actual names of the location. However, in other cases the names that we use in this book, and others in the series, are pseudonyms.

For most of the fieldsites there are also short films available which provide a brief introduction.[23] The fieldsites included in this volume are as follows (Fig. 1.3):

Figure 1.3 Map of the fieldsites of the ASSA Project (a small project in Trinidad is yet to be carried out). ASSA Project website, accessible at https://www.ucl.ac.uk/anthropology/assa/.

Bento, São Paulo, Brazil

Marília Duque, a Brazilian anthropologist, conducted her research in a district she calls Bento within São Paulo City, in Brazil. The fieldsite is a middle-class borough, with a large concentration of health services. It provides a variety of activities catering for older people and is well served by public transport, including underground trains. As a result the area is frequented by people from throughout São Paulo, meaning that the ethnography covered a wider population than those resident in Bento, including more from low-income areas. For 18 months Duque gave courses in WhatsApp and smartphone use and attended a large number of activities available for older people, including meditation, yoga and entrepreneurship courses/meet-ups.

Cuan, Ireland

Daniel (Danny) Miller, a British anthropologist, worked in a coastal town of around 10,000 people within an hour of Dublin, which he calls Cuan.[24] Originally a fishing village of around 2,300 people, the town has expanded through new estates since the 1970s. It is largely middle class, with typical occupations being in education, health, banking and the civil service, although there is some social housing. He focused on participants in their sixties and seventies and attended many retirement activities, among them learning the ukulele and participating in the Men's Shed. Most of Daniel's interviews were carried out in cafés or people's homes.

Dar al-Hawa, Al-Quds (East Jerusalem)

This ethnography is a joint project between Laila Abed Rabho, a Palestinian researcher, and Maya de Vries, an Israeli academic. Dar al-Hawa is a Palestinian community of around 13,000, which is today a neighbourhood in al-Quds. Prior to annexation by Israel, it was a village between the old city of Jerusalem and Bethlehem. This location remains very present in the daily lives of the people, influencing their relationship to various bureaucracies and to digital and health services. Laila and Maya focused upon the local seniors' club at the community centre and conducted many interviews and conversations both at the community centre and in people's homes. Maya also taught a course on smartphone use.

Kampala, Uganda

Charlotte Hawkins, a British anthropologist, carried out her research primarily in a neighbourhood of around 15,000 people in Kampala, here called 'Lusozi', which means 'hill' in Luganda. Her work was supported by a research assistant who grew up in the area, with the pseudonym Amor, who translated interviews into the many languages spoken in the area. The people who live in Lusozi come from all over the country and the region beyond, but as the majority of participants are from rural northern Uganda, Charlotte also spent time in some of their home villages near Gulu and Kitgum. To understand the use of smartphones specifically, Charlotte drew on methods such as surveys, as well as open-ended interviews and participation in community activities such as women's and savings groups.

Kyoto and Kōchi, Japan

Laura Haapio-Kirk, a British-Finnish anthropologist, worked primarily in two fieldsites. The first was the city centre of Kyoto, a city of 1.4 million people. The second was the Reihoku region of northern Kōchi Prefecture in southwest Japan. This area, like much of rural Japan, is suffering from depopulation and has a high proportion (40 per cent) of residents aged over 65. One of Laura's points of entry to the rural site was through volunteering for an annual health check for the over 75s run by a group of doctors primarily from Kyoto University; she followed this up with frequent visits to the area to gain a holistic understanding of people's lives. She was supported by a research assistant, Sasaki Lise.

NoLo, Milan, Italy

Shireen Walton, a British-Iranian anthropologist, carried out her research in a mixed-income neighbourhood in Milan. The area has recently been called NoLo (North of Loreto), following the idea of a social street in Italy.[25] It includes people from different regions of Italy, for instance Sicily, as well as from Egypt, Peru and the Philippines. She also worked with the Hazara community from Afghanistan. Shireen participated in activities ranging from a women's choir to a sewing group, a multicultural centre and Italian language classes. She lived in a diversely populated apartment block in the centre of the fieldsite, which became a hub of her urban digital ethnography.

Santiago, Chile

Alfonso Otaegui, an Argentinian anthropologist, conducted fieldwork among two populations within Santiago. The first were retired Chileans who attended the courses in smartphone use that he taught at a cultural centre for older adults. Long-term observation of participants allowed him to spot patterns and difficulties among older adults adopting new technologies. The second population were Peruvian migrants, who are very attached to their traditional expressions of Christianity and continue to honour their patron saints while abroad. Alfonso was able to observe how these late-middle-aged migrants stayed in touch with kin and friends across the diaspora through the use of WhatsApp groups and Facebook live broadcastings of parties and processions.

Shanghai, China

Xinyuan Wang, a Chinese anthropologist, conducted fieldwork in Shanghai, China's biggest metropolis, with more than 27 million inhabitants. She focused on a few 'mini' sites within Shanghai: a relatively low-rise residential compound in the city centre, a suburban area with crowded tower blocks of the type where most Shanghai residents live, a home for older residents in a medium-income suburb and a massive care centre for the elderly in a town adjacent to Shanghai. From early on in her fieldwork, Wang researched and then created a changing exhibition based on the family photo albums and oral history of the residential area in which she lived.

Thornhill, Dublin, Ireland

Pauline Garvey, an Irish anthropologist, lives with her family in Thornhill,[26] a middle-class coastal suburb on the northern side of Dublin city. Dublin city has a population of approximately half a million, while the core area of Thornhill includes approximately 20,000 people. Most people work in the city and travel on public transport from the suburb into the centre. Occupations include in the professions, banking, public sector and self-employed. Pauline joined various community groups, including a craft and coffee group, church groups and a walking club. Although she worked predominantly with retirees, her research also included participants in their forties and fifties. Because the sites of Cuan and Thornhill were studied in unison, there is more usage of the general terms 'Ireland' and 'Irish' than for other fieldsites, where generalisations at that level are mainly avoided. We also sometimes use 'Dublin' to stand for both fieldsites.

Yaoundé, Cameroon

Patrick Awondo, a Cameroonian anthropologist, conducted his research within Yaoundé, the capital city of Cameroon and home to 2.8 million people. His focus was on a middle-class district, here given the pseudonym 'Mfadena'. Most of the people in this area are senior civil servants working in central administration or in other public affairs such as education and culture. Many residents of the district are also involved in private businesses or work for private companies. They come from all over the country and include some expatriates. Two of Awondo's main entry points to the community were sports leisure groups and self-help groups, the latter known locally as *tontines*.

History of the smartphone

While this chapter cannot provide a history of all the precedents to the smartphone, it can at least offer a brief history of the smartphone itself, inasmuch as there is also a 'phone' in there somewhere. The mobile phone became an established consumer device during the 1990s. Increasing familiarity with the mobile phone in its earlier iterations did not, however, reduce the sense of awe and wonder when the iPhone was first presented to the world in 2007.[27] This may not have felt quite such a radical change in Japan, where internet-enabled feature phones known locally as *garakei* represented something of a bridge[28] and remain popular among older users (Fig. 1.4). The nearest candidate to a bridge otherwise may be the Blackberry, which became quite a popular device worldwide[29] before the emergence of the iPhone. Apart from that, this touchscreen, app-based device appeared to offer a whole new world.

The smartphone's own history has three defining moments.[30] The first was the arrival of the iPhone, because almost everything we think of as special about the smartphone was present in the initial offering that dazzled the world in 2007. The second was the rise of the Android phone, and especially the Samsung Galaxy. This development made smartphones a diverse species, shunting the iPhone into a minority presence. The third hugely important moment was the rise of cheaper smartphones, mostly

Figure 1.4 Example of a Japanese feature phone (*garakei*). Photo by Laura Haapio-Kirk.

from China. Huawei launched its first Android phone in 2009 and Xiaomi launched its first smartphone in 2011.[31]

There is relatively little significant difference, other than status, between the cheaper models and the more established smartphone brands. The rise of cheap handsets has allowed the smartphone to become a global presence no longer restricted to more affluent regions. Such an expansion was the prerequisite for this book, since much of the research included less affluent populations who may not possess top-end smartphones.[32] In the background are other multiple developments, ranging from assemblage and supply chains to the rise of third-party app development; these, however, lie beyond the scope of this project.[33]

If the 'computer in your pocket' analogy is the nearest thing to a precedent, then it is important to have some sense of the parallel development of the internet.[34] The World Wide Web was invented in 1989, with the first web page being served on the 'open internet' in 1991.[35] However, most people would have been unlikely to have come across this prior to the launch of Mosaic, the first web browser; this effectively democratised internet access by expanding access to web navigation to the ordinary user from 1993 onwards.[36] Much depended on the national context for such developments. Peters has recently argued[37] that while the US is supposed to be highly competitive, the internet was developed primarily through a mixture of state funding and a collaborative research environment. By contrast, while the Soviet Union was supposed to be a centralised state, the attempt to develop something like the internet failed due to diverse bureaucracies and institutions following their own narrow interests. In so doing they fragmented the potential internet into what became a rather competitive project.

The opposite lesson would emerge from a reflection upon the emergence of digital technologies in China.[38] Here state-sponsored development of new communication technologies was a clear and determined policy. It was based on an awareness that this would be vital to China's ability to leapfrog other nations in its quest for modernity.[39] While the state certainly watches over major Chinese tech companies such as Alibaba, ByteDance and Tencent, it also provides them with support in their quest to rival the likes of Google and Facebook.[40] As a result, our contemporary world contains only one clear regional division with respect to digital communication technologies: the one between China and the rest of the world. The population of the latter may be around four times the former, but, depending on what criteria are used for the calculation, the largest social media company in the world may be Tencent, rather than Facebook.[41]

The result is also evident in that six of the ten largest smartphone companies are Chinese.[42] They are Huawei, Xiaomi, Oppo, Vivo, Lenovo (which includes Motorola) and Tecno. Korea represents the only serious rival to China through Samsung, the dominant brand, as well as LG. The other companies are Apple in the US and Nokia HMD in Finland. The top three brands, Samsung, Apple and Huawei, have each managed to capture more than 10 per cent of the total market. Over the last few years Chinese companies such as Huawei and OnePlus have produced smartphones on a par with the premier brands of Samsung and Apple; other Chinese companies retain dominance of the cheaper smartphone market. In India there are now smartphones marketed for just a few pounds.[43] Versions of the iPhone 11 start around £679, while the 'Pro Max' version may cost more than £1,000.[44]

Equally important for anthropologists is the history of how populations have responded to prior technologies. What were the precedents that may help us account for the smartphone's appropriation, refusal and adaptation? The sociologist Claude Fischer[45] has studied the impact of landline phones in the US between 1900 and 1940, concluding that 'the telephone did not radically alter American ways of life; rather, Americans used it to more vigorously pursue their characteristic ways of life'. One of his most important observations is that those who marketed the phone to American households were quite slow to appreciate that it would largely be used for sociable conversation.[46] As a result, the people who really invented the landline were not so much the technical inventors or corporations. They were rather the consumers – especially rural consumers, who were the keenest to obtain the telephone and who best appreciated its potential.[47] Fischer discerns no obvious or major social or psychological impact. He concludes that 'the best estimate is that, on the whole, telephone calling solidified and deepened social relations',[48] rather than replacing face-to-face relationships. What the telephone appeared to engineer, then, was a general expansion of talk.[49]

Comparing the earliest to the latest developments, Fischer's observations remain quite close to the conclusions of a recent project in which several of us were involved called *Why We Post*.[50] This project examined the ways in which social media has been creatively reconfigured by populations of users. The evidence from *Why We Post* also revealed that much of social media employment has been quite conservative. Often these new media are used to repair something of the fragmentation and disorientation that are otherwise occurring in people's lives. The most obvious example is the way in which families who have been separated, for reasons ranging from war to the pursuit of new economic opportunities, use social media to try to reconstruct the close and continual family communications that they might otherwise have lost. This emphasis upon conservative usage helps to balance our natural fascination with whatever seems new and unprecedented.

These studies of telephones and then social media make a further point about the nature of causation. If the technologies were the primary cause of the way the devices are then used, we should be able easily to map subsequent usage against the specific technology intended for such usage.[51] Instead, the evidence from *Why We Post* was that genres of behaviour migrate quite easily from platform to platform. School banter moved readily from Blackberry to Facebook to Twitter, three completely different platforms.[52] If the genre of usage remained largely the same across platforms, then the properties of the platforms cannot be the primary explanation of their usage. The series *Why We Post* also uncovered a regional diversity of usage such that the main volume of results was called *How the World Changed Social Media*,[53] not how social media changed the world.

A similar point emerges if we consider the history, brief though it is, of an individual platform such as Facebook. At first its inventor tried to forbid usage to anyone not at Harvard; subsequent attempts were made to restrict the platform to university students. More recently, the *Why We Post* project documented that Facebook was declining in popularity among young people,[54] especially in more affluent markets such as the US.[55] There is simply no commercial logic that suggests that Facebook itself would want to lose its youthful image in those markets.[56] Again, therefore, other factors must lie behind these changes, beyond simply the interests of corporations.[57]

The landline is now fading into history, though it is still a presence for older people. Smartphones continue to be used alongside 'feature phones' and mobile phones, however, especially in some of the lower-income populations we studied. The history and understanding of these closer relatives is therefore also important, as they document trajectories that may continue to develop through smartphones. Not surprisingly, much of the associated literature on mobile phones has focused upon the specific consequences of mobility. Titles such as *Perpetual Contact* or *Personal, Portable, Pedestrian*[58] give some indication of these concerns. Of particular influence has been the ongoing work by Ling,[59] who has contributed a number of useful terms for analysing the mobile phone. Ling also pointed out how people were using these phones for micro-coordination such as 'midcourse adjustment', 'iterative coordination' and also the 'softening of schedules' – all of which illustrate the ways in which the mobile phone can be used to increase flexibility.[60] More recently Ling[61] has considered the ubiquity of the mobile phone and its subsequent 'taken for granted' status by looking at how it has become embedded alongside other 'taken for granted' technologies, for example the watch and the car. Also useful is the idea of technomobility introduced by Wallis.[62]

Studies of mobile phones have emphasised the intrusive nature of mobile telephony[63] and the dissolving of traditional boundaries between the public and the private sphere.[64] This in turn leads to a consideration of the etiquette that has developed around the use of the phone.[65] An initial focus was often on youth[66] – including the phone in relation to fashion, style and the body[67] – and also the impact of young people's use upon parenting. There is a wide literature that has emerged from communication studies and media studies. In addition, there has been increasing interest from development studies and sub-fields such as ICT4D (Information and Communications Technology for Development), concerned with the impact of mobile phones on world populations.[68]

Anthropology and other disciplines

Essential to anthropology has been the avoidance of perspectives that regard a device as a given object, which is then adapted or appropriated by local populations. In their study of how people used the internet in Trinidad, Miller and Slater[69] insisted that there is no such thing as the real or 'proper' internet. What we call the internet is simply the uses that any particular population has made of the possibilities of online. Trinidadian usage was not a distortion or localisation: it is simply another equal example of what the internet is. Similarly, populations are not fixed. A Trinidadian who uses the internet is no less Trinidadian, but does represent a change in the meaning of the term Trinidadian. The process is one of mutual change which respects the equality of all populations.

Anthropologists tend to eschew general debates about what a mobile phone is, preferring to delve more deeply into highly specific local circumstance and usage. For example, Archambault[70] studied mobile phones in a suburb of a provincial town in Mozambique. Her book does not start from UK or US debates concerning the impact of phones on privacy. Her primary interest is the mobile phone's capacity for subterfuge and secrecy, promoting and enhancing skills of concealment and exposure, as well as its consequences for new forms of intimacy and new ways of getting by financially. How, for example, to display social status without attracting envy? She also documents how the phone makes it easier to meet with a lover, but also creates new ways people can check up on each other, such that the phone is blamed for the exposure, not the infidelity itself. Respectability depends not upon what you do, but upon your ability to be discreet about it. In a situation of precarity, turning a blind eye may be essential to economic survival. Phones are also easy to steal and easy to sell, thus creating their own underground economy based on crime. The presence of the mobile phone raises all sorts of other questions about trust and intimacy that dominate local conversation about what a mobile phone is.

A second example is Tenhunen's study of mobile phones in a West Bengal village with around 2,400 inhabitants,[71] conducted between 1999 and 2013. Over time the phone takes its place among a wide variety of forms of co-presence. At first it is used for specific local tasks such as helping people to attend funerary rituals. Later it becomes part of wider political, social and economic change. The primary concern remains kinship – for example, telling one's relatives about work opportunities or organising healthcare. This includes radical changes: wives can remain in touch with natal families, for instance. The emancipatory impact is limited, however, as they may depend on husbands or in-laws to make calls. Here, as elsewhere, the primary drivers behind much of the uptake are not in the realms of politics or the economy, but rather the range of entertainment and fun that people can access through these new phones.[72] Lower castes who could not afford a television leapfrog such technologies, but they can also use phones to hear about factory employment. However, mobile phones do not have much impact upon the basic hierarchical social system.[73] While Tenhunen focuses upon a single village, her work is complemented by an excellent survey of the impact of phones on India more generally in *The Great Indian Phone Book*,[74] which scales this up to consider the implications for a vast population.

A third example of an ethnography, the method upon which this volume is based, is set in a small settlement within the Solomon Islands of the South Pacific. The study by Hobbis[75] strongly reinforces points made earlier in this chapter about smartphones being defined by usage, not by capabilities. The population that Hobbis studied make almost no use of texting and limit voice calls to around one call lasting only a minute or two per fortnight. Yet the smartphone has important consequences for the fundamental organisation of that society through kinship; it is used extensively in areas such as looking after children and within gender dynamics. This provides perhaps the extreme example of a smartphone that seems anything but a phone.

These three books are typical of the anthropological task: sustained and empathetic participation in people's lives that allows them to convey what it might be like to live in a village in India or to be a young man in urban Mozambique with a new mobile phone. There are, however, other complementary approaches – as found in an edited collection by Foster and Horst,[76] also focusing on the Pacific region. Within that collection Horst, for example, followed Digicel, the main supplier of mobile phones to the Pacific and the same company she had researched alongside Danny in Jamaica in 2005.[77] A study of Digicel's advertising reveals how the company tries to present itself as a friend or to embody the moral virtues of a good citizen. Jorgensen shows how the presence of phone masts planted by Digicel provokes anxieties about surveillance and control, but also allows local politicians to claim credit for 'development'.[78]

Other chapters in Foster and Horst's collection examine specific contexts. Lipset[79] presents a case in which the mobile phone, the first mechanism by which people can contact strangers, becomes a liberation from the constraints of traditional kinship. Wardlow[80] worked with women infected by HIV. Shunned by their kin, they use random calling to find supportive strangers who give them emotional and practical care; unlike family or boyfriends, these generous individuals do not ask for money, pigs or, in the latter case, sex in return. This edited collection shows how anthropologists may also study the relevant corporations and the wider political economy associated with the expansion of smartphone usage.[81] Horst has also argued for more ethnographic attention to the infrastructures behind mobile communication.[82]

Anthropology is just one of a constellation of disciplines making contributions to our understanding of mobile phones and smartphones. Summaries of the work on new media and personal relations include *Personal Connections in the Digital Age* and *Social Media and Personal Relationships*.[83] Sociologists focus more on individuals within networks. This approach is usefully summarised in *Networked*,[84] a volume which also employs the many excellent and ongoing surveys by the Pew Research Center in the US. Other influential approaches have been developed in computer and internet studies.[85] A major contribution has been that of communication studies, such as the volumes edited by Papacharissi,[86] which have introduced many of the more specialist terminologies. Since the smartphone has colonised every area of life, as noted above, there are now insights coming from every discipline, whether their primary concern is religion, crime or tourism. Often books with topics such as location technologies will feature contributions across various disciplines.[87] Anthropologists have contributed to many of these fields; for example Postill[88] has contributed to the study of digital politics. There are also nascent disciplines that study the unprecedented effects of digital communications, again with anthropological contributions such as Coleman's works on hackers and Anonymous.[89]

Externalities

The term we use for ethnography, 'holistic contextualisation', shows a desire to take into account everything that bears on the problem of understanding the global smartphone. It means that we cannot know in advance what the relevant context to our subject of study will be; an ethnography thus tries to consider all aspects of everyday life in case they turn out to be relevant. But there are many forces bearing on the smartphones that simply do not appear within an ethnography. Furthermore, this volume is only based on the ethnography of users; it does not contain equivalents to Horst and Foster's consideration of the companies and infrastructures. As Horst has also noted, there is always a still wider social and economic context which might be considered part of the smartphone's infrastructure, usually including state regulations.[90]

The ethnographic focus of our study therefore threatens to result in an absence of what economists call externalities. When the price of an object reflects the cost to the firm that produced it, but does not include the cost of dealing with the air pollution created by its manufacture, then air pollution has been made into an externality. What are the consequences of smartphones that are also less apparent within an ethnographic focus?

Fortunately there may be other studies that provide essential findings to complement those presented here. For example, Richard Maxwell and Toby Miller consider in their book *How Green is your Smartphone?*[91] the various ways in which smartphones may have a powerfully negative impact upon the environment and our welfare. There are wider ramifications of the components of smartphones, not only the politics of rare earths and the ecological consequences of the very materiality of the smartphone, but also the less tangible uses of energy involved, including the vast lattices of digital infrastructures involved in enabling global communication.[92] Smartphones have become part of the data collection that, in turn, feeds into artificial intelligence and other current developments. Their use in tracing people's interactions as a means of curbing the Covid-19 pandemic has made clear just how powerful they have become in this regard.

Other studies focus upon the wider political economy of smartphones and major corporations such as Apple, Facebook, Tencent or Samsung,[93] and extend these to a wider notion of 'platform capitalism'.[94] Several new studies have brought to attention groups otherwise ignored, such as ancillary occupations. Sarah Roberts's *Behind the Screen*[95] examined what she calls the 'obfuscated human labour' of content moderators, who have emerged as part of the corporations' response to the moral pressures for them to take responsibility for content.

A similar exposé is found in the book *Ghost Work* by Gray and Suri.[96] In chapter 9 we consider another major externality exposed by recent critical debates about the rise of surveillance capitalism[97] and the surveillance state.[98] Incorporating the recent experiences from the response to the Covid-19 pandemic, we will examine the fine line that exists between care and surveillance. The issue of surveillance introduces in turn an increasingly extensive literature around the development of Big Data and artificial intelligence (AI), both of which are also incorporated within the 'smart' elements of the smartphone.[99] The reason for calling these 'externalities' is precisely because they should not be external. They should actually be understood as integral to what smartphones are and the consequences they produce. However, they may not be apparent from the ethnographies that we carried out. This volume thus has to be complemented by those referenced above.

Conclusion

The Global Smartphone refers to our comparative study of smartphones based on 10 local ethnographies. Anthropologists tend to the plural. There are internet*s* and capitalism*s*, each of which may be quite different in specific contexts.[100] This volume is a composition based on scales – often trying to listen in to the singular tunes of an individual smartphone user and ensuring they are not drowned out by our interest in what is usual or widespread. Yet often tunes do not come from individuals, but from families or communities. We also listen to the different genres found in urban Ireland and in Santiago, and to the contrast between rural and urban Japan and low- and middle-income neighbourhoods in Kampala and Yaoundé. Where there is evidence for the smartphone as an instrument of global homogenisation or generalisation, this is acknowledged – for example, the increased role of visual forms in communication and care or the widespread use of step-counters. Even when it comes to theory in the final chapter, however, generalisation and abstraction are subject to nuance and variation that ultimately comes back to those families or individuals. Playing across such scales enables anthropology to face in both directions, giving equal respect to the parochial and the general.

Our 'smart from below' perspective focuses on how people craft smartphones and not just use them; they do far more than simply add content. They reject in-built components such as voice assistants, they reconfigure apps around certain tasks to suit their own routines and they acknowledge social etiquettes around what is acceptable and unacceptable usage. These selections then contribute to the creation of the smartphones we encounter. All of these activities take place in circumstances of constraint. Women or older people may be prevented from access or knowledge. Essential elements may be unaffordable. Corporations will push users in directions which create profits. Smartphones prove intrusive and some people talk of being addicted. This is rarely a volume that casts judgement, but it is one that seeks to acknowledge diversity. It is the diversity both of users and of these constraints that has created the global smartphone.

Notes

1. The continuity is represented by Laura Haapio-Kirk, Daniel Miller and Xinyuan Wang, all of whom worked on *Why We Post* prior to the current project. This study of social media around the world was also based at the Department of Anthropology, University College London. More information about the project can be found at the following URL: https://www.ucl.ac.uk/why-we-post/.
2. Statista 2019.
3. Mobile Internet Statistics 2020.
4. Clements 2014.
5. We wish to acknowledge that the term 'smart from below' is taken from a paper by Katrien Pype (2017). We believe Pype's work in Kinshasa to be among the best anthropological current work on smartphones and other digital technologies and cite several of her papers in the Bibliography to this volume.
6. *Telegraph* 2019.
7. This is not to say there are no helpful histories of the mobile phone. See, for example, Agar 2013.
8. Sarvas and Frohlich 2011.
9. For more complex, historically informed studies of visual (and visual-digital) communication genres see, for example, Mitchel 1992; Friedberg 2006; Dijck 2007; Mirzoeff 2015; Favero 2017.
10. Humphreys 2018, 29–49.
11. Modern mass media have been thought to facilitate a 'culture of narcissism', as claimed by cultural historian Christopher Lasch. In his seminal study of American postwar society, Lasch identified a 'pathological narcissism' that allegedly went hand in hand with the decline of the family: Lasch 1979. This kind of critique has been particularly levied at contemporary digital culture from within the field of Psychology; it is typified by preoccupations between the link between modern-day 'Selfie culture' and narcissism. See, for example, Weiser 2015; Sorokowski et al. 2015; Barry et al. 2017.
12. Frith 2015 and Greschke 2012.
13. Bogost 2020.
14. Cronin 2013.
15. Norman 2015.
16. In addition, a small follow-up study is planned in Trinidad. It has not yet commenced.
17. For example Istepanian et al. 2006 and Donner and Mechael 2013.
18. Schaffer et al. 2008.
19. Oudshoorn 2011.
20. Hingle et al. 2013.
21. European Commission 2020.
22. The team is currently writing a volume about our health-related projects and observation, but has already published a manual for the use of WhatsApp in relation to health. See Duque 2020.
23. Introductory films about some fieldsites can be found at https://www.youtube.com/playlist?list=PLm6rBY2z_0_jA3jTEJh5faHJoL0_-Ow7j.
24. The two Irish fieldsites had very similar populations and the two anthropologists came to very similar conclusions. For this reason we have combined them into a single 'fieldsite' of Dublin. Athough Cuan lies at some distance from Dublin city, Dublin is also a region which includes Cuan.
25. The Social Street was an idea born in Bologna in 2013, but is now common throughout Italy. It aims to promote greater socialising between residents of the area. See Social Street 2020. The website can be found at: http://www.socialstreet.it. The NoLo social street concept emerged in 2016 with the founding of the NoLo Social District Facebook Group.
26. See note 24 on combining the Irish fieldsites.
27. BBC News 2007.
28. Itō et al. 2005.
29. Sweeny 2009. Various personal assistant devices also served as precedents for elements of the smartphone, such as the Palm pilot and the Nokia communicator.
30. For a useful summary of this history see Woyke 2014.
31. Shirky 2015; Gupta and Dhillon 2014; Jia et al. 2018.

32. Or even smartphones at all. In Africa mobile phones remain more widely used than smartphones, although the latter are used by a growing minority of the population. See Xinhua 2019. For India see Counterpoint 2019.
33. For a discussion of the development of cheap smartphones in China, see Li Sun et al. 2010; Fu et al. 2018; Liu et al. 2015.
34. Of which there are many histories, for example Naughton 2000.
35. Web Foundation 2020.
36. Tagal 2008.
37. Peters 2016.
38. Shim and Shin 2016; Serger and Breidne 2007; Feigenbaum 2003; Plantin and de Seta 2019.
39. Hughes and Whacker 2003. For a quick summary see Keane 2020.
40. Jia and Winseck 2018.
41. Bhardwaj 2018.
42. Gadgets Now 2019.
43. Though claims of smartphones for a mere $4 need to be treated with caution. See Patil 2016.
44. Apple Inc. 2020.
45. Fischer 1992.
46. Fischer 1992, 85.
47. Fischer 1992, 119.
48. Fischer 1992, 266.
49. Fischer 1992, 268.
50. Miller et al. 2016.
51. Though technology can certainly be a major factor – see MacKenzie and Wajcman 1999.
52. Miller 2016, 183.
53. Miller et al. 2016.
54. Miller et al. 2016.
55. Al-Heeti 2019; Solon 2018.
56. Kirkpatrick 2010. The phenomenon is well referenced within business studies using the concept of 'crossing the chasm' – see Moore 1991.
57. For example, Miller argues that the increasing use of the term 'friending' is not a reflection of how we use Facebook; it is rather the fact that Facebook reflects longer term changes in the relationship between kinship and friendship. See Miller 2017.
58. Titles such as Katz and Aakhus 2002; with respect to Japan, Itō et al. 2005.
59. Ling 2004; Ling and Yuri 2002.
60. Ling 2004.
61. Ling and Yuri 2012.
62. Wallis 2013.
63. Ling 2004, 123–43.
64. Licoppe and Heurtin 2002.
65. Kim 2002.
66. Livingstone 2019.
67. See also Fortunati 2002; Fortunati, Katz and Ricini 2003.
68. See, for example, Ling 2004. For a summary see Green and Haddon 2009. For development studies see, for example, Donner 2015.
69. Miller and Slater 2000, 1–4.
70. Archambault 2017.
71. Tenhunen 2018.
72. *The Economist* 2019.
73. The conclusions are similar to Venkatraman working on the impact of social media at the other end of India in Chennai. See Venkatraman 2017.
74. Doron and Jeffrey 2013.
75. Hobbis 2020. For media practices and family dynamics more widely see Hjorth et al. 2020.
76. Foster and Horst 2018.
77. Digicel cut its teeth in bringing mobile phones to small island populations in the Caribbean, starting with Jamaica in 2001. The company then transferred this expertise to the Pacific, starting with Samoa in 2006 and reaching at one point 97 per cent of market share in Papua New Guinea.
78. Jorgensen 2018.

79. Lipset 2018.
80. Wardlow 2018.
81. For examples from Africa see Hell-Valle and Storm-Mathisen 2020.
82. Horst 2013.
83. Baym 2010.
84. Rainie and Wellman 2014.
85. Graham and Dutton 2019.
86. Papacharissi 2010; Papacharissi 2018.
87. For example Wilken, Goggin and Horst 2019.
88. Postill 2011; Postill 2018.
89. Coleman 2013; Coleman 2014.
90. Horst 2013.
91. Maxwell and Miller 2020.
92. Carroll 2020.
93. For example Garsten 1994; Kirkpatrick 2010.
94. Srnicek 2017.
95. Roberts 2019.
96. Gray and Suri 2019. These are not new concerns. For recent precedents in India see Xiang 2007. For much older precedents see Kriedte, Medick and Schlumbohm 1981.
97. Zuboff 2019.
98. Greenwald 2014.
99. boyd and Crawford 2012.
100. Miller 1997. See Bolter and Grusin 2003; Sarvas and Frohlich 2011; Dijck 2007; Bunz and Meikle 2017; Halavais 2017; Frith 2015; Duque 2020.

2

What people say about smartphones

Fieldsites: **Bento** – São Paulo, Brazil. **Dar al-Hawa** – Al-Quds (East Jerusalem). **Dublin** – Ireland. **Lusozi** – Kampala, Uganda. **Kyoto and Kōchi** – Japan. **NoLo** – Milan, Italy. **Santiago** – Chile. **Shanghai** – China. **Yaoundé** – Cameroon.

The definition of the telephone, at least until recently, would have been a device that is used for talking to or texting to another person. Yet by 2021 the smartphone has become hugely important as a device that we also talk *about*, rather than just talk through. This chapter will demonstrate just how significant these discourses about smartphones have become. Talking about smartphones has become a way of discussing a range of moral and other concerns people have about contemporary life. It follows that to understand the consequences of the smartphone, we need to address not just the technology and its usage, but also the role of the smartphone as an icon and idiom. Just as we now pay attention to the specifics of online language,[1] so should the discourses that surround smartphones be regarded as integral to what a smartphone is, alongside the material object. Our evidence will also suggest that these debates about the smartphone are rarely directed at its actual everyday use, the topic of the remaining chapters of this book. Rather, using smartphones to discuss morality is a quite separate function of the smartphone: the employment of smartphones to reflect a wider discourse about modern life. Discourses about smartphones therefore need to be acknowledged and considered in their own right, including major debates on topics such as addiction, fake news and surveillance.

Mostly, though not inevitably, the discourse tends to be negative when people are discussing the smartphone in general and more positive when people are asked about particular apps or uses, such as location finding or photography. In our Dublin fieldsites it was common for older people initially to claim that the only things they used their smartphones for were voice calling and text messaging. Yet when we carried out more systematic interviews based on inspecting their actual smartphones, it was apparent that these same individuals were often using around 25 or 30 different apps and functions. In part it seemed that participants were making these statements in order to distinguish themselves from young people who, they suggested, seemed to be constantly hooked on their devices. Having claimed this minimal use of the smartphone, they then quickly turned to problems they associated with the device, for instance the invasion of privacy or fake news, some of the topics explored in this chapter. This is why it is important to take these two forms of evidence separately: what people say about smartphones in general and the evidence of their everyday usage. Each is shown to have its own reasons and consequences.

An emphasis upon discourses is useful, however, for starting to explore some of what in the last chapter were termed 'externalities'. What people say about smartphones has clearly been influenced by larger forces such as governments, the media, commercial corporations and other institutions, such as religious ones.[2] One such force would be academia itself, including contributions by psychologists, political theorists and other disciplines involved in issues of policy and regulation.[3] For example, research participants might complain about commercial interests infiltrating their accounts, express anxiety regarding privacy and surveillance or comment on the health implications of possessing and using smartphones. They may also repeat stories from newspapers about the findings of a psychologist or claims by governments that people are in danger of becoming addicted to their smartphones. We rarely needed to extract such statements from research participants. As soon as you explain that you are studying smartphones, people often start to recite a whole litany of concerns and claims that they regularly read about, hear about and discuss among their relatives and friends.

The state and the media

Many governments now view the provision of basic digital infrastructure as essential to their people's welfare. They may consequently encourage the development of 4G or 5G, free Wi-Fi hotspots and other means of extending the network to include people in rural areas. For example, in the Japanese rural fieldsite, which suffers from depopulation, one of the recent concerns of the local government was the installation of high-speed broadband as part of revitalisation efforts to maintain the population and attract people to live there. Governments are well aware of the electoral benefits in providing what is increasingly regarded as a basic public utility to which people have rights.

At the same time the state may be viewed as one of the prime culprits when people express their anxiety about surveillance. In a recent report in Ireland, for instance, a government department sought tenders for media monitoring, including that of social media. They claimed this would help to inform them of citizens' opinions.[4] Although most social media content is relatively public data, such monitoring is widely regarded by research participants as a sign of the state's invasive inspection of private correspondence. This concern has grown alongside global controversies such as the Cambridge Analytica scandal,[5] where political parties were said to have harvested data about users for their own advantage, or the revelations made by Edward Snowden around state surveillance. Elections can equally become lightning rods for such fears about the rise of surveillance, as for demands around access to Wi-Fi or 4G. There are essentially two equally opposed trajectories, one considered positive and one negative, and many people appear to respond with profound ambivalence.

In Uganda debates about the moral consequences of smartphones have become an idiom for expressing intergenerational tensions. Older people in Lusozi complain that the knowledge they have gained from decades of personal experience is now undermined and disrespected by the younger 'dotcom' generation, influenced instead by various unedited claims to knowledge accessible through the internet and other global media. These complaints are then broadened into concerns about misinformation, 'Westernisation', the loss of traditional respect and togetherness, and various forms of harm being inflicted upon young people.

This narrative is both reflected by and fuelled by the government. In 2018 President Museveni (aged 76 in 2020) argued that social media promotes fake news, '*olugambo*' (gossip), accusations of witchcraft, pornography, addiction and general 'over the top' excess. On 1 July 2018 the Ugandan government introduced the 'Over the Top' (OTT) tax,[6] a directive to telecommunications companies to discourage 'olugambo'. This tax consisted of a charge of 200 UGX (£0.05) per day, or 6,000 UGX (£1.20) per month, for using various social media platforms including Facebook, WhatsApp, Instagram, Twitter, Skype and LinkedIn (Fig. 2.1).[7] When the OTT tax was announced, the speed with which the announcement spread to 15,000 WhatsApp users became news in itself. The tax prompted various petitions and campaigns, among them social

Figure 2.1 The Ugandan OTT tax on social media as shown on a mobile phone. The user has the option of paying OTT for their own number and also another number. Photo by Charlotte Hawkins.

media campaigns such as #ThisTaxMustGo. Protests, including those led by Bobi Wine, a musician turned opposition leader and proclaimed 'spokesman for Uganda's frustrated youth', dominated headlines in Uganda and attracted international attention. Bobi Wine (aged 38 in 2020) frequently draws on social media as a platform for the emancipation of the younger generation from the country's ageing president.

Political control over smartphone use may include attempts to restrict access to the internet. This occurs in Cameroon, where the government has cut off online access to a whole region of the country dominated by English-speaking Cameroonians. These regions are viewed as the opposition to the dominant French-speaking population and government.[8] As in Uganda, claims are made that such measures are required to protect people from 'bad technologies', a discourse promulgated by community leaders and journalists. In 2017 an article published under the title 'Cameroon: the mobile phone – beyond the value of use – death' summarised the risks of mobile phone use as 'the dangers incurred due to frequent exposure to radio frequencies', as well as 'hearing problems, cancer risks' and other risks that arise from speaking on the phone while driving.[9] Not only are physical dangers related to the phone presented as real and substantive risks, but the author also comments on the laziness of young people who rely on the phone to foster their social networks, all the while being subject to the dangers of street accidents, electrocution or fire.[10]

China has developed many additional facets of state control over the internet. Its central government actively shapes the infrastructure and rules of the country's information superhighways, as well as controlling access through a national search engine.[11] The internet is regulated through a 'three-layer-filtering' system: the Great Firewall, keyword blocking and manual censoring.[12] The Great Firewall blocks off blacklisted websites and social media services from mainland China, including Facebook, Twitter, Google and Wikipedia. The second filter, 'keyword blocking', can automatically censor sensitive material. The third content filter requires a huge investment in labour given the vast amount of information online. It is estimated that there are between 20,000 and 50,000 internet police and internet monitors nationwide, as well as between 250,000 and 300,000 paid propaganda posters (so-called *wu mao dang*). There may also be up to 1,000 in-house censors hired by an individual website for 'self-censorship'.[13] This is because Chinese internet companies need licences to operate and are required to police themselves, filtering out any illegal content, which ranges from pornography to politically sensitive material.

In many countries, the practices of the state are seen as centred on surveillance, something that is relatively common in Israel. A large segment of the Arab citizens and residents living in Israel know that security institutions are following them, including through what is said in mosques or schools. Israeli intelligence is a recognised and important body whose full capabilities are hidden, but the power of which is evident to the general public, especially Palestinian residents living in areas such as Dar al-Hawa. The internet and digital platforms have made tracking and monitoring easier because much information is openly online.[14] In 2014 many examples arose of young Palestinians posting statements or farewell messages on Facebook before carrying out attacks on Israelis. This escalated the surveillance and profiling of social media activities in particular,[15] leading in turn to a decline in this practice. The result is considerable sensitivity to the consequences of what is written more generally on social media. Subsequently, with the spread of Covid-19, Israel became one of the prime examples of the tensions between care and surveillance represented by the smartphone, as the smartphone's capacity for contact tracing became clear to all.

In many of our fieldsites we found that the press has become another major contributor to negative discourses surrounding smartphones. A factor may be the degree to which online media originally represented a serious competitor for newspapers and a challenge to their financial viability. In addition, there may be a sharp divide between older people, who tend to take their news from newspapers (increasingly including their online versions), and younger people, who are more likely to take their news from sources other than newspapers. As a result, it is possible that newspapers respond with more conservative and negative discourse about smartphones that appeals especially to these older readers. Newspapers generally regard themselves as incorporating traditional regulations and higher standards of reporting, as well as spearheading politically important traditions of critically challenging states and corporations increasingly believed to be engaged in illegitimate surveillance and data gathering.[16] The established press would mostly claim a contrast between their practices of stringent fact-checking and the lack of responsibility or commitment to integrity found in some online news; in some instances it is hard to trace information to any given source and these outlets may be seen as promulgating 'fake news'. The distinction between offline and online has blurred as the media increasingly include online formats. In China, for example, it is very rare to find any newspaper or magazine which does not also have a public WeChat account. But there remains a constant struggle for those who regard themselves as serious accredited journalists to remain distinguished from other kinds of posting.

In Italy there is now a clear convergence of negative and cautionary narratives surrounding smartphones from sources in academic literature and governmental organisations, as well as from NGOs. Children, teenagers and young adults, those regarded as 'digital natives', are deemed to be subject to addiction, becoming 'slaves' to the device'.[17] Psychiatry, in particular, is drawn upon by the media to present the case for smartphone addiction causing various psychiatric disorders among young people who are found to be in the 'high risk' group.[18] Both politicians and the media have fostered a national debate about smartphone addiction as a significant public concern.

In 2019 a bill was proposed to parliament in Italy to tackle 'widespread smartphone addiction', particularly among 15–20-year-olds, based on reports that this group 'consults their phone on average 75 times per day'.[19] Vittoria Casa, a politician in the Five Star Movement, stated that the problem is 'getting worse and worse and must be treated like an addiction … it's the same as gambling'.[20] The bill proposes courses in schools on the dangers of smartphone addiction, as well as campaigns to inform parents. There are also discussions about the possibility of establishing health centres, akin to rehabilitation centres, aimed at 're-educating' the young away from their smartphones, towards a more 'conscientious use of the internet and social networks'. The press in Italy report on 'no-mobile-phone phobia' to describe the anxiety caused by having no access to social networks or messaging apps. In another example the headline reads 'Italians, always more smartphone mad: 61 per cent use it in bed, 34 per cent at the table'.[21]

In September 2018 the daily Italian general interest newspaper *La Repubblica* highlighted the theme of smartphone use by older people. It cited a survey-based study that found that 76 per cent of senior Italians used smartphones regularly. The headline describes the over-55s as being 'inseparable' from the device; as the newspaper comments, 'Set aside the bowls and the playing cards, the over-55s spend their time on Facebook, Twitter and Instagram'.[22]

Citizenship and consensus

In China these criticisms of the effects of smartphones are also quite common, but there is a rather different relationship between the state, the media and the population. From the very beginning of internet development China has seen a profound alignment between the party-state, commercial companies and the media. As Xinyuan has previously noted,[23] the development of new media technologies has become one of the Chinese government's primary strategies as the country seeks technologically to 'leapfrog' the rest of the world; this ambition has increased in priority with the rise of Big Data and AI, both areas in which China is determined to lead the world.[24] The result is an overwhelming emphasis on the positive potential of new media and the importance of including all segments of the population in this drive to the future. For example, an article in the *People's Daily*,[25] the official newspaper of the Central Committee of the Communist Party of China, reads as follows:

As the power of the internet is increasingly integrated into the development of the times, society is constantly changing. Finding ways of helping people of all ages to adapt and embrace such changes is a difficult problem to be solved, and it is necessary to work together to solve it. It is necessary actively to face the ageing of the population and help the elderly to cross the digital divide, not leaving them behind in the internet era, and thus achieving the all-round development of society.

In 2014 several mainstream media reported, in a positive and encouraging tone, the story of a young man, Zhang Ming; he had tried to teach his parents how to use the new smartphones he bought for them, but found they had difficulty learning the skills and tended to forget his instructions.[26] After Zhang Ming returned to Beijing, his parents called him several times for advice on the use of WeChat. Finding himself too busy to deal with these requests, Zhang Ming hand-painted a nine-page 'manual' of WeChat instructions. The reason the media may have taken a particular interest in Zhang Ming is that this story creates an alignment between two popular themes: the drive to advance digital technologies and ancient Confucian ideas about respect for one's parents, expressed here as 'helping the elderly to cross the digital divide'.

Many older people in China subscribe to an ideal of citizenship that is strongly espoused by those brought up with Communist ideology. It is commonly said that the fate of the individual (*geren de mingyun*) is bundled with the fate of the state (*guojia de mingyun*). For older people it almost goes without saying that there is this intimate connection or 'fate community' (*mingyun gongtongti*) between individuals and the party-state. As a result, many people feel a personal responsibility as good citizens to support the state's drive to digital modernity. This results in a striking contrast between attitudes to smartphones held by older people in China and those in other countries. In China it may be younger people who complain that older people are always on their smartphones and ignoring face-to-face conversation. Most people believe that while some aspects of usage may become more difficult with age, it is never too late for people to adopt smartphones.

The sheer degree of direct political involvement in smartphone usage is exemplified by a rather mysterious app called *XueXi Qiangguo*. The name of the app literally means 'Study makes a powerful country', but there is also an implicit pun that carries the sense of 'learning from President Xi'.[27] In a few months this became the most downloaded item on Apple's domestic app store. The presence of this app soon became a good indication that an individual either was, or aspired to be, a Communist Party member. The app serves as a news aggregation platform for articles, short video clips and documentaries about President Xi

Figures 2.2a and 2.2b WeChat stickers of Karl Marx as a superhero and a diligent reader, sent to researcher Xinyuan Wang by a research participant. Screengrab by Xinyuan Wang.

Jinping's political philosophy. 'Study points' are earned by users who log on to the app, read articles, make comments every day and participate in multiple-choice tests about the Party's policies. According to recent state media reports, Party cadres are required to use the app every day and to accumulate high scores. More generally, Party propaganda now incorporates genres that are smartphone-friendly, such as these WeChat stickers depicting Karl Marx (Figs 2.2a and 2.2b).

The case of China has been considered here to make a contrast to the prior discussion. But a final example, that of Japan, may serve as a bridge between this evidence for the influence of the media and the state and later sections, which will focus more on ordinary people's discourses about smartphones. In Japan a strict etiquette exists surrounding the use of mobile phones in public spaces. Making phone calls on public transport, for example, has long been frowned upon.[28] While Japanese trains are often full of people focused intently on their phones, anyone starting to have a conversation on their phone would quickly receive glares from fellow passengers. In trains between Osaka and Kyoto there are designated seating areas for the elderly or people with disabilities, complete with signs telling people that they cannot use their phones at all in these areas, in case they are too absorbed to notice that someone might need a seat. Throughout trains in Kyoto signs tell people to keep their phone on silent mode (known as 'manner mode'), forming part of a wider trend of etiquette posters on Japanese trains associated with slogans such as 'Good Manners, Good Life'. These signs represent a very public, state-organised discourse about how individuals should behave in public[29] – a consensus central to Japanese life more generally.

These examples all concern national responses. However, there are also important consequences for smartphones that arise from international relations. The most recent example of this has been the impact of deteriorating US–China relations on one of the largest of the smartphones companies, Huawai. Political tensions have also created problems for certain apps such as TikTok, the development of 5G and the sourcing of smartphone components.

Commerce: the smartphone and app industries

The other massive influence over discourses around smartphones, other than the state and the media, comes from the relevant commercial forces. Smartphones remain strongly associated with brands. The promotion of the world's most popular smartphones, the iPhone and the Samsung Galaxy, is pervasive and presumably in many cases persuasive. Many studies are now available of how these companies, along with more recent entrants into the market such as Xiaomi, have developed their own commercial strategies.[30] Campaigns tend to be pitched to the younger user, revealed by a decade of advertising the best 'selfie' smartphone. Relatively little advertising for digital technologies is pitched to older users, despite the economic power of the 'silver yen'[31] in Japan or the equivalent term to highlight the affluence of older people in other countries. This may be in part because it was common to find that older people, especially those aged under 70, feel uncomfortable at being targeted as a 'senior user'. However, there is an incipient range of technologies available for seniors, including simpler smartphones such as those marketed by Doro.

In 2019 the world's smartphone users downloaded more than 200 billion mobile apps. They spent an average of more than US$21 per connected smartphone per year on apps and app-related purchases. Gaming apps account for more than 20 per cent of downloads.[32] This is notwithstanding the fact that most people expect apps to be free,[33] with many of our research participants across the fieldsites declaring adamantly that they would never pay for an app. They may accept various indirect costs, however, including the presence of advertising. An ad-based revenue strategy may result in higher financial benefits for the company than charging for apps.[34] Increasingly, however, the primary cost is that of consumer privacy, with smartphones becoming a major conduit for the collection of data as an increasingly important component of contemporary commerce. Many participants are aware that the terms and conditions for using an app provide for the collection of data that seems to go well beyond the actual requirements for that app to function. Their acceptance does not indicate that they are happy about such conditions; it is rather an acceptance of this as a requirement for use of those apps.

The region where commerce has been most fully incorporated into smartphone use is China, where payment through apps has almost replaced the use of either cash or bank cards. Even older people, often initially reluctant, are now accepting that a social media app (WeChat) will be combined with crucial personal ID information,[35] including bank details. In addition 'freemium', the combination of 'free' and 'premium', has become the dominant business model among smartphone app developers.[36] The app and basic functions are free, but people may pay for premium functions and services. China has also seen a rise in free apps providing 'paid content'.

One striking feature, given the size of these industries, is that overall levels of advertising may be low compared to other consumer products.[37] One possible explanation for the comparative lack of overt or direct advertising is simply that the commercial world has far more effective tools for influencing the general public. There is considerable involvement of commercial companies in sponsorship of sporting events, for example. They also have access to more direct marketing through messaging the phone itself.

Yet perhaps their most important weapon in such campaigns comes from the impact of other players in their field – those that make life increasingly difficult for people who fail to employ their smartphones. A key factor that drives people to smartphone use is their growing inability to perform daily tasks cheaply and efficiently without the smartphone; in Dublin, for instance, banking and booking flights are both increasingly done online, through smartphones. Perhaps the single most important factor is the side-effect of the relentless drive towards cost-cutting by public utilities, government agencies, retail and banking sectors, all of whom are trying to replace customer service and call centres with online-only access to their facilities. As will be explored in chapter 7, the decline in offline access creates a formidable digital divide, leaving most people with no choice. They either learn to employ online technologies or they become effectively incapacitated. A recent example was that in order to obtain financial support from the Brazilian government during the Covid-19 pandemic, it was essential to be able to receive a temporary password code through one's own smartphone.[38] These examples illustrate that the very same governments who blame smartphones for every form of addiction and harm are themselves making it impossible for ordinary people to avoid regular usage. Under these conditions, smartphones and platform companies do not really need to spend money on advertising other than with respect to the choice of brand; the job of promoting smartphones is being done more effectively by others.

People's discourse and ambivalence

The evidence so far has suggested a series of complex contradictions in the involvement of states and the media, at least outside of China. Governments have to provide online access as a public resource, but also feel responsible for dealing with a litany of harmful effects claimed mainly to affect young people. At the same time the relentless digitising of infrastructure forces people to depend upon online access. Traditional media such as newspapers, which started as rivals to online media for advertising revenue, are increasingly recognising their own future to be online. In some of our fieldsites almost all online engagement is through smartphones as few people possess computers or tablets.

Turning to the discourses of ordinary people, this book is also subject to bias. Our primary demographic was older people who, outside of China, tend to reiterate negative appraisals of the impact of smartphones. Some older people see themselves as victims of surveillance and data extraction, but may otherwise consider that they are relatively inviolate from many of the other forms of harm. These, they claim, will mainly affect the young people they seek to criticise.

In Santiago, older adults appear to be in a battle – often with themselves – against the stigma of old age. On the one hand they are expressing the extent to which they feel such new digital technologies are very much 'not for them', while at the same time indicating how much they want to learn about and use smartphones. They spend much time blaming smartphones for various kinds of antisocial behaviour, for example claiming that 'people on the Santiago metro are staring down at their phones, not engaging with the real world and other people'. Yet, increasingly, they express such sentiments by sharing nostalgic memes on their own smartphones, as in the examples below (Figs 2.3, 2.4 and 2.5).[39]

The emphasis on nostalgia may be also prompted by this sense of loss of respect for their knowledge. An older man in Bento was very

Figure 2.3 Meme saying: 'Don't complain about homework. This was my "Google"', widely circulated online in Santiago. Screengrab by Alfonso Otaegui.

Figure 2.4 Meme saying: 'This was the WhatsApp of my childhood', also circulated online in Santiago. Screengrab by Alfonso Otaegui.

Figure 2.5 Meme saying: 'I'm so thankful for having lived my childhood before technology invaded our lives.' This meme is also widely circulated online in Santiago. Screengrab by Alfonso Otaegui.

proud of having memorised all the streets of São Paulo, all of which was made redundant by Waze,[40] GPS technology and Google Maps. Memes shared in NoLo show similar concern with the lack of sociability on the Milan metro and again people respond with nostalgic memes (Figs 2.6 and 2.7).

On the other hand, these same research participants in Milan simultaneously acknowledge the benefits of smartphone use. 'It serves me' (*mi serve*) is a widely heard phrase that refers to the utility of smartphones, whether for getting up in the morning, planning the day or facilitating long-distance/transnational family communication. Yet they may then immediately turn back to the negative when they state that the smartphone 'robs them' of their time, their attention or their offline presence with others.

Not surprisingly, faced with these contradictions, popular discourses often resort to humour or irony. For example Anna, a schoolteacher in NoLo, often talks about the weather. She turns the smartphone into her personal weatherman, a useful and friendly presence in her daily life. After all, forecasting the weather helps her to plan everything from her classes with the children to what clothes and shoes to wear that day. Anna also sheds light on this idea of the smartphone as something that robs people of their time. Sitting at the kitchen table or on the sofa in the evenings in her small but homely apartment, she will often browse

Figure 2.6   The Milan metro. Photo by Shireen Walton.

Figure 2.7   A typical kind of meta-commentary on the ubiquity of smartphone usage today, shared on WhatsApp and other social media platforms via smartphones. Screengrab by Shireen Walton.

Facebook or WhatsApp for substantial periods of time. She then regards this as getting 'trapped' in the smartphone and feels ashamed of wasting her time in this way.

Yet Anna clearly also adores her smartphone and its endless possibilities for searching information and connecting with her family. In reality, she actually needs things that 'rob' her of time. Since an upsetting separation from her husband a few years ago, Anna spends many evenings in the cold winters of Milan knitting in front of the television. 'I like it [knitting] because it distracts the mind, which is very important,' she explains. Knitting takes Anna's mind off things. Crucially, the act itself appears morally and socially acceptable to Anna, a fairly pious woman who is dedicated to her family and to keeping up domestic activities such as cooking. It is the smartphone that is most effective in keeping her connected, bringing her contact with her family, her children in Milan and a host of relatives with whom she discusses food recipes and chats through to WhatsApp groups that make her feel less bored and lonely. On the other hand the smartphone is a relatively unfamiliar modern object. So, comparing these two activities, although the smartphone knits her family and friends together it has not – at least as yet – accrued the positive moral connotations possessed by knitting, a traditional motherly or grandmotherly craft that creates clothing for her children and grandchildren.

Similarly, the following short film concerns Deirdre in Ireland. Her smartphone informs her that she has been spending sometimes six or seven hours a day on her smartphone, which might seem like an obvious case of addiction. However, as this film shows, she tries instead to empathise with her own predicament and to explain the experience in terms of her extremely difficult circumstances at that time (Fig. 2.8).

Figure 2.8   Film: *Deirdre*. Available at http://bit.ly/DEirdre.

As well as developing a discourse about smartphones, these devices are in every region also the basis for new forms of discourse. Chapter 8 will consider in more detail the new modes of communication that integrate visual and textual communication. As McIntosh[41] describes for Kenya, but which also relates to most regions, text messaging led to all sorts of new and synthetic genres of language and communication that are themselves often highly expressive of local idioms. These in turn enable people to express this ambivalence and other local feelings about the way in which life has changed with mobile phones.

In Japan, ambivalence also reflects the particularities of local history. Many Japanese participants saw technological advancement as the key to their country's growth during the 1980s 'bubble era', represented by international companies such as Sony. The generation who were of working age during this period cite their experience of it as a reason why they are generally positive and open towards new devices such as smartphones today. Younger generations who grew up after the bubble had burst may feel that the smartphone is yet another way to be overworked or to experience the pressures of social life more generally. Ambivalence is often present within discussions, for example that with Ishikawa san from Kyoto:

> I think 70 per cent of the time I look at my phone for no reason. It's like a drug. My daughter is like that. For her, it's normal to have this device around, so the smartphone doesn't look like a robot or a device anymore, it's just another being. But I don't think it's a bad thing.

Ishikawa san started the discussion by stating that she does not use her smartphone much and is not particularly attached to it. However, by the end of the conversation she had come to realise just how many tasks involve her smartphone, and how much she relies on it. She then framed this as addiction:

> Maybe I don't acknowledge or realise that I'm addicted to my phone and I actually am. I mean I start my day with my phone, with the alarm!

Sato san shared this ambivalence. She explained that her smartphone is at the centre of her life, yet this means that she sometimes feels that she neglects her duties outside of the social realm of the smartphone.

For me, smartphones are something I need, and I don't think this is necessarily a good thing, but I find myself too concentrated on my smartphone and my chores get a little ... not well done, which is bad. It's also the centre of my friendship.

Across most of our fieldsites ambivalence lies at the heart of popular discussion. These laments, which tend to combine all sorts of complaints, are very common across the fieldsites. Older people in Bento describe how smartphones can be used to stay cognitively active, improve mental health, connect with families and fight loneliness.[42] But at the same time Olivio complains that:

The smartphone became our second brain. You can do everything using your smartphone. What I don't like is to see how people can become dependent. Especially teenagers who can commit suicide if someone said something they didn't like on social media, or when they suffer bullying. There are many inappropriate things, like putting an end to a relationship or dismissing someone using the internet.

Earlier in this chapter cases of state involvement in online access in Cameroon and Uganda were noted. The ambivalence that many people express about smartphones may in part derive from wider discourses about the future of these states. While people over 50 in Yaoundé complain about what they regard as the negative consequences of smartphone use, they also feel the need to embrace these icons of 'modernity' and 'openness' and to benefit from the increase in personal capacity that accompany them. Their ambivalence is partly resolved, here as elsewhere, by projecting their criticisms upon the young. Older people see smartphones as abetting the manipulation of new generations by unscrupulous intellectual elites, and as key factors in what a schoolteacher called the 'loss of African cultural specificities'.

In Kampala, ambivalence is encompassed within the everyday term 'dotcom', which is used, often humorously, to refer to modern developments. In particular, dotcom refers to 'Westernisation' through media internet exposure, something that applies especially to the younger 'dotcom generation' or 'the children of dotcom' who have grown up with this. Some view dotcom with suspicion, believing it to be having a detrimental impact on respect between the generations. They see dotcom as something that is out of control, leading the younger generation to vices such as addiction and pornography, or simply to being cut off from the world immediately around them. Nafula, who is responsible for her two teenage grandsons, believes that dotcom 'spoils kids' and avoids it for herself.

> My life has no dotcom because where I am, I'm free. I don't care about TV, I don't care about this … maybe what I really want is a radio. That is good for me. … Searching is not in my mind. Me, I receive and call. I don't want to know. … This dotcom is something which can spoil if you are not strict with your children.

Yet simultaneously older people in Kampala may also see dotcom as something they want to learn from their children. Some participants express gratitude for the way that their children teach them to use smartphones, seeing it as a mark of respect. Dotcom can therefore be seen to invert the hierarchy between older and younger generations, through which children should be learning from their parents, while also supporting values of togetherness and respect for elders. These examples show the ambivalence about smartphones, the potential of which is bound up in broader political and intergenerational tensions.

The unambivalent

In such an extensive research project the evidence around any topic will likely be highly diverse. Ambivalence may be typical, but not universal. The point is made through these two final examples.

Karima, from Alexandria in Egypt, but now living in NoLo, appears to feel no guilt about the amount of time she spends on her smartphone – an activity regarded above all as an opportunity to be online with her family and friends in Alexandria and in Milan. Karima, along with a number of female Egyptian research participants, saw no moral distinction between online and offline communication. Given the way that her smartphone facilitated her social communication and helped her to overcome previous difficulties – such as running out of battery, a problem with her previous model – the device was an unalloyed blessing. What mattered far more to Karima was being able to remain in constant social communication online and offline with her Egyptian women friends, the core of her social network in Milan (Fig. 2.9).

Olivia, from Dublin, is particularly forthright in her opinions. Her worries centre around effects regarding radiofrequency exposure. She found that in the 'Settings' and 'Legal' sections of her phone, the

Figure 2.9 'La Festa del Pane', or international bread festival, is one of several community events in NoLo. Photo by Shireen Walton.

manufacturer recommends carrying the device at least 5 mm away from the body. The same section recommends using a hands-free option, such as the earphones supplied with the phone, which worried her even more. Olivia began to look seriously at any information she could find on the ill effects of radiation and radiofrequency exposure. She looked first at the leaflets displayed in the doctor's surgery, then read books on the subject and spoke to friends. Finally, she asked her GP if there was anything to worry about, but was not really reassured when he claimed there was not.

After finding a press release from the World Health Organization (WHO) that tentatively recognises such exposure as a possible carcinogen, Olivia began to leaflet her workplace and local school. She now scans her environment, mentally plotting the locations of local mobile phone masts. She finds that responses to her campaigning tend to cluster around two poles: people are either in complete agreement or they do not agree at all. There seems to be no middle ground in this debate.

Fake news

As noted above, the media is given to many forms of negative portrayal of smartphones. One of the most important of these, surveillance, will be discussed in chapter 9. Another that is frequently cited, probably because it reflects the newspapers' own concerns, is that of 'fake news'. The term is unfortunate inasmuch as it contains the misleading connotation that traditional media reportage could be relied upon as 'true news'. In Brazil older people are regarded as one of the main groups which promulgate this practice of fake news, a claim that has also been made in the US.[43] Such an accusation adds to other ways in which older people are stigmatised.[44] In turn, older people respond with a variety of strategies. Some remonstrate directly against their peers who have shared news without checking, while others try to be patient and wait until there is some clarification. As one respondent puts it, 'I wait. In some minutes someone will comment on it, and I will know if it is fake or not'. According to one survey,[45] 79 per cent of Brazilians now regard WhatsApp as their primary source of information.

The issue of fake news has come to the fore following the highly divisive 2018 election campaign and subsequent government of Jair Bolsonaro. Reuters claims that roughly one million WhatsApp groups were opened to promote political candidates.[46] The same phenomenon has also led to the rise of fact-checking groups, such as that of Projeto Comprova in Brazil, which received 67,000 messages to just one of the fact-check groups created at that time. Fake news is also closely related to the spread of malicious cyberscams. Here again WhatsApp was found responsible for disseminating 64 per cent of these links.[47]

People do not necessarily take such discussions at face value. In Italy fake news has become a massive feature of media coverage and popular conversation. An example, locally relevant to NoLo given the high proportion of migrant groups, occurred in July 2018. A public Facebook post gained significant attention across Italian social media, and eventually the world. An individual had posted an image on Facebook ostensibly depicting thousands of people at a heavily crowded port; small boats were shown carrying people, whose vast numbers overflowed onto the surrounding docks. The image's caption read: 'The Port of Libya … they never let you see these images … they are all ready to set sail to Italy'. Its intention was to make Italians nervous and angry at this imminent 'invasion' of supposed migrants. The image was initially highly successful and widely shared across the internet. The context is of widespread anti-immigrant rhetoric espoused above all by Matteo Salvini, a far-right Italian politician and former Interior Minister. Within a few hours, however, the image was revealed to be a picture of a Pink Floyd concert held in Venice in 1989 (Fig. 2.10).

Across NoLo the image first appeared on individuals' social media in its subsequent 'revealed' form as a fake news hoax, showing the 'absurdity'

Porto Libico..NON TE LE FARANNO MAI
VEDERE QUESTE IMMAGINI..SONO PRONTI
TUTTI A SALPARE IN.ITALIA

Figure 2.10   A widely shared social media post that falsely depicted Libyan migrants as being ready to 'set sail to Italy'. It was later revealed to be from a Pink Floyd concert in 1989. Screengrab by Shireen Walton.

of fake news. In NoLo, where there is a strong liberal presence, the image was transformed via social media into a resource for people expressing their active opposition to racism and xenophobia. Many people also recognised the historical roots of the way in which smartphones are used to spread disinformation. After all, a number of the country's established media had been used to promote anxiety about topics, including immigration, in previous decades, for instance during the Berlusconi period. Some people, whose parents had lived through and witnessed the propaganda of the fascist era, had challenged the suggestion that there used to be a historical period dominated by 'true news'.

Academic studies of these discourses

This volume will provide little evidence to help assess most of the claims that are made about the smartphone in the dominant discourses.[48] For reasons given in the introduction to this chapter, we have treated discourse as a quite separate property of smartphones rather than as evidence for actual usage. There is little in our ethnographies that can contribute to a discussion of whether or not these claims are true. We prefer to concentrate on that for which we have abundant evidence, which is the use of these claims in contributing to moral debates within our respective fieldsites. However, given the huge significance and consequences of these discourses, we briefly provide a guide to academic works that do make such claims of assessment.

A useful starting point is the history of these discourses, the early period of which was well researched by Adam Burgess.[49] He examined the various public fears and anxieties about possible threats to health believed to be posed by early mobile phones, as well as exploring the sources of these fears and why they became prominent among certain populations. Such discussions feed into a long-standing academic debate about how populations perceive risk, revealing the considerable time depth to many of these common fears as well as the ways in which these anxieties have managed such continuity in content – even when the devices onto which such concerns are projected have themselves changed beyond recognition.

Probably the most extensive academic debates about smartphones focus on their political impact, something also commonly discussed by our research participants. While there are some balanced discussions on the topic,[50] this is a field that also tends to produce highly polarised debates. For example, there has been a discussion for some years now surrounding the idea that new media has created either a filter bubble or an echo chamber. These terms suggest that social media and smartphones have narrowed our exposure to political discourses to only those that reinforce our opinions, preventing us from viewing counter arguments. On the one hand, many books now suggest that filter bubbles have markedly increased, to a perhaps disastrous extent.[51] At the same time other books seek to debunk such claims, suggesting that the evidence leads to entirely different conclusions.[52]

A more difficult debate concerns the concept of addiction, because it is less clear what the term refers to. The word addiction is frequently employed by users, often about themselves; there is a string of self-help books aimed at helping addicts.[53] But what does 'smartphone addiction' really mean? Clearly no one is staring at a blank screen. They are looking at some genre of content. Such addiction might then refer to anything from playing card games, obsessive interests in news about Trump, following celebrities on Instagram or schoolchildren wanting to know what their friends are saying about them. Each is a quite specific interest with its own causes and consequences. The results would best be described not as addiction to smartphones, but rather as evidence that smartphones help to facilitate addiction to such specific content or practices. Similarly, as Sutton has shown, the idea of 'detoxing' from smartphones covers a wide variety of intentions and understandings of addiction.[54]

The most important aspect of such addiction to content, and perhaps the most historical,[55] is being addicted to knowing what other people are saying about you. This is the usual reason why a teenager consults their smartphone under their pillow at 3 a.m. No-one regards this as a healthy practice and teachers sometimes suggest that it reflects children's lack of self-confidence.[56] But it would be strange to see such concerns as unnatural, or merely about the smartphone itself. It seems the opposite to the common claim that smartphones lead us to be concerned with screens rather than with people. It also seems reasonable to pair the topic with something else that looms large in the discussion by users, which is smartphones as a response to boredom.[57]

Unfortunately, these phenomena are rarely separated out from what should be considered addiction to smartphones, where there is a constant desire to view any content irrespective of genre. This form of addiction may be implied by wider discussions around *attention* – the criticism that smartphones have led people to be both inattentive or to find difficulty retaining attention to the non-screen world. These concerns seem to dovetail with the massive contemporary interest in notions of mindfulness and wellbeing that are intended to promote attentive slow thinking, being present in the moment and care – although, somewhat ironically, it is now very common to access mindfulness through smartphone apps such as Headspace.[58]

Even if it is difficult to tease out the substance of the phrase 'smartphone addiction', it is surely appropriate to be concerned with the degree to which smartphones facilitate addiction more generally. For example, Albarrán-Torres and Goggin[59] discuss the rise of smartphone betting. On the one hand there are commercial companies such as the Dublin-based Paddy Power, keen to increase their profits by extending betting to mobile apps. On the other hand there are demands for state intervention and regulation, arising from a fear that smartphones make addiction to gambling more likely. The evidence is that users employ smartphones to bypass traditional commercial forces such as the bookies. Instead they bet against one another, partly to develop an online sociality of betting. In turn, these developments then lead the commercial companies to produce 'social betting' apps; these first appeared in 2013 and are now an important component of smartphone betting. At least in this case there is a clear argument as to precisely how smartphones might be related to addiction.

The third example has been chosen because it is the single most common critique that emerges from our fieldwork. This is the idea that smartphones are particularly harmful for young people. A recent paper in *Scientific American*[60] summarises some of the more clinical and scientific attempts to assess the consequences of social media on young people from the perspective of disciplines such as psychology. Overall, while the results of the initial work tended to be extremely negative, more recent studies focus on simultaneous negative and positive consequences. The effects are viewed as generally mild overall, though this is not the case for all young people. The academic discussion around children and parenting in the digital age is worth highlighting because it is the field of enquiry that has generated perhaps the most exemplary and sustained attempt within social science to observe, analyse, draw conclusions, create appropriate policy and provide sensible, informed advice for populations of increasingly anxious parents. Through a series of studies that incorporate ethnography in conjunction with a range of other methodologies, researchers have done much to ensure that children's use of the internet and smartphones is understood in a broader context.

For example, boyd[61] has argued that parents have developed their complaints about children's reliance on online communication and content during the very same period in which they have become increasingly restrictive about their children playing in public with other children. Research in the US by Clark[62] has shown how these intergenerational conflicts about children's behaviour are closely related to wider issues of class. Lim examines what she terms transcendent parenting in Singapore.[63] *Primus Inter Pares* is the work of Sonia Livingstone,[64] who has engaged in a wide range of research projects on the topic. These range from more ethnographic engagements, as in her recent book *The Class*, to large-scale comparative surveys across Europe. The result is a judicious, balanced presentation of findings: it acknowledges potential harms, but is also cautious about the many projected anxieties that parents tend to presume about their children's behaviour. Livingstone's work is also impressive because she spans highly academic work with considerable engagement with policy responses. Furthermore, through recent initiatives such as the blog 'Parenting for a Digital Future',[65] Livingstone and her associates have created their own avenues of digital engagement that can be accessed directly by parents. Their resources allow parents to be better informed in their day-to-day decision making with regard to their children's online lives.

It is important to end on this more positive note about the potential contribution of research, since so many of these debates tend to be highly opinionated and performative interventions that might obscure more than they illuminate and stoke anxiety rather than provide assistance. By contrast, this final example shows that it is possible to have a well-informed, well-reasoned and balanced assessment that pertains to one of the most common critical discourses about the consequences of smartphones. As a team we also benefited directly from their initiatives. When you are engaged for 16 months in fieldwork on the topic of smartphones, you encounter many anxious parents asking for advice. It really helps to have a place to which research participants can be directed, where you are confident they will find useful and sensible advice.

Conclusion

This chapter started with a point that is perhaps surprising: when people are talking about smartphones in very general terms, they rarely address their own actual usage of smartphones. Their response is often very different in interviews about specific uses: a story about how someone used Google Maps for a hospital appointment, or how often they listen to music, or how they struggle with their online banking. This distinction was merely the first of several contradictions. The discussion around the impact of the state started with the observation that states commonly both promise better access, but are also considered a source of surveillance. Many other examples suggest that smartphones are viewed as having simultaneous beneficial and negative consequences.

These evident contradictions have tended to create a profound ambivalence among the general population. While this was already a conclusion drawn from our ethnographic research, the strength of this conclusion has grown as a result of the Covid-19 pandemic. A key part of the response to the pandemic concerned the potential of contact-tracing smartphone apps. These made the way in which smartphones may extend surveillance and intrusion abundantly clear. Yet the very same capability was proposed as a technological solution to the pandemic and thereby an expression of care. From March 2020 onwards we witnessed a remarkable heterogeneity in the global response, ranging from countries such as South Korea, where the government became more popular for having favoured public knowledge over individual privacy, through to Republicans in the US who tend to privilege individual privacy over the state's desire for data about health. This diversity of attitude is because the balance between care and surveillance is essentially a moral, not a technological issue. The example then clarifies the main assertion of this chapter, which is that these discourses about the smartphone are generally used to debate much wider ethical and political concerns.

These discourses about smartphones are often generated by the interests of the various groups involved. Governments may condemn smartphone usage that has become critical of their rule. The more established media may be responding to the demographic of their readers and threats to their financial viability. Experienced journalists also worry about the long-term effects on the quality and adjudication of reporting, and the threat smartphone usage poses to critical journalism. Commercial forces are mainly concerned with sustaining profits. There are also reasons why older people, who see how the smartphone undermines the respect young people once had for their knowledge, accumulated over many years, respond by arguing that the smartphone is evidence of the increasing shallowness of youth. Overriding many such concerns, and often directly opposed to the expressed desire to limit smartphone usage, is the constant temptation for governments and companies to try to save money by going digital.

Different regions have developed quite different alignments between the state and the citizen. In China, it is older people who believe that being a good citizen means helping the state to leapfrog the rest of the world by developing digital capabilities. In Japan, the state is mainly expressing a traditional concern with social harmony and the avoidance of discord. In Cameroon, the use of smartphones is a sign of modernity and development in both official and popular debate. Ordinary people themselves constantly fluctuate between negative and positive assessments, sometimes in the same sentence. When individuals claim that the smartphone is both a blessing and a curse, this is neither hypocrisy nor ignorance. It is perhaps the only reasonable response to a situation in which, as every chapter in this book will confirm, these devices simultaneously bring huge benefits and create new problems. At the end of the volume we will return to what is currently the most conspicuous example of this balance between care and surveillance: the response to the Covid-19 pandemic.

The discourses about the consequences of smartphones have then their own results. For anthropology, some of the most important pertain to the impact on social relations. An example was the use of this discourse within intergenerational tensions. Many of the moral debates presented here show the smartphone as an idiom through which some older people might discuss and condemn younger people's behaviour.

In a later chapter we will see how younger people can at times exploit the difficulties some older people have with smartphones to make equally cutting remarks about them. Intergenerational tensions concern not only who respects whom; they also raise complex questions of dependence, autonomy, dignity and inequalities based on age. The point made by this chapter is that much of this conflict is based not on the smartphones themselves, but rather on what people say about them. At this point, however, we will turn to the much larger topic of the effect on our social relations of what people do with their smartphones, as opposed to what they say about them.

Notes

1. McCulloch 2019.
2. For an example of religious influence see Pype 2016.
3. Typical examples of these kinds of academic discussion would include Deursen et al. 2015. Also see Elhai et al. 2020.
4. Edwards 2018. Again, in a radio interview on Ireland's national television and radio broadcaster (RTÉ News at One, 15 January 2020), David Cochrane, online editor for the *Irish Times,* noted that 66 per cent of the Irish population have a Facebook account and over half that number use it daily. He said that Facebook represents one of the primary sources for election candidates to reach voters. He also noted that the number of Facebook users dropped in 2018, a fall that he linked to privacy concerns. Since that year, however, the number of users on Facebook has climbed again. See News at One 2020.
5. Cambridge Analytica, a company owned by the hedge fund billionaire Robert Mercer, used personal information taken from Facebook without authorisation in early 2014 to build a profile of 50 million US voters, in order to direct personalised political advertisements to them. See Cadwalladr and Graham-Harrison 2018.
6. Mugerwa and Malaba 2018.
7. Boylan 2018.
8. Al Jazeera 2017.
9. Bikoko 2017.
10. Bikoko 2017.
11. Jiang 2012.
12. Wang 2016, 129–30.
13. Chen and Ang 2011.
14. More generally see Morozov 2012.
15. Hirshauga and Sheizaf 2017.
16. *The Guardian* [Editorial] 2013. Legal actions have also been taken to challenge mass surveillance programmes and existing surveillance laws. The joined cases *Digital Rights Ireland and Seitlinger and Others v Ireland* led to the invalidation of the Data Retention Directive by the Court of Justice of the European Union. See Court of Justice of the European Union (2014).
17. De Pasquale et al. 2017.
18. Servidio 2019.
19. *The Local* 2019.
20. Scancarello 2020.
21. Merola 2018.
22. *Wired Italy* 2019. *Wired* cites the original survey, conducted by Ipsos and promoted by Amplifon, which is entitled 'Smart Ageing: Technology has no age'.
23. Wang 2016, 25.
24. Hughes and Whacker 2003.
25. Fan 2018 and Sina Technology Comprehensive 2019.
26. Luo 2014.

27. Huang 2019.
28. Ito 2005.
29. In Brazil there is also a law that says that headphones are mandatory when using musical devices on public transport, including mobile phones. See Prefeitura de São Paulo (São Paulo City Hall) 2013.
30. For example Shirky 2015.
31. Long 2012.
32. Data from App Annie reports combined with a report from Ericsson, see Kemp 2020. See Tiongson 2015.
33. See Tiongson 2015.
34. Petsas et al. 2013.
35. In China there is a 'real name' policy that is compulsory when it comes to social media use.
36. Kumar 2014.
37. This situation is not uniform; there is more advertising in Brazil, for example.
38. Governo Federal (Brazilian Government) 2020.
39. Similar memes are shared in Bento.
40. Waze is a travel and navigation app. Developed in Israel, it was launched in Brazil in 2012. See Grupo Casa 2012.
41. See McIntosh 2010.
42. Vieira 2019 or de Sousa Pinto 2018.
43. Guess et al. 2019 claims that people over 65 shared nearly seven times the amounts of articles from fake domains compared to a younger cohort. Though applied to Brazil, the source of the evidence was from a US study.
44. Monnerat 2019.
45. DataSenado 2019.
46. Reuters Institute and Oxford Internet Institute 2019.
47. Simoni 2019.
48. For an example of a book that claims to assess the overall impact of smartphones see Carrier 2018.
49. Burgess 2004.
50. By now there is considerable general coverage of how social media in particular has impacted upon politics. Two examples are Bruns et al. 2018 and Margetts et al. 2016.
51. Examples include Pariser 2012 or, for the more general threat to democracy, see McNamee 2019.
52. Examples include the mathematician David Sumpter's book; see Sumpter 2018 or Bruns 2019.
53. For example Price 2018 and Burke 2019.
54. See Sutton 2020.
55. Standage 2013.
56. As was found in Miller's previous work in schools, as part of his study of social media in an English village. See Miller 2016, 123–36.
57. Jovicic, under review.
58. Headspace 2020. This comes from an English-American healthcare company specialising in meditation through this app.
59. Albarrán-Torres and Goggin 2017.
60. Denworth 2019.
61. boyd 2014.
62. Clark 2013.
63. Lim 2020.
64. For example, Livingstone 2009 and Livingstone and Sefton-Green 2016.
65. See 'Parenting for a Digital Future' blog.

3
The smartphone in context

Fieldsites: **Bento** – São Paulo, Brazil. **Dar al-Hawa** – Al-Quds (East Jerusalem). **Dublin** – Ireland. **Lusozi** – Kampala, Uganda. **Kyoto and Kōchi** – Japan. **NoLo** – Milan, Italy. **Santiago** – Chile. **Shanghai** – China. **Yaoundé** – Cameroon.

Smartphones as objects

Before considering the use of the smartphone for communication, we should acknowledge its presence as a material object. Its tangibility may matter more to some populations than to others, and for different reasons. For example, most Italians are well aware of their reputation for style.[1] It is perhaps no coincidence that some of the most interesting work on mobile phones as fashion items comes from the Italian sociologist Leopoldina Fortunati.[2] NoLo lies within Milan, a city whose economy is closely bound to fashion and style. Style can apply to content such as the screen, external appearance, additions or associated accessories. In other regions the material aspect of the smartphone that matters is cost – not just of the handset, but also of being able to afford data or Wi-Fi access.

Eleanora is a widow who lives alone in Milan. An active grandmother, she collects her two grandchildren from nursery every day and looks after them until their parents return from work at around 7 p.m., coordinating these activities with the parents through her smartphone. Eleanora's smartphone is also a visual shrine to her grandchildren; its wallpaper is a photograph of them on holiday and other photographs of them are taped to the back of her smartphone. The collage she has created resembles her fridge door, replete with old photographs and memories rendered as fridge magnets. The fridge from which she feeds her grandchildren and the smartphone from which she connects with her family have become sites where she can see them, even when they are not physically present.

In Japan, the mobile phone case and any associated 'charms' dangling from it can often be an expression of personal aesthetics. For example Midori san, a woman in her sixties from Kyoto, is a professional singer and dresses glamorously. She chose a phone case that she felt expressed her fun and feminine personality – a plastic Daisy the Duck case with a lipstick and high heeled shoe charms attached (Fig. 3.1). More frequently, people chose cases of a 'serious' notebook design, typically made from leather, that served also as protection for their phone screen. They often had business cards tucked into an inside pocket. Another woman in her early sixties explained that she would never buy clothes that were brightly coloured or 'young' looking as she felt they were inappropriate for her age. Her smartphone case was an extension of her age-appropriate aesthetic, being a plain notebook design.

This concern with appearing not to be too flamboyant, especially as one gets older, was shared by many participants, both female and male. The image below shows a Facebook post of Sawada san, a Kyoto Buddhist priest in his sixties, in which he explains to his friends why he was now using a new red smartphone case that he considered 'flashy' (Fig. 3.2). He writes how the case had previously belonged to his wife,

Figure 3.1   A professional singer in her sixties who uses mobile phone charms to match a particular 'look'. Photo by Laura Haapio-Kirk.

Figure 3.2   A red phone case which a Buddhist priest felt was inappropriate. He explained that it had previously been used by his wife. Photo by Laura Haapio-Kirk.

but he was now using it to replace his old blue one, which had worn out. He tried to make the new one look less 'flashy' with the addition of book binding tape, and he asked for understanding from anyone who saw him with it.

For Onono, a policeman in Lusozi, the decoration on his smartphones relates to his Christian faith. He has Jesus as his background image 'for protection'; 'if you have any problem, put this on and keep the light on,' he adds. He selected this wallpaper from the Google Play Store. At night, he chooses an image of Jesus on the Cross for special protection. Onono will change these images constantly to reflect the season, for example at Christmas or Easter. If he has bad dreams or receives bad news, for example about a death, he also places the image on his bed.

Individuals can come up with quite playful and inventive possibilities. Elisa, who lives in NoLo in Milan, experimented with combining her smartphone with a traditional receiver designed for a landline (Fig. 3.3). For her this provided a tangible link between the ability to talk for an 'unlimited time' through WhatsApp and the feel of the more familiar landline.

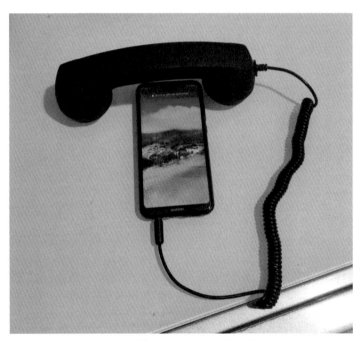

Figure 3.3   This device, halfway between a landline and an internet-enabled smartphone, was assembled by research participant Elisa. Photo by Shireen Walton.

Each of these examples represents a kind of aesthetic domestication of the smartphone,[3] making it more like a fashion accessory. The other way in which a smartphone impinges on us as an object, which can be a burdensome one, is its need to be placed somewhere when in use or merely being carried. For Dina, who now lives in NoLo, but is originally from Egypt, a daily task involves holding her smartphone for speaking to relatives or friends while simultaneously attending to her 4-year-old son or when walking and pushing a shopping trolley. Like several other women in NoLo who wear a hijab, Dina has got into the habit of tucking the smartphone into her headscarf. She can then breastfeed her baby or use a sewing machine while speaking on her smartphone.

Smartphones and status

Yoko was the only student who brought an iPhone to Marília's WhatsApp course in Bento, near São Paulo, where it was conspicuously displayed on the table. Because of its design, reputation and expense, her iPhone felt like some form of status symbol – which posed a problem for Yoko, who felt that she then had to be especially proficient in using it to match its prestige. Her solution was to reassure the others that she had not *purchased* this smartphone. As with most iPhones possessed by older people in Bento, it had been inherited from her children.

In Yaoundé most people have two smartphones. This is partly because there are multiple networks, which differ in quality across the regions of Cameroon. A high school teacher in Yaoundé explains:

> You see in some neighbourhoods you merely find the two main companies – Orange and MTN. You are then obliged to have a third one, which could be Nextel or even Camtel. Some of my friends have two or three SIM cards. They think this is a good solution.

Owning multiple smartphones may signify affluence and command respect, but only if one of them is a high-status brand. Even these may not signify much about older owners, however, as they have mainly inherited their smartphones from children who replaced them when taking out a new contract, as Yoko did. It is common in Yaoundé for the smartphone to be on display for all to see. Older people with limited mobility put it on a table or on some other surface that is easily in reach. Younger people care more about appearance and often print their own image onto the phone cover. Carrying them around in their hands and pockets may have encouraged the very regular pickpocketing of mobile phones that occurs in the city.

Older inherited smartphones may not be reliable, as discovered by Marie, a widow and former teacher in Yaoundé. She has nine children and has already been gifted five smartphones, but they seem to become redundant very quickly, especially when her grandchildren borrow and then break them. This becomes quite an impediment to her attempts to learn to use the smartphone. It is also a nuisance since by the time the smartphone stops functioning Marie has come to rely on several of its functions, ranging from the alarm, vital to her complex regimes of medication, through to WhatsApp, Skype and the phone's ability to retain pictures.

Status is not the only social dimension that may be reflected by phones. In Japan, where mobile phones have been internet-enabled since the early twenty-first century, 'feature phones' (*garakei*), with their distinctive flip style, remain popular, especially among older research participants.[4] Smartphone adoption was often gendered, with female participants tending to have stronger social networks of friends and family outside work. They were also more inclined to upgrade to smartphones and attend smartphone classes. By contrast male participants were more likely to retain *garakei* or even their landline as their primary mode of communication. One Kyoto man in his sixties explained that he kept his *garakei* even though he now also had a smartphone. His reason was that it contained all his work contacts which were important to his professional identity.

The cost of smartphones

While smartphones are becoming more ubiquitous, a significant digital divide remains between those who can afford them and those who cannot. In Lusozi the majority of research participants still had mobile phones rather than smartphones, as seen in this pie chart based on 204 individuals with an average age of 51 (Fig. 3.4). Of the 19 respondents who had no phone at all, 15 had had phones stolen and 4 have never owned a phone.

Not being able to afford a smartphone matters, especially as the phone is the primary point of internet access;[5] only 3 per cent of households in Lusozi have a working computer.[6] A growing minority can now afford cheap smartphones, however, such as one that sells for around £12.50 from the Chinese company Tecno; these thereby become

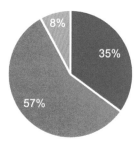

Phone ownership among research participants in Kampala, Uganda

8%

35%

57%

- Had a working smartphone
- Had a working mobile phone
- Had no phone of their own

Figure 3.4 Survey undertaken in the field by researcher Charlotte Hawkins. The percentages are based on 204 particpants.

the cheapest form of internet access.[7] But there are other subsequent costs, and these cheaper models may also have limited data storage for messages and images. Variations in smartphone possession were found to be related to wider inequalities of age, gender[8] and whether people live in cities or rural areas.[9]

According to a global survey,[10] 19.1 million Cameroonians (76 per cent of the population) have a mobile phone subscription. Thanks to the fall in the price of smartphones, from an average of £54 in 2014 to £36 in 2018, most of the middle class now possess one, although they can be second-hand or of poor quality. Durability thus becomes a key concern, with people occasionally referring to their phones as 'throronko', meaning an unreliable brand. Owners sometimes find they end up spending more money because they bought a cheap handset. These basic smartphones do provide access to WhatsApp and Google for information, however, as well as YouTube. This access is important, as older people spend considerable time watching videos on their phones.

As in Lusozi, people who have only a cheap smartphone brand may encounter problems of RAM or storage, limiting access to the latest software and apps. A university lecturer who owned an iPhone noted that by the time he had installed his favourite apps (Facebook, WhatsApp, an Instagram photo collage application, Photogrid, LinkedIn Messaging, Gmail and Yahoo, among others), he had no more storage space left.

Cost as a barrier to access was not limited to these two fieldsites. In Brazil there was a considerable expansion in the use of smartphones among people with lower incomes following the launch of Motorola's Moto G in 2013. In Japan many research participants gave the expense of 'pay monthly' plans as their reason for not having smartphones. A government bill aimed at lowering these fees[11] resulted in long queues at many mobile phone shops in Kyoto, which subsequently advertised heavily discounted smartphone deals. This government pressure to make smartphones more accessible coincided with a push towards greater surveillance through a new digital social security and tax number system ('my number'), which ties together an individual's financial, medical and social security data. Finally, there are fieldsites such as Dublin in which premium iPhone or Samsung Galaxy smartphones are ubiquitous. Since almost everyone can afford these smartphones, there is little sense of status competition.

Problems of access

Obtaining a smartphone may then open up further divides in the ability to afford access to data and the internet. A daily gigabyte of data in Cameroon requires a weekly plan costing £14. Typically, participants with low incomes spent around £3.50 and middle-class participants around £10 a month on access. In fieldsites such as Santiago Wi-Fi is accessible free of charge in many places, for example metro stations, public libraries and squares. While older participants in Dublin could comfortably afford monthly payment plans, many had only a vague idea of the difference between data and Wi-Fi. When asked, they often said they would not download a film over Wi-Fi at home because they assumed that must incur an additional cost. Here an age-based digital divide has more to do with knowledge than income.

In Lusozi another reason some participants had no phones was the prohibitive costs of repair or replacement. Locally the cost of calls is known as airtime. In a survey of 50 mobile phone owners, 74 per cent topped up 'airtime' daily, buying the smallest available bundles for between 20 and 40 pence as and when needed; only one respondent topped up their phone monthly for the equivalent of £20. This suggests that phone calls and the internet would have become inaccessible to most smartphone owners at least once a day. Given these limitations, phones are often used on a rationed basis. WhatsApp groups might be avoided in favour of less data-intensive messaging. Instagram and YouTube were rarely used, and data was switched on only when actually being used.

At the time of fieldwork, telecom providers in Uganda had stopped the sale of airtime scratch cards – previously the predominant method of topping up, especially in rural areas. This was considered a challenge by 31 of Charlotte's 50 participants, provoking comments such as 'it's hard to get airtime, especially at night' and 'it's terrible, you have to move long distances to search for airtime'. Relatives in villages now had to travel to the nearest trading centre every time they wanted to top up their phones.

In Lusozi most participants were connected to electricity at home. Those without could use charging stations in shops, phone repair stores or internet cafés for the equivalent of 11 pence; others preferred to charge their phone at their workplaces. The relative longevity of older mobile phone batteries was often lauded. One older man replaced his smartphone with a 'small phone', explaining that he preferred not having to worry about charging. In northern Uganda solar panels are the primary source of electricity, and people in a village wait their turn to charge up their phones. For example, a woman whose mobile phone was mainly for keeping in touch with her children and relatives might spend between 2 and 3 hours charging her phone from the solar panel she bought for £11 for this sole purpose.

Problems of access also applied to the middle class of Yaoundé. Many participants love to play games on apps, but downloading these requires a stable connection and patience. The Central African region, in which Cameroon is situated, has the world's lowest level of internet coverage: 25 per cent internet penetration in January 2018.[12] Cameroon is in advance of this regional average, but behind most of the rest of Africa.[13] App stores such as Google Play and the Apple Store require customer accounts which some smartphone users may not know how to operate, especially given that sometimes they discover that Cameroon is not within the 'authorised zone' for a particular app.[14] The procedures seem 'too technical, too long, too demanding', requiring credit card numbers, Apple IDs and other things they may not have. At road intersections and in the marketplace one can find 'downloaders' (*graveurs*) who fix phones that are broken. They can also create fake accounts that make it appear that the owner is in France rather than Cameroon, in order to gain wider access.

Problems of access may also be a reflection of disability. Laila Abed Rabho from Dar al-Hawa, one of the authors of this volume, has been blind from childhood due to an eye disease. She remembers little from the time before she lost her eyesight. Laila learned quite quickly to read and write using Braille; she graduated from university and went on to obtain her PhD. Until a year ago Laila had a 'stupid', simple mobile phone with no internet connection. She taught herself how to use the buttons and could send text messages and ring people, but she could not tell who was calling her. It was always a guess. Despite this, the phone made her life easier. When Laila left the house, she could still communicate with her family if required, or order a taxi through calling rather than texting. But that was pretty much the phone's limit.

Laila bought her first smartphone a year before she became a researcher in this project. She chose an iPhone because she was informed that these have excellent built-in sound software for blind people. She used both her own money and a government grant provided to blind people in Israel. The grant money meant she was also entitled to free guidance from a specialist on the use of digital devices by people with disabilities. He came to her house about eight times for two-hour sessions. This is not easy training, because you have to wait very patiently for the voice to say out loud what Laila is clicking on, which activates the accompanying voice. There is a gap of a few seconds between what appears on the iPhone screen and what the voice manages to describe. It is very difficult to get used to this pause in what others experience as an instantly responsive device.

At some points the voice just becomes unbearable for Laila. She turns it off many times through a double-click on the iPhone screen play button. The reading is not always smooth and often includes material that is not important to Laila, such as irrelevant signs within her Gmail. Yet the iPhone has changed her life, making it so much easier to be in touch with both fellow professionals and other blind people. Whether on messenger or email, the voice software that reads and writes is pretty good. Laila's difficulties come from the visual side of the smartphone. Obviously she cannot see images; as she observes, 'I don't have Instagram, I don't want to open an account, for what?' Yet her WhatsApp group includes people who are not completely blind and who often do share images. This is frustrating as she cannot see them and has to ask the group to describe them. 'I also hear the Quran and search for materials on Google more easily, like news. But I don't know how to take pictures in my iPhone.' Another thing that has helped Laila greatly is the dictionary: as she explains, 'if I want to search for the meaning of a word, English or Arabic, I use the iPhone'.

In this short film Laila talks about her experience using her smartphone while researching with Maya for this book (Fig. 3.5).

Figure 3.5 Film: *Laila's smartphone*. Available at http://bit.ly/lailasmartphone.

> If I accidentally click on the wrong button or too many times and then all of a sudden my click brings up iCloud instead of my Gmail, and then my sister-in-law needs to help me and press the right button. Sometimes the WhatsApp is disappearing, and I cannot find it, and I have to let someone fix it.

Clearly for Laila what matters most is being able to proceed with her research and other activities by herself, and not having to resort to the assistance of others every time some function of her iPhone needs to be reset or a stage overcome that cannot be achieved without seeing the screen. Laila finds apps easier to use than the device itself. This is because it is the main smartphone interface where she feels if she touches the screen and gets something wrong, she cannot retrieve the place she had been at without help. Nevertheless, where previously she was mainly using her computer, by this stage she loves her frustrating smartphone.

There is a reason why this section on cost and access follows the discussion of the smartphone as a material object. It is often cost that brings the material qualities of smartphones to the fore. In Hobbis's[15] study of smartphones in the Lau Lagoon region of the Solomon Islands, the key is in understanding the micro SD card. In his book *After Access*, Donner notes the impact of how people pay for their smartphone use, including the rise of a 'metered mindset' when people 'remain aware of the incremental costs of using their devices'.[16] In Lusozi, for example, many participants explained how they adapted their social media use to limit data costs, while airtime vendors are now a very visible feature of the urban landscape in Kampala and Yaoundé. Listening to the issues faced by Laila, one is made conscious of the problems of getting one click wrong and the underlying technologies of the interface. Often materiality is relative to what people cannot help but notice. These discussions also impact our conception of the smartphone itself – no longer a platform of unlimited possibilities, but rather a tool based on careful prioritisation in order to achieve specific communicative ends.

Screen Ecology

Both of the next two sections derive from the earlier description of ethnography as holistic contextualisation. The first, 'Screen Ecology', situates the smartphone in its relation to other screens, such as tablets, computers and smart televisions. The second, 'Social Ecology', demonstrates that the owner is also not isolated as an individual. The smartphone may be shared by several different people. In most fieldsites people have access to a variety of screens, as shown by the infographics for 30 research participants in NoLo and 146 people surveyed in Kyoto and Kōchi Prefecture, Japan (Figs 3.6 and 3.7).

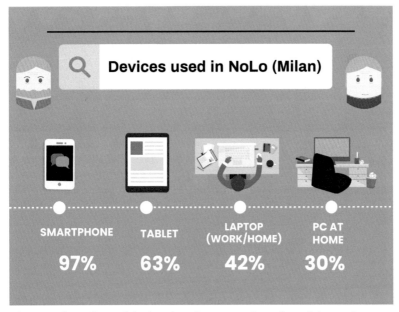

Figure 3.6 Infographic showing the proportion of participants in NoLo who use different devices, based on a survey of 30 people aged 45–75 conducted by Shireen Walton.

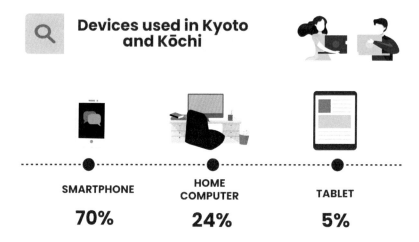

Figure 3.7 Infographic showing the proportion of particpants in the Japan fieldsite (Kyoto and Kōchi Prefecture) who use different devices, based on a survey of 146 people carried out by Laura Haapio-Kirk.

Most participants in the Dublin fieldsites have a tablet, either a laptop or desktop, and increasingly a smart television in addition to their smartphones. The most common reason people give for selecting one or other is screen size. While some elderly participants have eyesight problems, making the size of the screen an issue, it seems that established routine is just as important. People claim that a smartphone is not suitable for watching a television programme, but in practice they do use smartphones to watch programmes on YouTube. Smartphones also have clear advantages of mobility and may be the only screen available to watch important sports events when outside the home.

In Dublin, the iPad had been a revelation for older people. Even those in their eighties and nineties, who had previously resisted using any form of computer, were soon contacting relatives and making photo albums on their tablets. By 2019, however, tablets were losing out to the new larger smartphones because of the latter's mobility.[17] Maia now facetimes people with her iPhone, though she keeps the iPad for her creative writing. Some people, however, went in the opposite direction of increasing iPad use. Eamon, for example, uses his iPad both as a camera and a phone, preferring the larger icons and screen for his ageing fingers. His iPad comes with him by train or car. To achieve his one-size-fits-all commitment to the iPad, Eamon has also scaled his television viewing, including Netflix, down from the television to the tablet. For other Dublin participants, it is the laptop that dominates screen use. They bank online, shop online and generally access websites more easily through their laptop rather than smartphone apps. A woman in her forties finds phone apps clunky and inconvenient compared to the equivalent websites, and so avoids almost all apps.

For still others in Dublin, there has been some migration to the smart television – used not only for streaming television programmes but also other content that would benefit from a large screen, such as videos and photos from holidays or weddings. In summary, it is possible for an individual to focus upon just one device such as a tablet or laptop as almost their sole point of online access. But this is rare and most people move between one screen and another depending upon the task involved. With the rise of cloud computing and automatic data synchronisation, people seamlessly employ the smartphone when out and about, the tablet when going to bed, the laptop for a sustained piece of writing and the television for a family Skype session. Finally, not every device has a screen, and the landline remains important to some older people in Ireland – mainly because they still have parents alive, perhaps in their nineties, who will never use anything else. By contrast in Bento, as everyday communication has migrated to WhatsApp, people have been deactivating their landlines, having become fed up with endless cold calling from people trying to sell them 'stuff'.

It has been noted above that people in Yaoundé often have two smartphones corresponding to the two major phone networks, but they may also separate out their phones by usage. Retired people, for example, often reserve one smartphone for WhatsApp and one for other social networks such as Facebook. Almost everyone has a good-sized television and older people have radios, but very few have a laptop or a desktop computer. If present, laptops are just used for storing or transferring files such as photos or videos, in a similar way to traditional photo albums. The screen that dominates is the television, regarded as a family device rather than merely an individual screen. Generally once people have woken up, after prayer or perhaps sports, it seems an automatic reflex for them to turn on the television while having breakfast.

In Yaoundé David and Essy follow this routine of turning on the television after going to church at 6.00 a.m. Otherwise, screens may be distributed around the family. The couple use a tablet – a present from one of their sons, who is a doctor – mainly for playing games such as Zuma and Solitaire, while their youngest son uses it for racing games and online shopping. Each family member consults their own smartphone. Televisions may enter into this distribution as one son watches a second television. The smartphone is most commonly used to share videos and images; as with the television, these are often commented upon by friends and relatives. It is therefore common to overhear someone saying 'Oh, have you seen this…?' or 'What do you think of that video a friend just sent?' In a room with around 10 people, there is then a constant passing around of smartphones and chatting about their contents.

Screen Ecology is not just distributed around the family. The next example from Shanghai demonstrates how it can have quite a profound impact on the very nature of households and families. Mr Huang is used to his wife scolding him when he tries to finish reading a WeChat article on his smartphone after she has called him for dinner, although they agree it is all right to have the television news on during the meal (Fig. 3.8). The news mentions a flower show, so Mrs Huang checks the weather while her husband uses the *Gao De* app (a Chinese mapping and navigation app). This tells the couple that it will take them two hours by underground and bus to get there.

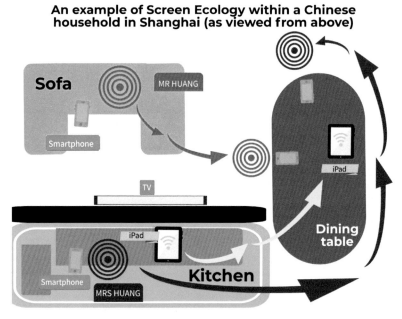

Figure 3.8 The dining area in Mr and Mrs Huang's home in Shanghai, as re-created by Xinyuan Wang. The illustration shows how different screens are placed around the house.

Just then the iPad, resting on a kitchen surface, rings. 'It must be Xiaotao!' Mrs Huang exclaims in delight. She fetches the iPad and places it on the dining table so they can speak to their grandson in Beijing, where her son-in-law works. They are only visited every three months, but her daughter gave Mrs Huang the iPad so they could WeChat on a bigger portable screen. From time to time Mr Huang takes photos of Mrs Huang happily talking with Xiaotao, and then sends them to the family WeChat group. While they are still talking Xiaotao's *Nainai* (the child's grandmother on his father's side) has replied to these photos with WeChat cute stickers saying 'nice shot'. Since she was visiting Xiaotao, she can then post photos of the video call from the other end. Mrs Huang in turn shares the photos with her WeChat group 'Sisters', which includes her three close female friends.

There is nothing exceptional about this dinner scenario. It involved at least eight screens in three locations within a single hour, with the images supporting moments of intergenerational bonding for a retired couple. An examination of the positioning of these screens reveals how they simultaneously reinforce the sense of their domestic surroundings, as well as incorporating more distant kin. In the past, placing family photographs around the rooms would have served the same purpose; now, thanks to these screens, it feels like those images have come alive. This domestic Screen Ecology is quite sophisticated. The bedroom includes another television, along with a laptop and desktop computer inherited from their daughter and used mainly by Mr Huang, but also a place to nap for their cat (Fig. 3.9).

In the early afternoon, if it is a nice day, the couple will sit on the balcony for a cup of tea, each of them on their own smartphone. The iPad means that Mrs Huang can either watch soap operas there or take the device into the kitchen to continue watching while she cooks. Cooking also involves making use of the 'Go to Kitchen' (*Xia Chu Fang*) app for her video-illustrated recipes and *iQiyi* – one of the largest online video sites in the world, used for six billion hours each month and often dubbed 'the Netflix of China'. After dinner, the couple prefer to use the desktop computer to play Chinese chess, do online shopping and check the stock market. One problem with Mrs Huang's smartphone is that it tempted her to check how her stocks were doing every few minutes, which led to her deleting the app altogether. 'My mind was controlled by this app; it's like an addiction and very unhealthy. I felt less happy those days,' she commented. While the bedroom does contain a television, the most active screens are the couple's two smartphones, including their allocated half-hour of bedtime reading interspersed with viewing friends' WeChat

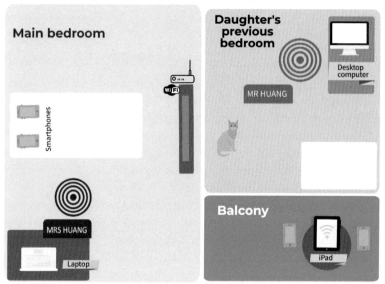

The Huangs' house plan, showing two bedrooms

Figure 3.9 The Huangs' house plan, showing the two bedrooms. Plan re-created by Xinyuan Wang, based on ethnographic research with the couple.

profiles (for Mrs Huang) and listening to history podcasts on the *Ximalaya FM* app (for Mr Huang).

The result exemplifies the contemporary experience of polymedia[18] – the experience of living in an environment where most people have multiple media that complement one another at their fingertips. Each medium develops its own 'ecological' niche within the whole. The prior example from Yaoundé presented the 'always on' television that stands for the family as a whole, even as individuals are simultaneously looking at their smartphones. This makes a nonsense of treating smartphones in isolation, because the definition and experience of what they are is relative to the alternatives that simultaneously present themselves and the development of people's ideas as to which is appropriate for each task.

Social Ecology

Just as smartphones only make sense relative to other screens, their owners need to be considered relative to other people, which is the principal point of what we are calling Social Ecology.[19] The clearest example comes from Lusozi where only four people out of the 50 phone survey respondents were the sole users of any particular phone, with the rest citing an average of three other people who shared access to their device. Such sharing might include their children, siblings, partners, neighbours and friends, who use the devices to play games, take photos, call friends and play music. Sometimes respondents said they would refuse to lend their phone if they felt it was being 'misused' – for example, if someone else borrowed airtime from the phone, took up too much storage or wanted to make calls after midnight.

The cost of making calls is commonly circulated between family and friends, with 33 (66 per cent) of respondents reporting that they had shared airtime in the previous six months and 30 (60 per cent) reporting having received some. 'Beeping' is also common; this is when someone makes a call, then lets the phone ring once or twice before cancelling, in the hope that the other person will call them back and incur the cost. All these practices may enable resources to be distributed across social networks, which in turn consolidates such social interdependence.

Nakito and her son jointly own and run a hair salon in Lusozi (Fig. 3.10). While she has her own 'small phone' for work calls, she does

Figure 3.10 Nakito with her son and grandson in their salon. Photo by Charlotte Hawkins.

not have enough money to buy her own smartphone, so she shares one with her son. On alternate weeks they take turns as the main phone holder, updating the password and background photo. That way, both Nakito and her son have periods of independent ownership, but can also use the phone at any time with permission from the current owner. Within the phone, there are certain apps that only one or other uses, for instance an app called 'Love Quotes', which Nakito's son uses to choose messages for his girlfriend. He is also the one who knows how to load music on the phone from a memory card which they update regularly, especially when they hear something new they like on the radio. During the week Nakito looks for her own music on the phone, as she prefers Baganda[20] songs. They have the same photos, mostly those Nakito has taken of her young grandchildren 'to keep the memories', especially on special occasions such as their birthdays.

Other instances of sharing are neither mutual nor egalitarian. Burrell,[21] working elsewhere in Uganda, explores how sharing can also be used to reinforce social hierarchies. For example Acen, a single mother in Lusozi, has heard of the internet but does not know what it is. She regularly hears from other people that through the internet you can get to know what is happening outside of Uganda. Without education, stable employment or support from the father of her children, she struggles to pay rent and school fees and cannot afford her own phone. Once or twice a month, Acen used to load some airtime onto a neighbour's phone in order to communicate with her relatives at home in her village. The neighbours usually had to show her how to dial and make the call. Normally, she was calling to check on her relatives, to learn whether anyone was sick and to find out if everything was still stable. If Acen's relatives needed to talk to her, they could also call the neighbour's number. Last time she heard from her relatives, they told her that her mother was sick. Acen would have preferred to go back to the village in person to check on her, but she failed to raise the money for transport, so she sent them money instead.

At the time of the interview, Acen was waiting for an update on her mother's health because she had not been able to call her relatives since then. She has faced some challenges when asking the neighbours to use the phone. Sometimes when they have seen her coming, she has overheard them complain that 'she's coming to disturb us', leaving Acen now afraid to ask. She tried to ask a second neighbour instead, but she 'refused totally there and then'; the woman claimed that her phone did not have any battery and that she was always out. These experiences have left Acen feeling 'totally helpless', but she is determined to be strong as she is now both 'the mother and father' of her children.

Often people in Lusozi said they had bought a phone for older relatives in the village to keep in contact with them as a means of caring for them at a distance. The situation was similar for the Palestinian population of Dar al-Hawa, where around one-third of research participants received their smartphone from a family member rather than buying their own. Although Laila and Maya worked with single and widowed women, none of the main group of research participants lived alone. There is always the expectation that an individual will live with a member of their nuclear family, whether that is parents, children or siblings, which in turn has repercussions for how the smartphone is used in daily life. Since most older people are living in family homes, they are regularly involved in looking after grandchildren. These days that often means that the smartphone is being shared as the grandchildren watch kids' shows on the grandparents' smartphones. This is not always what the grandparent wants, but children in most regions have an impressive ability to cajole adults into letting them borrow their smartphones.

In regions where the extended family has largely disappeared, it can still be the case that the prime user of a smartphone is a couple rather than an individual. In Dublin a man who on principle had no phone at all would sometimes give his wife's phone number as a contact to other people, or ask her to bring up Google Maps on her phone when he was driving. However, he would in turn do the couple's banking on the home computer, as well as other tasks she was not fond of. Traditions of gender roles are often key. If the smartphone is conceptualised as a device for keeping in touch with family and friends, use of it may fall within the established role for the wife. Couples may then see their phones as interchangeable, knowing each other's passwords and answering either phone if they happen to be nearest the device at the time, though this is not typical. A woman in the Dublin fieldsite also admits to playing up her lack of skills as a way of flattering her children's skills:

> Sometimes we don't even mind playing a little bit naive to allow them to do it, so, you can say to them 'You do that because you're better', like a little bit of a game playing roles.

This sharing within couples is examined in more detail in Figure 3.11. It is an infographic based on interviews with 12 Shanghai couples, segmented by age (Fig. 3.11). Currently the number of apps used declines with age. The couple in their forties tend to have in common apps such

Use of smartphone apps among 12 couples in different age brackets in Shanghai

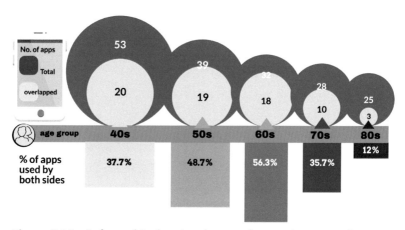

Figure 3.11 Infographic showing the use of smartphone apps between 12 couples in different age brackets in the Shanghai fieldsite. Survey undertaken by Xinyuan Wang.

as Dianping, a popular review and advice app (dubbed the 'Chinese Yelp'), since this age group is more likely to try out new restaurants or visit unfamiliar areas. They also share functional apps for payments and trip information. People in their fifties and sixties tend to have entertainment apps in common, such as videos and games, as they share retired leisure time. The survey suggests that couples over 70 have fewer apps in common, but the interviews reveal that the opposite is true. These older couples have become so interdependent and do so much together that they are commonly sharing the smartphones themselves, so that there is no point in replicating an app on both smartphones. For example, the taxi-hailing app Didi is usually found on only one smartphone between a couple in their seventies. 'Taobao' or 'Pinduoduo', both online shopping apps, are only downloaded to the wife's smartphone, as traditionally it is the wife who takes care of the shopping. In this way, the smartphone has become a means by which people express their changing ideas of what it means to be a couple.

Social Ecology may refer to intergenerational links as well as couples. Surveying older people's smartphones often revealed ringtones or games downloaded by children or grandchildren who had borrowed the phones. In Santiago, a student of Alfonso's smartphone classes often got angry with her grandson for taking her smartphone without permission. At one point she asked Alfonso to delete a video her grandson had downloaded which she found 'disgusting', as well as some games and apps that had already cost her roughly £60 – all were games and content related to girls. The problem is that her grandson has already registered his fingerprint with the phone and knows her password. During his smartphone class Alfonso advised her on how to change the security on her phone.

Networks

Both Screen Ecology and Social Ecology demonstrate why it is misleading to study smartphones merely as the relationship between a device and its individual owner. Even the household is insufficient as the basic unit, as the Huang family use their screens to link with other families who in turn have an array of screens. Smartphones often thereby link networks rather than just individuals. Before this project, an influential book by two senior social scientists called *Networked*[22] suggested that, partly as a result of the rise of the internet and new communication technologies, we should think less in terms of people living in groups or within a community of proximity and more in terms of networks where the individual is now the hub. The main evidence from the *Why We Post* project, which studied social media, turned out to be opposed to this argument. Instead its researchers found social media was often used to repair and retain traditional groups such as the family and community.[23]

When it comes to the smartphone, which does so much more than social media, it is very hard to fit our findings within any particular trajectory of sociality. It can be argued, however, that it is easier now to reconcile these apparently opposite trends – the shift to networks or the repair of groups. Both seem to be true simultaneously. On the one hand, the smartphone is constantly used as a hub that connects individuals through networks – linking us to our friends and relatives, for example, wherever they may be. On the other hand, in fieldsites such as Milan and Dublin, Facebook was found to be a major community site for posting about local activities such as community breakfasts, events at the local allotments, sports or 'Tidy Towns'.[24] The section on Screen Ecology showed how smartphones consolidate and expand families rather than replacing groups. We are more reconciled to the concept of networks when these, as in the Chinese case, refer to smartphones connecting networks, rather than just connecting an individual to a network.

Similar points apply if we turn from the way that smartphones network people to the way they act as a control hub for networking things. For decades we have also been told that we are just about to experience something called the 'Internet of Things'[25] as a significant change in our lives, leading also to new concerns with the potential security issues.[26] Many of the claims made for the 'Internet of Things' reflect the hype of commercial desires. Yet even if this is still far from flourishing, in Dublin we can start to observe the first shoots. There were rare occasions of people using their smartphones to control domestic devices, such as turning on the heating systems in preparation for when they get home or being able to check security systems while abroad. The first video doorbells are being installed. It is now reasonably common for people in Dublin to Bluetooth link their smartphone to their car, to permit long conversations while driving. For example, one man routinely reports to his sister at length while driving home from visiting their elderly father. Phone assistants are not as yet used much to link with other devices. Some speak to Siri, but while many people in Dublin have Alexa, it is almost invariably relegated in practice to a voice-activated radio. This is all quite limited, but it does suggest that the smartphone is destined to grow in importance as a kind of remote-control hub for interacting with other technologies as a network.

Conclusion

Using the word 'context' in this chapter's title is not intended to make the chapter sound like just a scene-setting background or introduction to this book before getting to the meat in the sandwich. In fact the materials presented here, as in chapter 2, describe important components of what smartphones are and the consequences of their popularity. This chapter helps us to understand how smartphones have an impact as material objects. Their value may be used to express status or expose us to theft. Their costs may be a significant burden for people with low incomes. For some owners, smartphones transform their relationship to sibling screen technologies such as laptops, tablets and television.

When smartphones are shared, they help to constitute relationships within couples and to others. These are mutual relationships, which explains why our term for ethnography, holistic contextualisation, is so important. The phrase clearly implies that context is reciprocal. It is not just that people use smartphones. Smartphones in turn shape relationships. Relations between husbands and wives may be the context for understanding smartphone use, but smartphones have become part of everyday life that forms the context for how couples now relate to each other. In the case of the Huang family in Shanghai, the point was not simply that many different screens were being employed. It was rather that through understanding the internal relationship between the devices, we came to appreciate both the nature of domestic relationships and changes in the way the family operates, as the screens enable them to incorporate relatives outside the immediate household.

Previously, the walls of the domestic space created a divide between a household living within and an external family living elsewhere. Only the photographs on display were evidence of this wider sphere. Today, through using these multiple screens as polymedia, wider kin can appear within the domestic space, popping up regularly on WeChat links or video links. We can then ask whether this incorporation of the wider family is something new – or a return to the more traditional extended family that people experienced in China before they moved into the cities? The point is that Screen Ecology contributes as much to explaining family dynamics as Social Ecology does.

Similarly, issues of cost and access are not simply the economic aspect of smartphones – nor are their implications limited to understanding who gets to use what. Both reflect and impact upon wider relationships of inequality and power. A person who depends upon their neighbours for charging or gaining access to a phone opens themselves up to insult and humiliation. Digital divides can become chasms. On one side are those who have access and can thereby become part of global communications – something the smartphone achieves by bringing migrant families together irrespective of where they live. On the other side are those who cannot afford access or lack the requisite knowledge and skills. They do not remain static in the face of technology, but are rather transformed into a digitally illiterate underclass compared to their peers. By contrast, in Bento this situation may in fact create positive networks of care, as people have to seek help from their friends, creating the solidarity of interdependence. Thinking in terms of context is especially important for a device such as the smartphone – one that can become a kind of control hub from which a proliferation of other technologies and other people are organised. Few objects can have ever been quite so embedded in our everyday lives and relationships as the smartphone is now. So context truly matters.

Notes

1. In Nicolescu's work on social media in South Italy, he showed how style and appearance are seen almost as a civic duty in respect to the reputation of Italians, rather than just an individualistic project. See Nicolescu 2016, 121–48.
2. Fortunati 2013.
3. The term 'domestication' also references a much wider theory of media domestication, for which see Silverstone and Morley 1992, 16–22.
4. Holroyd 2017.
5. National Information Technology Authority (NITA) 2018.
6. National Information Technology Authority (NITA) 2018.
7. See Deloitte 2016, 4.
8. According to a survey by Charlotte of 50 respondents between September and December 2018, with an average of 5.6 people in their households, 1 male and 0.65 females had mobile phones, with an average of 0.9 males vs 0.6 females with a smartphone. Smartphone owners in the household were an average of 31 years old, as opposed to mobile phone owners, who were 38 years old.
9. In Kampala there is better access to telecommunication, electricity and internet infrastructure than in rural areas. See Namatovu and Saebo 2015, 38.
10. WeAreSocial 2018.
11. Kyodo News Agency 2019.
12. WeAreSocial 2018.
13. For surveys of internet use in Cameroon in 2018 see WeAreSocial 2018 and the second slide in the article by Mumbere 2018.
14. Certain apps from the Apple Store are not available in all countries, with the store imposing geographic restrictions. If a user is not from an 'available country', they will not be able to download and/or access the particular game or app in question. This tends to frustrate users in these regions.
15. Hobbis 2020.
16. Donner 2015, 215.
17. Spadafora 2018.
18. Madianou and Miller 2012, 125–39.
19. We recognise that this is our own specific use of the term Social Ecology and do not intend a reference to other uses of the same term (for example, see Ling 2012).
20. The Baganda are an ethnic group native to Buganda, the largest traditional kingdom in present-day Uganda. Baganda also refers to a music culture developed by the people of Uganda.
21. Burrell's ethnography in rural southwestern Uganda (2010).
22. Rainie and Wellman 2014.
23. Miller et al. 2016, 181–92.
24. This is an annual national competition in Ireland, where the most beautiful towns taking part are given an award for their attempt to improve the quality of life for local residents.
25. There is, for example, a journal called *IEEE Internet of Things Journal*. It can be accessed at: https://ieeexplore.ieee.org/xpl/RecentIssue.jsp?punumber=6488907. See IEEE 2020.
26. Li et al. 2017.

4
From apps to everyday life

Fieldsites: **Bento** – São Paulo, Brazil. **Dar al-Hawa** – Al-Quds (East Jerusalem). **Dublin** – Ireland. **Lusozi** – Kampala, Uganda. **Kyoto and Kōchi** – Japan. **NoLo**   Milan, Italy. **Santiago** – Chile. **Shanghai** – China. **Yaoundé** – Cameroon.

Introduction: not starting with apps

The temptation when examining a smartphone is to imagine that it is, in essence, an app machine, reducible to the various apps of which it is comprised. Apps, in turn, could be understood as the mechanisms that align the smartphone to some particular purpose associated with that app. If so, answering the question 'what is an app?' would go a long way towards answering one of our project's main questions: what is a smartphone? However, ethnography approaches this question from observable use, not potential use. On that basis this chapter will unfold as an account of people orientated not to apps, but rather to tasks. To help them accomplish these, they may employ combinations of appropriate bits of various apps.[1]

This chapter will also consider clearly relevant externalities. These are the work of developers, who may not appear within the ethnography themselves, but whose creation of apps and envisaging of tasks is very apparent in the smartphone itself. There will also be continuity with our consideration of smartphones as material objects presented in chapter 3. Apps are also objects; exposed on a smartphone's screen in the form of icons, they spring to life when touched. As icons, apps can be relocated to different screens and organised within folders according to particular interests, function or frequency of use. Apps also vary hugely in the degree to which they have become general or single purpose elements within the smartphone – properties that will be addressed in this chapter.

If this chapter progresses from apps to the tasks of everyday life, then this trajectory reflects one of our main methodologies – one designed to elicit a comprehensive coverage of apps. But because this methodology took the form of stories about their usage, the apps were instantly transformed from isolated technologies within the smartphone to being viewed within the context of everyday life. After following that trajectory, this chapter will shift still further towards apps understood within ordinary usage by focusing on one particular genre of deployment: that in relation to health. This will lead to a further conclusion, reinforcing our perspective of understanding apps from deployment, not from technological properties. The apps involved in health were mostly not those designed for that purpose. We are not trying to suggest that usage is in opposition to design. Rather we reveal that many apps – as well as the handsets of smartphones – have been designed in so open a fashion that only through the creative imagination of users do the possibilities within an app emerge into the light.

The app interviews

When planning the ethnography, the team agreed that each of us would interview at least 25 participants about their smartphone apps. The interviews were to be carried out in quite a specific manner. Instead of talking to people about their smartphone use in an abstract or general way, we would ask them to open up their smartphones so that we could see every single app present on every single screen, and then discuss these one by one (Fig. 4.1). In the case of Android phones, this might include both home screens and the background app drawer.

This method was essential because when going through the apps in this systematic fashion, participants would frequently remark that they had quite forgotten that they used this or that app. As with so much of everyday life, such things are quickly taken for granted and memory requires prompting. Often pointing to an app icon on the smartphone screen stimulated a detailed story or discussion that would not have otherwise arisen. Without actually inspecting the smartphone app by app, we would have been unlikely to have garnered this more comprehensive encounter with the sheer range of tasks that now involve smartphones. This method also gave us a sense of how many apps were being used and which these were, as shown in these two summaries.

The chart (Fig. 4.2) shown on p. 82 is based on information from 30 participants in Shanghai with an average age of 59. It suggests that

Figure 4.1   A typical Samsung Galaxy screen, showing different apps.
Photo by Daniel Miller.

the older people are, the fewer apps they use. But it is possible that, as the currently middle-aged group age further, these differences will diminish. Taking this group as a whole, the average number of apps in use is 24.5. On page 82 the first chart relates this to age and gender, while the second (Fig. 4.3) indicates which apps are most common.

In the case of Dublin, there were 57 such interviews across the two fieldsites, with participants' ages ranging from their forties to their eighties. As with Shanghai, the survey only included apps that were *used*, excluding those that are just present on the smartphone. Before considering apps per se, there are the inbuilt functions that almost everyone uses. These would include the camera, the clock/alarm feature, the torch, the voice phone and text messaging. Next come the apps that at least 80 per cent of users would employ, including WhatsApp, an email app such as Gmail, a calendar app and a browser app such as Chrome or Safari.

Apps used by more than 50 per cent but less than 80 per cent of participants included transport apps such as Dublin Bus and Irish Rail,

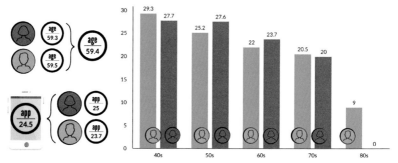

Figure 4.2 The average number of apps in different age and gender groups in the Shanghai fieldsite. Survey of research participants undertaken by Xinyuan Wang in 2018.

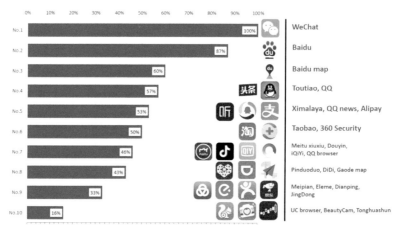

Figure 4.3 The chart above represents the 10 most commonly used apps among 30 of Xinyuan Wang's research participants in the Shanghai fieldsite.[2]

news apps such as RTÉ news, Journal.ie, the BBC, *The Irish Times*, the *Independent* or the *Guardian*, weather apps such as Met Eireann or YR, photo apps such as Gallery or Google Photos, radio apps such as RTÉ radio, airline apps such as RyanAir or Aer Lingus, webcam apps such as Skype or Facetime, music apps such as Spotify or iTunes and map apps such as Google Maps, as well as Facebook, Facebook Messenger and YouTube. Examples of less commonly used apps are included in this infographic (Fig. 4.4).

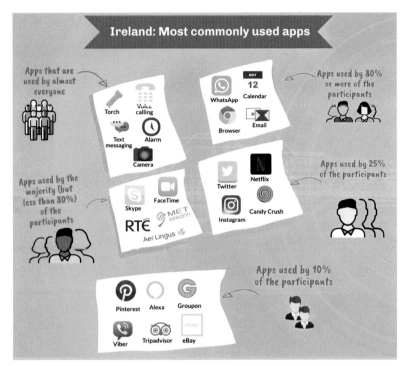

Figure 4.4 A selection of the most commonly used apps in the Irish fieldsites, based on 57 interviews. Please note that the illustration is not comprehensive. Graphic created by Georgiana Murariu.

Typically, an older user might employ between 25 and 30 functions and apps on their smartphones. Younger people in the same fieldsites might reach up to 100 actively used apps and the proportion of young people using many of the apps listed above would be much higher.

We might have been expected to present more such quantified results. However, having carried out this exercise, the team felt that the results are better understood as visualisations of what are essentially qualitative observations. The real benefit from these interviews was the in-depth stories and accounts of usage they produced. As quantified figures, these are likely to be misleading. Reasons for this include the problem of defining exactly what it means to say an app is used or not used. An app may have been put there by one's children, for example. In many cases, it was downloaded, used once and then never again; but then again it may have been used two or three times, with people tending to be vague about this. They might say they did not use it and then suddenly remember occasions when they did. Secondly, an app can be just one of several forms of access. If a person has an app for Tripadvisor on their phone, but then declares that actually they mainly access Tripadvisor directly through their browser, rather than through the app, should they be recorded as using Tripadvisor or not?

Thirdly, there are all the complications documented in chapter 3 under the headings of 'Social Ecology' and 'Screen Ecology'. If a person does not have banking apps on their phone because their partner does the banking, is this a vicarious presence of banking apps on their phone? If they use the apps on an iPad instead of their phone, is this a sort of smartphone access? Or should it be considered irrelevant to the way they use their smartphones? This is why, both in this instance and throughout this volume, we tend to be circumspect in the presentation of quantified results, often placing greater trust in their qualitative aspects. It is the details of how they are employed and the consequences of that employment that matter – not trying to define what exactly counts as use and what does not, or calculating the exact percentage of use among what are in any case quite small and specific groups of people.

Scalable Solutionism

All of this begs the question of what an app really is. The word 'app' is itself in some ways misleading, since it groups together some very different beasts. Often an app is better understood as a term for a zoo, rather than an individual creature. A recent book called *Appified*[3] helps to clarify this point. Each of its 30 chapters is dedicated to a different app and named after it. One of these chapters is called 'Is it Tuesday?'.[4] This app is an intentional joke, as the only thing the app does is answer the question of whether today is Tuesday (Fig. 4.5). So far we have found it to be accurate. The app reveals the way in which we use humour and irony to address our perception of this new app culture – a perception best summarised by the phrase 'there's an app for that'. As the author of that chapter observes, if to the hammer everything looks like a nail, then to the app developer everything looks like a problem that can be solved by an app. These observations lead the author to a discussion of topics such as micro-functionality and solutionism.

At the opposite end of the spectrum is a Chinese app discussed in another chapter of the same book.[5] From social media to paying water bills, WeChat is the ultimate Swiss Army penknife of the app world, featuring many more functions than, for example, Facebook. The chapter's author argues that there are specific reasons why it tends to be messenger

Figure 4.5   A screengrab of the 'Is it Tuesday?' app for iPhones. The screen shows how many times the user has checked whether today is Tuesday, as well as how many checks have been done globally on that day. Screenshot taken by Georgiana Murariu.

apps, based around texting, that tend to develop the incremental functionality that lies behind their success. The app develops an underlying infrastructure that can then be turned into pretty much anything that users might require from their smartphones, ranging from a way to pay for goods, to a means of obtaining a doctor's appointment, to possibly hundreds of other applications which might, in other contexts, have each represented a stand-alone app.

In the more technologically focused literature, the emphasis will tend to be on the term affordance, meaning that the design of an app will lead to a propensity for people to use it in some particular manner. From an ethnographic perspective, however, an app is only what users *actually do with it*. So it is equally important to observe the opposite process, starting from complexity and going back to simplicity: the wine drinker who has a Swiss Army penknife, but only ever uses the corkscrew. For example, when Alfonso was teaching smartphone use to older Chileans in Santiago, he found that for some students YouTube turned out to be, in terms of their daily use of it, reduced to one single functionality – that of playing music. During her research in Brazil, Marília found that one of her research participants in Bento uses Facebook solely as a means to remember birthdays.

We will use the term 'Scalable Solutionism' to refer to both of these arguments. Firstly these two apps – 'Is it Tuesday?' and WeChat – represent two extremes: the app with a single micro-function and the app that aspires to be useful for everything. Most apps lie somewhere between these two. This is one form of Scalable Solutionism. The term also applies to an equally important factor discussed in the paragraph above. This suggests that an app is not only what its designers created and intended. Mostly an individual user has a problem they want to solve or a particular task in mind. For them the app is simply that part of its potential that is relevant to their concerns, and which may become for them the totality of that app.

Choosing the word 'solutionism' is significant because the term signifies another important consequence of the rise of app culture. An example comes from a study of digital infrastructure and start-ups by anthropologist Katrien Pype. She works in Kinshasa, in the Democratic Republic of the Congo, where she observes more than just how people make and use apps. Pype argues that the rise of digital technologies has had a more general impact on the way people consider the world around them:[6] it has in fact promoted a concept of solutionism. In Kinshasa there is now a whole discourse about 'solvability', a concept that resonates with the narratives of development aid, but is now also internalised as the promise of digital solutions to various aspects of the urban predicament. This would be a local example of what Morozov in his book *To Save Everything, Click Here* has exposed as a global tendency towards such a technological solutionism.[7]

The implication follows chapter 2 in acknowledging that an important consequence of smartphones sometimes lies in the field of discourse rather than usage. Developments in the field of digital technology are creating a whole new language and set of expectations. People now tend to see the world more through the lens of solving problems and the imagination of '*vivre mieux*' (living a better life). This is embedded in official discourses about technology and the idea of 'smart cities'. Yet this utopian vision, when situated in the Democratic Republic of the Congo, was very far from people's actual experience of poor infrastructure and restricted internet access. When it comes to finding actual solutions people turn to 'smart from below', a concept also developed by Pype.

How the world changed the app

The obvious result of a 'smart from below' perspective would be that apps are deployed according to local and cultural concerns. The comparative book from the *Why We Post* project was called *How the World Changed Social Media*.[8] The title followed the observation that far from social media homogenising the world, each fieldsite saw quite different uses of the same platform. That project in turn followed an earlier publication, *Tales from Facebook*,[9] which examined what Facebook looks like when it is studied through its use by people in Trinidad. Facebook came to incorporate many quite specific features of Trinidadian society. As a result, unless one is able to understand terms such as *commess* and *bacchanal*, which refer to the particular ways in which gossip and scandal operate in Trinidad, it is not possible to understand properly what Trinidadian Facebook is.

The differences are still clearer through a direct comparison of the visuals posted on Facebook in Trinidad and England.[10] The *Why We Post* project showed similar diversity across all the fieldsites studied. For example, low-income Brazilians might post pictures of themselves near swimming pools and gyms to show their aspirations, while low-income Chileans followed the opposite ethos of seeing Facebook as a place to show the reality of their actual lives.[11] Acknowledging the importance of cultural differences in the use of social media will also apply to the use of smartphones. It is not just that in one region people use WhatsApp and in another they use WeChat. There are regional differences in the way a single platform such as Facebook is used, such that ethnographers are dealing with multiple Facebooks rather than just one.

One facet of this diversity derives not from differences in the usage of a single app, but from the way that people combine apps in carrying out tasks. A shift away from a focus on apps is particularly important when working with older people. We can demonstrate this with the example of Fernanda from Bento. Extremely organised and in charge of both family and her own business finances, she puts her entire 'to do' list into her calendar, including bills to be paid. Most of the bills come by email. When the right day comes, she accesses her bank app to make the payment and shares the receipts, such as with her landlord, using WhatsApp.

None of this is unexpected. Others, however, bypass the obvious app for achieving these tasks and go about them in a more roundabout manner. For example Susana, a migrant to Santiago from Venezuela, does not use her banking app. When she wants to pay a bill, she will google her bank's name, go to its website and make the payment. Ernestina manages to complicate things still further. She needs to forward a bill through her email to her sister, but she does not know how to do this. What she actually does is take a screenshot of the bill from within her email app. She then goes to the gallery app and shares this screengrab with her sister through WhatsApp. It was often the fact that they did not understand how to use one app 'properly' that led older people to be highly creative in adapting others. They make it clear why merely listing the affordances of individual apps would not get us very far. These people are not concerned with apps but with *paying bills*. In order to accomplish this task, they use combinations of apps in ways that may not have been originally envisaged by app developers. However, these examples are exceptional. To make this point more generally, the following section delves into a whole genre of app usage.

Health beyond solutionism

This research project started with a commitment to helping to facilitate initiatives within the scope of mHealth. As such, it followed the literature on mHealth, which is mainly concerned with the current development of bespoke apps for smartphones.[12] Typically, these might be symptom-checking apps, apps for rehabilitation exercises or apps for better sleeping or fitness. In other words, mHealth is a prime example of solutionism, based on the hope that for every health problem there is a potential app that might at least contribute towards a solution. The results of this part of the project will be published elsewhere, but the original intention was quickly undermined by our findings as ethnographers.

We decided to focus upon mHealth since the age group we were targeting are themselves increasingly affected by health issues. It was soon apparent, however, that this did not mean there was much uptake of mHealth apps of any kind. For example, in a survey undertaken by Alfonso at a cultural centre for older adults in Santiago, he found that out of 64 participants, 52 of them (81 per cent) did not use bespoke mobile health apps. Of those who did use apps related to health, none were the kind of biomedical apps usually considered as mHealth. In several of the fieldsites, the health-related apps that were evident would have been almost entirely what could be considered as a kind of 'soft health', such as step counting, meditation and diet-related apps. In Dar al-Hawa, Laila and Maya similarly found that none of the 27 women interviewed (all aged over 40) used an mHealth app, though they had vaguely heard of them. For example, Hala knew of an app for a local health clinic and used her smartphone to make contact, but she felt no reason to download their app because she lived nearby.

Yet, if anything, these older people were using smartphone apps *even more* for health than we had envisaged. It was just that they did not follow the solutionist route of finding the app designed for a specific health problem. Instead, they adapted, combined and made relevant the apps that they used for other purposes. In Yaoundé, 19 out of 65 research participants (29 per cent) said they often used health-related applications. This might be a bespoke app if this was pre-installed, such as a step-counter. But mostly it was the use of the ubiquitous smartphone apps. In general, uses of the smartphone for health fall into three categories: nutrition-related, sports- and fitness-related, and medical task-related, such as apps for sleep tracking or medication. A common usage of smartphones is where people google and carry out YouTube searches for medicinal plants and other health-related information. The variety of plants researched ranged from a woman who used the 'King of Herbs' plant for skin problems, as well as citronella and palm oil for stomach aches, to another who used guava leaves as a cure for thyroid problems. A former administrative executive regularly asks his WhatsApp group for general information and advice on rheumatism and prostate cancer, two of the most common health problems within this population.

This short film (Fig. 4.6) illustrates the use of the smartphone in sourcing traditional medicines.

Figure 4.6   Film: *Healthcare in Yaoundé*. Available at http://bit.ly/healthcareyaounde.

Similarly, the main 'health' app used by Dar al-Hawa's seniors is their WhatsApp group, where people forward messages about proper diet and sports for older people. Messages also circulate about diabetes, of which there is a high incidence among the Arab population[13] in al-Quds. There might be more of these messages before one of the various holidays, since these are times when people traditionally tend to eat sweets and festive foods.

A genre of bespoke health-related apps that *are* used in some fieldsites are those provided by medical insurance companies. These may, for example, allow photographs of invoices to be sent through the app which speeds up the payment of claims, a facility found in use in Dublin. By comparison, people in Bento generally struggle with apps that allow them, for example, to make appointments; they find ways around them through using configurations of other apps. In Bento, Sandra uses the Unified Health System (SUS) and tried to use the Agenda Fácil app – which, in addition to making appointments on the public network, also generates a digital version of the national health card. However, because she finds it difficult to use this app, Sandra takes a picture of her card and saves the image to a folder in Google Drive. When she goes to appointments, she opens the file and shows the card on her smartphone screen.

It may not even be apps that are the key to deploying the smartphone for health purposes. In Lusozi the crucial impact of phones in relation to health comes through the movement of money. Voice calling and mobile money are the most ubiquitous uses of mobile phones, often connecting people to their relatives in distant villages. A phone call will be made to 'check' on these relatives or to request financial assistance, which can then be followed with a transfer via mobile money.

Across two separate surveys, Charlotte asked people about their last three phone calls: who these were with, the purpose of the call and the duration of the call. A total of 195 respondents offered replies about 585 phone calls. A breakdown of the main aims of these calls is given in the chart featured below (Fig. 4.7). Many calls were for seeking or sending 'help', sometimes in the form of money or sometimes food; people explained that 'he wanted me to help him' or 'she called to send money home to my sister'. Many calls in the first two categories of checking on relatives or sending money were in relation to health. In addition, 16 per cent of phone calls were made solely for health purposes. This includes 60 calls for updates on family health: 'she was sick so I called to find out how she woke up' or 'he was telling me about his father's sickness'. A further

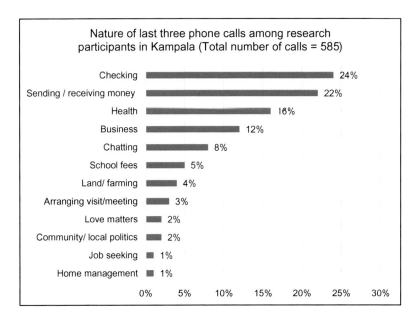

Figure 4.7 Chart of the last three phone calls among research participants in Godown, Kampala. Survey undertaken by Charlotte Hawkins.

23 calls were made directly to health professionals, including 15 doctors and 8 nurses. These were described as 'doing follow-up', 'calling to confirm the medication' or 'to know if my health was improving'.

Mobile money is often lauded as an example of adapting technology to requirements 'from below', offering financial flexibility and connection.[14] In Lusozi this practice is now embedded in daily life, for example to support elderly parents' health at a distance. Almost all the research participants here use mobile money. With 33 mobile money vendors in Lusozi alone, it is the most conveniently accessible form of financial transfer and banking. People sending money take cash to an agent, who arranges the transfer to the recipient's phone number via their mobile. Charlotte also asked participants about the last three times they had sent or received mobile money. Of 130 recorded remittances, 37 (28 per cent) were for 'help', which could include anything from money for upkeep, food, 'pocket money' or gifts. This was followed by 32 remittances (25 per cent) for health purposes, which could include hospital bills, medicine, transport to hospital and surgery costs.

Figure 4.8   Film: *Mobile money in Uganda*. Available at http://bit.ly/mobilemoneyuganda.

The ways in which mobile money is used are explained in the short film featured here (Fig. 4.8).

The ethnographic evidence complicates assumptions that dotcom technologies are facilitating increasing individualism and selfishness. Instead, this shows how dotcom can also facilitate family obligations and respect, even at a distance. One woman explained that she is the only person providing for her parents at home with money. Recently her mother had had a stomach ulcer, so she sent her money to go to hospital. A village elder noted that 'life's easier now with phones', as they are able to communicate family problems to relatives in the city who can 'mobilise' necessary funds.

WhatsApp has also been commonly adopted for health purposes in Kampala. Large WhatsApp groups for sharing information are common in neighbourhoods, or among service professions such as nurses. During a recent outbreak of cholera in Lusozi, the Ministry of Health sent not only radio and television announcements but also text messages to people in affected areas; these were then circulated on WhatsApp. At the government hospital all of the staff are part of a WhatsApp group where announcements are made. Each department also has its own WhatsApp group, through which people can let one another know if they are going to have to miss work or provide updates about patients and medical supplies. They can even circulate information with networks through WhatsApp groups outside the hospital. As one electrician explained, 'in that group of mine, one is a teacher, one is a doctor ... so any information one of them gets, I have to get it here'. A woman explained that she had 'learned so much' about health from WhatsApp – for example how to check for breast cancer, as well as nutritional information.

Our project was focused upon older people rather than the elderly. As a result a common health issue comprised looking after frail parents, who might be in their nineties.[15] Scrolling through the phone history of Frances from Ireland, it is evident that around 80 per cent of all voice calls and text messages received and sent related to organising care for her frail father. Since a recent fall, he is more or less bed-bound and needs to be changed, washed and cared for there. The state provides Frances with 10 hours a week of care, but this has become her full-time work for the rest of the time – so much for her dreams of retirement. In the last month alone, she has sent 270 texts concerned with her father's care. This need percolates through her smartphone. For example, Frances has a voice recorder that records all her phone calls; she can then use this to produce evidence of her conversations with the care authorities when they dispute her claims about promised care. She also carries a power bank everywhere to ensure her phone never runs out of battery. Of her four WhatsApp groups, two are family groups concerned with organising her father's care and two concern sailing. Her father has a Doro brand phone (a simple phone devised for use by the elderly) so that he can talk to his sister. Given that the sister in question is in the early stages of Alzheimer's, these may be long conversations that drift in many directions. (Another participant, Stephanie, purchased the same phone for her 89-year-old mother-in-law.) The result is a smartphone almost entirely used for health purposes – but none of this involves mHealth-related apps. Instead it is all about Frances's creativity in engineering her everyday use of a smartphone into an effective tool for looking after her father.

Finally, the evidence from Japan points to possible future uses of smartphones in relation to a wider health technology. Technology here is at the forefront of the country's strategy for coping with a rapidly ageing population and a shortfall of care workers. An ecosystem of devices, including wearables, alarms and motion sensors, has been designed to prolong the time that elderly people can continue living at home, reducing the burden on both families and local healthcare institutions. Costs for care technologies are reimbursable by the state's long-term care insurance system after a complex, needs-based assessment.[16] The policy is to augment face-to-face care with technology rather than replace it.

Kawamura san, an 85-year-old man living alone in rural Kōchi, had a motion sensor installed on the ceiling above his bed; this could alert local care services to any unusual immobility. He also had an emergency alarm button which he could press if he fell. Kawamura san was still strong enough to chop wood to heat his evening baths in his outside bathing house, but regular visits from a local member of the social welfare office were integral to ensure that he was getting the support necessary to continue his independent living for as long as possible. In a similar manner Toriyama san, a Kyoto man who lived with his 78-year-old mother, explained that she had been able to come off blood pressure medication completely with the help of Japanese traditional medicine (*kampo*) and daily home monitoring via a blood pressure device. Through a combination of self-monitoring and her son's daily encouragement of walks and a healthy diet, she was helped to make positive changes leading to better health.

Apps and screens

Through these various examples of usage, the ethnographies show why the focus is on tasks rather than on apps per se. But there is a highly significant property of the smartphone that underlies this shift in emphasis. The design of the smartphone facilitates this reconfiguring of apps in combination, as people can now shift easily between apps that appear as adjacent icons on the screens of their smartphones. This makes it important to understand how people organise those screens and the app icons they contain, usually as part of their general customisation of smartphones after purchase. The organisation of apps is often central to the way owners turn their smartphones into a kind of control hub that can marshal associated apps together, making them especially convenient. In effect there are two such control hubs: the remote-control hub discussed in chapter 3 in relation to the 'Internet of Things' and also a more internally focused control hub, considered here, which is based on the organisation of apps within the smartphone.

Creating this control hub is not always a simple operation. One of the complications introduced in the discussion of Screen Ecology was the fact that some people are working with apps spread across different devices, including tablets and laptops, as well as smartphones. There are three main observable changes. The first is deleting unused apps and gathering the most commonly used apps on the initial screen. The second change is nesting apps around particular functions (Fig. 4.9). Many users now have a news icon, which includes all those apps that have to do with news, as well as a sports icon, travel icon or finance icon used in similar

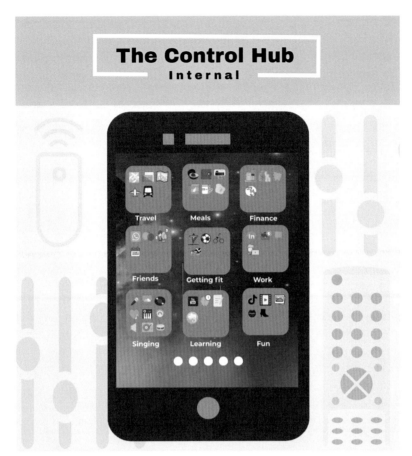

Figure 4.9 Example of how the process of nesting icons helps to make an organised smartphone into a kind of control hub. Visual created by Georgiana Murariu.

ways. The third change is simply placing apps in juxtaposition to each other, usually because they are often used together. There are exceptions, however. Alessandra from NoLo has carefully organised the apps on her phone into alphabetical order, while in Milan Bruno, a retired architect originally from Sardinia, colour coordinates his apps.

When it comes to older people, their screens may require some interpretation. In general, researchers found that the older people with whom we mainly worked were less knowledgeable about how to organise their screens than younger people; as they gained experience, however, they quickly caught up. Some older people have a single app on each screen because no one has shown them how to consolidate apps. In her smartphone class in Dar al-Hawa, Maya dedicated an entire lesson to learning about the multiple screens on smartphones and what a 'home screen' is. The students often had difficulty searching for particular apps during a class. An app that most participants wanted special assistance with downloading was Waze, but it seemed they needed to know their way around the smartphone before they could use it to navigate around the city. During an interview in Bento, Rita revealed that she only recognises half the apps on her home screens and uses only 23 of the 45 apps on her smartphone.[17] As an accountant, Eduardo has 104 apps of which he uses 70, meaning that there are 34 he either does not use or does not know how to use. In proportional terms he is outdone by Iara, however, who knows very little about almost two-thirds (35 out of 55) of the apps on her phone.

By contrast Esteban, a Peruvian migrant and successful businessman now living in Santiago, is very careful with the organisation of his three home screen pages. Apps appear strictly according to the frequency of usage, with a decline in usage or redundancy condemning an app to the less frequently used screen to the right. The third page is a kind of condemned cell, containing apps destined to be deleted. Esteban also organises folders of apps according to uses. As he discussed his smartphone, he explained the apps within his 'travel/taxi' folder, including Booking.com, Latam (an airline app), Tripadvisor, Airbnb, Despegar (a travel agency), Hoteles.com, Latam Play (an in-flight entertainment app) and Wallet. However, he also took the opportunity to move Cabify from there to his 'Maps' folder, which also contained Google Earth, Apple Maps, Google Maps, Waze and Uber. He highlights in his home screen his 'most important app', that of the ATP Tour – Esteban plans to travel next year to Europe and follow the tennis tournaments. He also notes his 'most important' folder, that dedicated to 'Music', which includes Panamericana (Peruvian radio), Peru Radio, A la carta (a television app), Spotify, Music Player, Radio Union and Oasis FM (a Chilean radio station with music and very little news; Esteban says he does not want to hear 'depressing stuff'). He also has a couple of health apps, one of which is to remind him to take his medication (he suffers from a heart condition). Esteban has also set up an iPhone facility for making an emergency call, which only requires him to press the volume and 'unlock' button.

Many participants could be considered either 'housekeepers' or 'hoarders'. Housekeepers take control of the organisation of their apps, keeping their smartphones tidy, while hoarders may be overwhelmed by the proliferation of apps on their phone and lose control. Our interviews also provided a sense of apps as a sign of people's attitude to time. Some mention apps that are downloaded because they envisage a future need, while others download apps only when they have to. Some delete an app as soon as they no longer need it, but others retain the app just in case they need it again.

Where do apps come from?

Apps do not appear out of the ether. They are created and owned by companies that mostly seek profits. In the app business there is a distinction between clients and users, with clients being those who pay for the development and maintenance of the apps. A user may also pay for an app, but nowadays the most commonly used apps are free for users. Examples of ubiquitously used free apps include LINE, WeChat, Facebook, Messenger and WhatsApp,[18] as well as apps related to the Google suite such as Google Drive. They are not exactly 'free', of course, as users exchange their privacy for the service.[19] The mandatory terms of use set this deal when users first engage with those apps.[20] For just a brief moment the business concerns of those behind the app are revealed, but these terms and conditions are so extensive that almost no one reads them.[21] In a way there is very little incentive to do so, as these are not negotiable: if you do not accept the terms and conditions, you cannot use the app. Across our fieldsites most users show almost no awareness of or interest in who owns what app or platform, or whether they are using apps that derive from the same commercial source. For them Facebook is a separate platform; they do not greatly care that the same company also owns Instagram and WhatsApp. They just want to download the chosen app and be free to use it as they wish.

Most of our participants are thoroughly opposed to paying for an app. They are more willing to do so in China, but may be influenced by factors such as trust in the developer[22] and online social conformity as well as social identity.[23] Mostly, as elsewhere, users in China are unaware of the identity of developers. They simply search or look for recommendations. Their focus is upon the app's usefulness, rather than its trademark.

'I care about the egg, not the chicken,' remarks Weiwei, a retired taxi driver in Shanghai. In all the fieldsites, most people cannot tell who owns most of the apps they are using, except the best-known ones such as Facebook and WeChat. A market report from Google shows that even though the App Store remains a popular way to find new apps, a quarter of customers discover new apps through search.[24] For example, Mrs Qian in Shanghai installed the restaurant reservation managing app 'Meiweibuyongdeng' when she was waiting to be seated outside a popular restaurant; now notifications regarding table availability are sent to her smartphone. Similarly, someone may discover a bike-sharing app as a by-product of a search on Google Maps.

Some companies strive to combat this indifference, however, and try to keep people within their own corporate universe. Perhaps the best-known example with respect to smartphones is Apple, who do have some control of which apps are used on iPhones through their App Store; they also tend to sync data automatically to other Apple devices owned by the same users, be they iPads or Mac computers. However, the company's control starts with the 'App Review' – a restrictive set of guidelines that developers have to comply with if their apps are to be published in the App Store. Guidelines for Apple's App Store include matters such as design, broken links, data extraction, use and protection;[25] much of this is justified in terms of security.[26] By contrast, Android employs an open-source approach, although recently Google has also started reviewing new developers.[27] For established developers, the process remains quite easy and quick.[28]

Early in the chapter the concept of Scalable Solutionism was used to consider the range of apps. In the industry, however, 'scalability' is used to refer to the app's ability to scale up the number of users or user requests[29] – the term describes an app's ability to create, or adjust to, growing demand. 'Growth', in turn, may refer to an expansion in the numbers of users or an expansion of the app's functions. The two are obviously related, as new features may bring new users and also keep the app relevant. Equally, users may start using a feature in ways that developers did not anticipate, prompting the latter to respond to the expansion. These dynamics were very evident in the development of Facebook, for example. As noted in chapter 1, Facebook was originally limited just to those studying at Harvard and then, in turn, to other university students. Mark Zuckerberg was subsequently very successful in capitalising on the platform's hugely expanded use for attracting advertising. But this arose mainly thanks to people's desire for sociability and because they simply ignored his initial attempts to restrict how his creation was used.

Facebook subsequently grew in more than just numbers and profitability. The platform became increasingly complex, with an impressive number of features being released each year. Since 2007 Facebook has launched the Marketplace feature, the Facebook Application Developer, the False News Story Flag, Facebook Reactions and Instant Articles,[30] to mention just a few. Sometimes a new feature is released in one market and then scaled to others. Facebook Dating, for instance, is a match-maker functionality that was first tested in Colombia and then released to Argentina, Canada, Thailand and Mexico; further expansion is planned.[31] The company may also co-opt usage that it has observed in one local market and scale this to others. Such was the case of the 'Safety Check' feature released by Facebook in 2014. It followed observations by the company's Japanese engineers of how the platform was used by coastal communities during the 2011 tsunami.[32]

Apps are not simply a given technology that subsequently may or may not be used as predicted. By now a circularity and constant to-ing and fro-ing exist between users and developers. In later chapters we will explore several instances of companies developing their apps as a result of their own studies of usage, for example creating a 'kinship'-oriented app in China. By not focusing so much on apps or platforms, the dis-cussion of usage in this chapter can transcend the concept of the app itself. Changes are already afoot that seem to imply a future in which smartphones are organised through alternatives to conventional apps, such as the mini-programme.

WeChat – the dominant social media platform within China, owned by the conglomerate Tencent – has become a kind of app store within an app store. Starting in 2017, WeChat introduced a new feature that allows mini-programmes to be employed inside the platform. Within a year *Xiao cheng xu*, as these mini-programmes are known, had been adopted by 72 per cent of WeChat users.[33] The use of mini-programmes cuts down on the use of phone memory. It allows users to access apps without installing them and can provide coupons, discounts and easier communi-cation with other WeChat users. There are four main categories: games, news, utility and e-commerce.[34] To give two examples, 'Jump Jump' (*Tiao Yi Tiao*) is a mobile game mini-programme which reached 400 million players in its first three days, partly because it scored players relative to their WeChat friends.[35] At the same time, a utility bill mini-programme introduced by WeChat in March 2019 reached 147 million monthly active users in just three months.[36] Mini-programmes then quickly spread into other areas, for instance apps for local public transport. As a result, within two years the number of mini-programmes available within WeChat is around half the number of apps available in Apple's App Store, since mini-programmes have attracted a hoard of developers who covet WeChat's massive user base. All of this builds on other properties that make We-Chat 'super-sticky' – an app that no one wants to leave.[37]

Outside of China, Apple and Google have also developed their equivalents to WeChat mini-programmes. These include the Health app and Wallet app from Apple and Google. Google promotes greater integration of its apps within the Google suite with slogans such as 'One account, all of Google',[38] and these have achieved considerable success. But phone corporations have not been so successful at pushing their own applications. Most smartphones come with pre-installed apps, for example, but these are sometimes spectacularly unsuccessful. Most of the users we observed do not use these pre-installed apps and would delete them if they were allowed to. A particularly conspicuous failure is Bixby, Samsung's voice assistant, generally regarded as a considerable nuisance.

Conclusion

The end result of these complex interactions are the actual smartphones that were inspected during the interviews with which this chapter began. As stories are told about where apps came from, why they remain on someone's phone and whether they are used, the incredibly messy nature of everyday life comes to the fore. As previously noted, apps may be downloaded by others. When Carla from Bento lent her phone to her granddaughter, it returned with nine extra apps including a meditation app, a delivery app, a banking app and a language learning app.

Pressure to have an app may come from institutions other than companies. The governments of both Brazil and Chile are currently working towards digitising state-provided services in a bid to offer these as paperless,[39] as is also found in other fieldsites. This means that in order to access state services, certain apps may, in effect, be mandatory. Most of our older informants disliked the proliferation of apps that result from these processes. One of the reasons people use ubiquitous apps for health purposes, rather than bespoke apps, is to control the number of apps on their smartphones. Still many people at every age tend to end up with an array of apps on their smartphones that they either do not use at all or use only once or twice; this can sometimes amount to half the apps on their smartphone.[40]

Users can be interested in and talk at length about particular apps. They are happy to have conversations about whether Waze is better or worse than Google Maps for navigation, or about which is the most accurate weather app. They may delight in telling one another about a new app they have discovered for a specific interest, such as recognising bird songs or plant names. But once we enter into the messy world of doing stuff with smartphones, such as in relation to health, we see all sorts of creative reconfigurations of information, images, insurance forms and monitoring, and the consequent involvement of other activities. It can then turn out that sending money may be more significant for health outcomes than any apps created specifically in relation to health. In the ideal world of mHealth there would be an increasing range of clear bespoke apps that assist with specific health and welfare requirements. But in the ethnography of how smartphones are used for health, we are more likely to find a combination of apps, none specifically designed for health purposes, but which are put together to help care for a frail elderly parent, whether living in the same house in Dublin or in a remote village in Uganda.

If the proliferation of specialist apps has had an impact, it may be less through the way that they are used and more in the approach to tasks which they imply. The discourse and mindset that this involves has been called 'solutionism'. This chapter also recognises that there is no simple division between the world of users and the world of developers. It is important to understand how companies such as Apple, Google and Tencent develop strategies for keeping people engaged with the apps they developed. For example, in recent development the unit of the app is being superseded through the creation of mini-programmes by Tencent. All of this may help to explain why this volume has taken the smartphone to be so much more than an app machine.

Notes

1. This could be considered as analogous to the artful alignment of infrastructures among teams of scientists. See Vertesi 2014.
2. Results of Xinyuan Wang's survey in her fieldsite of Shanghai, showing the most used apps on research participants' smartphones.

Penetration rate	App name	App function
100%	WeChat	All-in-one (social media)
87%	Baidu	Search engine
60%	Baidu Map	Map
57%	Toutiao/QQ	News/social media
53%	Ximalaya/Tencent News/Alipay	iPod/news/payment
50%	360 weishi/Taobao	Security/shopping
46%	Meituxiuxiu/QQ mobile browser/iQyi	Photo editing/mobile browser/long video
43%	Pinduoduo/DiDi/Gaode map	Shopping/taxi hailing/map
35%	Meipian/Elema/Dianping/JingDong	Blog/food delivery/review & advice/shopping
15%	UC browser/beautyCam/tonghuashun	Browser/camera/stock market

3. Morris and Murray 2018.
4. Morris 2018.
5. Brunton 2018.
6. Pype 2017.
7. Morozov 2013.
8. Miller 2016.
9. Miller 2011.
10. Miller and Sinanan 2017.
11. Spyer 2017, 63–82; Haynes 2016, 63–87.
12. Istepanian et al. 2006; Donner and Mechael 2013.
13. See Taub Center 2017.
14. Kusimba et al. 2016, 266; Maurer 2012, 589.
15. Platforms such as WhatsApp are increasingly employed to gather together 'care collectives' to help look after frail parents. See Ahlin 2018.
16. Yong and Saito 2012.
17. Here 'home screens' refers to the screens that are open on Android smartphones, as opposed to the ones that require the user to go to the background app drawers.
18. WhatsApp used to have an annual subscription fee of around 69 cents/pence, but this policy was abandoned in 2016. See BBC 2016.
19. Couldry and Mejias 2019.
20. Nissenbaum 2010.
21. Duque Pereira 2018.
22. Ku et al. 2017.
23. Wu et al. 2017.
24. See Tiongson 2015.
25. See Apple Inc. 2020.
26. Leswing 2019.
27. Samat 2019.
28. Mohan 2019.
29. Williams and Smith 2005.
30. Boyd 2019.
31. Lavado 2019.
32. Kedmey 2014.
33. See Parulis Cook 2019.
34. Lui 2019.
35. Jao 2018.
36. Lui 2019.
37. An important discussion of WeChat may be found in Chen et al. 2018. This includes other possible terms such as 'super-app' and 'mega-platform'.
38. This slogan is usually displayed on the sign-in page when the user is signed out of their Google account. It is available via accounts.google.com › ServiceLogin, if one has a Google account.
39. Otaegui 2019. For Brazil see Governo Do Brazil (Government of Brazil) 2020.
40. One source of evidence for this claim is that we have been asking 28 students to carry out the same style of interviews as part of a lecture course we are running at the Department of Anthropology at UCL.

5
Perpetual Opportunism

Fieldsites: **Bento** – São Paulo, Brazil. **Dar al-Hawa** – Al-Quds (East Jerusalem). **Dublin** – Ireland. **Lusozi** – Kampala, Uganda. **Kyoto and Kōchi** – Japan. **NoLo** – Milan, Italy. **Santiago** – Chile. **Shanghai** – China. **Yaoundé** – Cameroon.

Chapter 4 traced a journey from treating the app as the fundamental unit of the smartphone to the messier world of everyday life. Smartphones are associated with a form of solutionism, but the concept is not as simple as 'for every problem there is an app'. Instead, it starts by considering tasks that vary across these fieldsites, then involves individual users in finding combinations of apps and functions that work for them. In the conclusion of chapter 4, we suggested that smartphone design is crucial to this process. With icons placed in close proximity, it is easy to flit between them. This point is equally relevant to the discussion in chapter 3 about how the smartphone is becoming a kind of remote-control hub – potentially useful for organising external objects such as an incipient 'Internet of Things', but at the moment mainly focused on people's social relations.

These internal properties of the smartphone can now be linked to its most obvious external attribute. As a mobile phone, it is small enough to be carried in a pocket or handbag and is therefore easily present from the moment a person wakes up to when they go to sleep. We call this mobility, but its most important property may perhaps be the opposite. It is not just that of being carried to different places, but of being constantly in the very same place – next to our bodies and therefore always immediately present. This will prove foundational for several of our main conclusions. One of these is the notion of a 'Transportal Home', to be discussed in chapter 9. The concept that dominates this chapter is called 'Perpetual Opportunism'.

The term 'Perpetual Opportunism' builds upon a previous legacy. One of the influential academic books about the mobile phone was called *Perpetual Contact*,[1] referring to the way in which the mobile phone allowed us to be constantly available to another person. For example, smartphones are starting to replace the red button supplied to elderly people to summon emergency help in the event of a fall. This example shows the security in perpetual contact, but there are also burdens. Teenagers may be forced always to remain in potential contact with parents, or find themselves unable to escape the fear that someone they thought was their friend might be saying nasty things about them online.

A second, related development discussed by Ling[2] examined the way in which mobile phones have changed our relationship to space and time, for example bringing us the capacity for micro-coordination. Previously we had to make a plan to meet with other people and stick to it, since they could not know that we wanted to change that plan. With mobile phones, we could make an initially vague agreement about time and place, which would become more specific nearer to the event. For example, we had all planned to meet at a certain pub one evening. Then the first person to arrive may find it is too crowded, so they will use WhatsApp to direct everyone to a different pub.

Perpetual contact pertained to a mobile phone – a device that was still primarily used for conversations or texting between people. Yet none of the fields of usage that comprise the content of this chapter are mainly about the smartphone as a phone. They are about the use of smartphones for entertainment, travel, gaining information and taking photographs. So perpetual contact is no longer adequate as the theorisation of the smartphone.

What emerges instead as the key quality is *opportunism*. Simply because the smartphone is always with us, it creates the possibility of being opportunistic as a constant. But what matters is the evidence that this possibility is appreciated by users, who in turn develop a more opportunistic attitude to their everyday lives. The first example in this chapter shows how the smartphone has transformed photography. Here the crucial difference comes with the ever-present possibility of taking and then instantly sharing a picture. By the same token, we can look at the latest news or listen to some music simply because we are waiting in a queue or otherwise feeling bored. Perpetual Opportunism thus goes well beyond perpetual contact in changing our relation to movement and travel.

As noted in chapter 1, it is impossible to be comprehensive in discussing smartphones because they now engage every part of our lives. This chapter therefore concentrates on just four disparate examples, chosen to examine both the diversity and commonality in smartphone usage across these fieldsites. In each of these instances, however, the discussion is not limited to the impact of Perpetual Opportunism. This is always addressed within its more general context, which will include whatever other factors appear relevant to those genres of usage.

Opportunistic photography

The interviews about the apps on people's smartphones confirmed that the camera is one of the most commonly used functions across these fieldsites. But, as noted in chapter 1, calling the device on smartphones a 'camera' is misleading; it implies that smartphone photography is merely a mobile version of previous photography. Certainly the camera on a smartphone takes pictures, but a more careful examination suggests that both the smartphone camera and the photographs it captures are better appreciated by emphasising the contrasts with former photography rather than the continuities. One evident difference is simply scale. The sheer numbers of images that people take, circulate, exhibit and store once they possess smartphones are incomparably greater than analogue or even digital photography.[3]

There are also entirely new genres of photography, such as the 'functional photography' mentioned in chapter 1. These are the photographs that people now routinely take of an item that they might later want to buy, the opening hours on a shop window or a notice board flyer for a yoga class in a town hall.[4] Functional photography is a prime example of the point about the smartphone as a control hub: taking these pictures is commonly a first step towards organising one's time and tasks. It is equally an example of how these new uses of the smartphone take advantage of the proximity between icons on the smartphone screen. We first take the picture and then immediately use it in relation to some other apps, for instance the calendar or a social media platform, to let someone else know about this event.[5] Changes in photography as a practice thus map onto changes in how we access and share information, relate to location, use calendars and link the digital to human memory.[6]

Smartphone photography is also indebted to other new technologies, including the reduced size of the jpeg format and the massive storage available on devices. Beyond the phones themselves now lies the widespread availability of cloud storage, which means that suddenly the 'cost' of taking and storing a photo has radically diminished. Digital cameras may also now come with Bluetooth, Wi-Fi and GPS for geotagging and immediate sharing. Yet smartphone photography is a special case within digital photography. The crucial difference is that of opportunism, which has had profound effects on what photography is. Early photography was strongly associated with the idea of long-term storage and the archive: it was a means of capturing and retaining images of people, places and things.[7] Photography was about endurance or permanence. In stark contrast, contemporary smartphone photography has become one of the key manifestations of transience, the ultimate expression of which is found in the title of Snapchat[8] for one of its main platforms. This means that for the first time a photographic image can be used as part of the conversation; it then becomes just as fleeting and transient as oral communication. Most photographs are shared on WhatsApp, Instagram or Facebook with the idea that they will be seen for a day or two and then be replaced by others. This shifts the meaning of photography 180 degrees from its origins as an enduring archive. Where once the primary purpose was permanence, it has now become transience. Representation and archiving remain as purposes within photography, but are more often secondary rather than dominant.

Not everywhere has shifted to the same degree or in the same way. Older people are more likely to combine traditional and new possibilities. An individual might be creative precisely through the way they combine smartphone photography with analogue photography. An example can be seen in this short film from Dublin (Fig. 5.1).

The production of images is now an integral part of the everyday retirees' environment in Yaoundé, where people may try and transform every memorable moment into an image. In effect they use the smartphone to keep the 'trace' of the small events that punctuate their

Figure 5.1 Film: *Photography in retirement.* Available at http://bit.ly/retirementphotography.

daily experience. One of the major components of older people's use of smartphones is the time they spend looking at pictures and videos: those sent to the family WhatsApp groups by their children and grandchildren, and those from extended families and friends. If younger people have seen an intensification of the way in which they take photographs, it is equally important that older people have in turn accelerated the way they consume photography. This becomes clear through comparing the way in which they constantly look through social media images with the much more occasional previous use of photo albums, framing practices and displays within the home.[9]

While most viewing is transient, older people may continue to be concerned with archiving. A role that has changed to accommodate these developments is that of the person who becomes responsible for organising and keeping images in the family archives. For instance, Roger from Bento has become the 'memory keeper' of his family. He is now responsible for organising these photos in Google Drive with folders addressing the main events of the family, but he also archives them by name, year or family member. In addition, he may improve upon the images using filters from two apps on his smartphone and computer. When someone wants a specific picture, he is the one they will turn to.

Not only has the smartphone camera consolidated the shift from analogue to digital in transforming our relationship to the image, but it may also have had a commensurate impact on how we see the world around us.[10] Taking a picture puts a frame around some of the things we see, separating them out from that which is not framed.[11] Framing through photography is at least analogous with framing as a separation of art from non-art, or the sacred from the mundane. Framing then helps account for what otherwise might be a conundrum. While many images are shared, people also take vast numbers of images that will never be viewed. So why do they take these pictures in the first place? To understand smartphone photography requires paying attention not only to how people consume images, but also to the question of why they take images knowing they will not be consumed.

As we go about our daily life, most things are experienced as mundane and passed by. Now and then, however, we see something we want to acknowledge as standing out.[12] This may be planned and ritualised, as in a child's birthday party or as a tourist's view, or it may simply be something unanticipated that strikes us in the landscape. Merely having the facility of Perpetual Opportunism, enabling us to take a picture whenever we see something that stands out, may result in us looking at the world differently. These images are taken on a whim: we may or may not then share them through social media. Rather as in the realm of art,[13] taking a picture has the effect of instantly putting a frame around something; it is thereby marked, at least in a small way, as something transcendent of the mundane. We may never look at the image again nor share it with anyone else. But for that moment we felt we could not pass by that butterfly, or funny-looking stone, or our friend's expression, without paying homage to it by adding a frame through taking its picture.

As a Japanese woman, Sawada san, explained while walking around a Kyoto temple and observing a garden through a perfectly framed window, 'Japanese people love to frame things; it's in our culture'.[14] The traditional Japanese garden is designed to be observed from framed viewing spots in temples and homes, rather than walked within. The smartphone has democratised this act of framing from artists and garden designers to something that many people enact every day. Whether we choose to take a photo of a meal or a tree in bloom, the act of framing a subject is a mode of paying attention, while also claiming a presence;[15] not just 'I was here', but 'I am here, right now, having this experience in real time'.[16] Smartphone photography may then have rather more affinities with art and religion (and perhaps mindfulness) as an act of at least transient sacralisation than we might have expected.

The one instance where the affinity of photography with religion is clear and explicit is among the Peruvian migrants in Santiago. Much of the fieldwork took place among a Catholic brotherhood dedicated to an annual religious procession based on an icon of the Lord of Miracles. So many people now take pictures throughout the eight-hour procession that nowadays the entire event is illuminated by the bright screens of smartphones pointed at the sacred image. Similarly, any *pollada* (a fundraising chicken dinner) is also festooned with the flashes from smartphone cameras. National Peruvian Day celebrations are also registered by hundreds of phones. Almost any event, religious or otherwise, ends up being broadcast through Skype or Facebook (Fig. 5.2).

Religion is more of a constraint in Dar al-Hawa. Here, being the subject of a photograph can be a threat to personal family honour, especially for women. An image of a woman smoking a cigarette or not wearing a hijab or can cause serious problems if it is shared,[17] although as women marry, have children and age, the social norms of modesty and proper behaviour may become less of a pressure. For these older people, taking photographs or being photographed is more used as an evidence of vitality and a productive life that a woman can share with her friends or show off to her family.

Figure 5.2   Peruvian migrants broadcasting the Lord of Miracles in Santiago, Chile. Photo by Alfonso Otaegui.

Most members of the community centre partake in group travel outside of Dar al-Hawa several times a year. Field trips are organised and subsidised by the community centre. On one such trip to Acre, a town in the northwest of Israel, those with smartphones took photographs continually (Fig. 5.3), while those without were equally included in the shots. The coordinator also arranged group photos. All the images were relayed back via WhatsApp to those older people in Dar al-Hawa who had not been able to come, whether through illness or having had other commitments. The images evoked positive responses such as 'take care of yourself', 'oh, the sea is beautiful', enabling those who remained behind at least a form of presence within this community activity.

This sense of being present is complemented by the more traditional use of photography as memory – a concept that has not entirely disappeared. Hibat's WhatsApp profile picture features a black and white photograph of a young girl, also in black clothing with a white collared buttoned shirt. The image reminds her and tells others of how she looked as a young girl and her fond memories of that period of her life. It is also a reminder to herself when she is looking at her photos to see what she did and where she was. There are other similar pictures in her Facebook albums. As a result, perhaps the most vibrant classes in teaching smartphone use were those dedicated to photography, particularly when teaching people how to take a selfie.

Figure 5.3 A picture taken from a boat on a fieldtrip to Acre. Photo by Maya de Vries.

These uses of photography suggest that the idea of craft is no longer limited to the taking of pictures. It applies also to their transformation, sharing and consumption. Examples here are the technologies of selection, editing and posting of images on Instagram. The use of Instagram across and within our fieldsites varied. Some people spent time crafting 'artistic' pictures such as flower arrangement displays in Japan; others used it more like Facebook, as a platform on which to share their daily lives. In sharing photos of events they have attended or images of their friends and family, these research participants are acknowledging their appreciation of Instagram as a democratised craft with which almost anyone in their family could engage. The viewing of photographs on Instagram in most fieldsites far outweighed the time spent on uploading images. Photography can also be used simply to facilitate familial relationships, as Komatsu san from Osaka explained. On Instagram she most enjoyed following her daughter-in-law, who would regularly post photos of her grandchildren. They lived on the other side of the city and Komatsu san saw them face-to-face only about once a month.

One example of this development of craft is the use of smartphone photography to take portraits. People in Shanghai regard the camera as part of the phone's hardware rather than as an app. A survey of 200 people aged between 50 and 80 confirmed that for most people it is now their *only* camera. As a result, the quality of the camera has become a key selling point for more high-end phone brands such as Huawei and OPPO. For older people in Shanghai this sense of reverence for photography evolved quite naturally from a previous era, when photographs were expensive and reserved for very special moments. Cameras had to be borrowed from shops, as people could not afford them.

Photography is no longer rare or expensive, but the taking of a picture of a meal before eating it may still feel like a ritual gesture, an enduring legacy of that former time. In some places the photography of a meal prior to eating it has become an essential routine, to the extent that one cook in Xinyuan's fieldsite was quite offended when her son and fiancée ate their meal without first taking a picture. Things may be shared constantly on WeChat, but that still leaves a question of whether something is worth taking a picture of in the first place, and indeed subsequently sharing. The whole process has become an assessment of value. Holiday pictures may be common subjects on social media partly because of the expense that the holiday involves.

Mr Shou always mentions the phrase 'sense of ritual' (*yi shi gan*) to highlight the significance of each photography session. He sees his not-for-profit photography project among the elderly as a mark of respect. As he puts it:

> Many people have passed away without a proper photograph. Every person deserves a proper portrait photo in his or her life. What I want to do is not just take photos, but keep the great memory of the person. I take it with great respect, and people can also feel the sense of ritual. Life needs a sense of ritual, don't you think so?

Mr Hu, aged 88, regards smartphone photography as a serious and specialist hobby. He has filled up one and a half screens of his latest OPPO smartphone simply with photo apps. Mr Hu also has a cabinet full of elaborate and expensive camera equipment, such as a long infrared Nikon lens. They take up quite a bit of room in his small apartment (Figs 5.4a and 5.4b), but he does not mind that they are now gathering dust, given the new capacities he has discovered in smartphone photography.

The survey of apps used in Shanghai presented in the previous chapters shows two photo editing apps among the top 10 downloaded apps. As Mrs Huahua put it, the key is the powerful 'retouch' (*mei yan*) feature which renders immediate enhancements (Figs 5.5a and 5.5b).

Figures 5.4a and 5.4b Mr Hu's array of purpose-built camera lenses (Fig. 5.4a); Mr Hu in his studio flat (Fig. 5.4b). Photos by Xinyuan Wang.

Figures 5.5a and 5.5b The subject's natural appearance (Fig. 5.5a); the subject's appearance after on-screen manipulation, with wrinkles removed, skin smoothed and whitened, nose given higher bridge and corners of mouth adjusted (Fig. 5.5b). 'Washington Chinese Culture Festival 2015' by S. Pakhrin, licensed under CC BY 2.0.

She describes this as a 'safe and free plastic surgery without the pain and cost'. Wrinkles, pimples, scars, dark circles and age spots can all be eliminated with one button using 'auto-beautification' (*yi jian mei rong*). The app can also be used for 'digital make-up' such as adding lip colour, blusher, false eyelashes and eyeshadow, and drawing and adjusting the shape of eyebrows.

People in Shanghai would not usually dismiss Mrs Huahua's self-edited images simply as 'fakes' because, in her own way, she has adapted the camera to become an artisanal craft, as Mr Hu has done. As a result Mrs Huahua is judged on how well she has adjusted the image to create an idealised picture. After all, altering one's appearance has long been established as an everyday practice in the 'analogue' world. Cosmetics or flattering clothing are regarded not as fakes, but rather as examples of a domestic craft that accords with social convention. People might be condemned for doing these things badly, but not for attempting them in the first place. As Mrs Huahua states:

> It is not that I am particularly narcissistic – I am just trying to comply with the social norms of good WeChat photos.

Meanwhile a third research participant, Mr Li, is rather annoyed that they do not seem to have an app designed to disguise his hair loss. In Shanghai[18] concern with surface appearance is not denigrated as superficial. It is rather viewed as the place where people can demonstrate their aesthetic skill, providing evidence of who they are and what their capabilities may be.

The examples above serve to illustrate a contrast with the earlier example of how older people in Yaoundé consider photography. In Yaoundé they are much more likely to see this entire activity as the creation of a fake appearance. This may be partly because they are not so invested in this development of 'crafting the image'; their focus is rather orientated to new ways of *consuming* photographs. If anything, older people in Yaoundé may feel that the rise of the smartphone as a technology for taking pictures is a new source of anxiety. Mr Etou, a retired mechanical technician (Fig. 5.6), reports the following:

> You get into the camera, you operate it thinking you're taking pictures, but you're doing videos. It's really a problem. With age, we have more shakes. It's hard to keep the focus on what you want fixed without moving. But as soon as you move, the image is cheated. For

weeks when I had my first smartphone, I could not make a single clear picture, despite the assistance of children and small children. I ended up giving up. The other problem is to organise these photos and videos. Sometimes you try to do that and you lose it and you do not find it anymore. That's really annoying. You want to break your phone. It's really annoying.

Older people in Yaoundé are also sensitive to the images of themselves that circulate online. Without the facility for altering the images, they find themselves reminded more constantly of how much they have aged, which may not correspond to their sense of themselves as still relatively young. Some older people may refuse to show their photo gallery to others, seeing this as an intrusion on their privacy. They feel disturbed and may not appreciate other people's insistence on taking

Figure 5.6 Mr Etou, one of Patrick Awondo's research participants in Yaoundé. Photo by Patrick Awondo.

their image. Even in Shanghai, some older people refuse to allow others to circulate their image. For older people the tragedy is that it is one's *actual appearance* that feels alien or 'fake', not the images that people take of it.

A different problem with beautifying filters was noted by Fujiwara san, a woman from Kyoto. She suggested that creating such heavily edited and idealised images sets an unnaturally high standard that unfiltered photos cannot achieve. If everyone is using filters, then you feel that you must too, resulting in a sense of competition. It is not so bad if the use is simply seen as fun, as in the selfie below, sent to Laura during the Covid-19 pandemic (Fig. 5.7). Here the make-up filter was just a thing to play with on a bus trip, as friends experiemented to see whether they could add eye shadow and recognise each other even with face masks on.

Figure 5.7 Make-up filters work even with face masks on. Photo taken by anonymous research participants.

Maps/movement/travel

Apps used for travel and transport are often a major component of the smartphone. In this section we will start by considering uses that relate to local transport, including the way Uber adds an element of Perpetual Opportunism. The review then shifts to a consideration of holidays and travel abroad. With regard to local transport, this infographic (Fig. 5.8) shows the percentage of research participants in NoLo who use various transport-related apps. There was a sense across several fieldsites that this particular usage of apps is likely to increase.

For older people in Bento, two main apps act in combination with WhatsApp to support the expansion of sociability in old age: Google Maps and Uber. Together they create an urban mobility package that gives older people the autonomy to consume the city with their friends. This

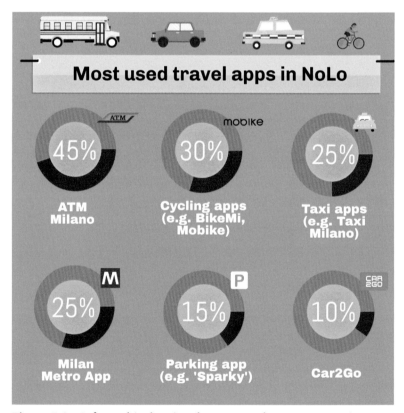

Figure 5.8 Infographic showing the most used transport apps in NoLo, based on research by Shireen Walton.

is the time of life when people are abandoning their car and discovering public transport, now free for people over 60. Many of them only started to take the bus or the underground after retirement. For information about public transport, they use the Moovit app as well as Google Maps. As Fernanda declares:

> I'm loving it! I look at the app and just go to the bus stop – often I don't have to wait more than five minutes before the bus goes by.

Uber works as a complement to public transport. The app is usually launched in the evenings. It has become the Perpetual Opportunism that gives older people the freedom both to stay out and if they wish to drink alcohol when meeting friends. Mauro, a dance teacher, no longer accepts excuses from his female students not to go out to dance at night, when previously they may have been wary of doing so for safety reasons. 'Call an Uber, for Christ's sake,' he says. In Santiago, Ernestina's husband has been diagnosed with Alzheimer's and is now banned from driving. Her own driving licence has expired, since she has a problem with her eyes, so she has quickly become dependent on Uber's taxi service. Uber is also an example of a smartphone app being used in the manner described as a 'control hub' in chapter 3, but this time in the labour market, where it is networked to a sophisticated organisation and auditing of work.

Some older people in Santiago are reluctant to use GPS because they fear that their location is then being tracked – as indeed it would be by Google, if this has not been deactivated. They would prefer to memorise ways of travel, but they do appreciate an app that tells them when the bus is coming so that they do not have to wait so long. Younger users, such as the Peruvian participants, are mostly concerned with the advantages of Waze over Google Maps because it seems better informed about local traffic conditions. Santiago, as with so many cities, is getting increasingly congested. A glance at the iPhone of Federico shows just how important travel apps have become for this Peruvian entrepreneur living in Chile. Whole sections of the phone's front screen are devoted exclusively to travel apps such as Airbnb and the local airline LATAM; another section is more focused on local transport (Fig. 5.9). These are by no means all the relevant apps on the screens of Federico's phone (Fig. 5.10). He also has Flightradar24, which shows him the location of commercial aircraft anywhere in the world, in real-time. Sometimes, when he is waiting at the airport, he takes a look at the app to see where his plane is, or even looks at other flights just out of curiosity.

Figure 5.9 The travel/taxis folder on Federico's phone. Photo by Alfonso Otaegui.

Figure 5.10 The maps folder on Federico's phone. Photo by Alfonso Otaegui.

In Shanghai, Baidu and Gao De are the two most popular map apps. In practice, older people made less use of these maps than younger people, as they are less likely to visit new areas and have more confidence in navigating without smartphone maps. 'The map of my neighbourhood is printed in my mind. My mind works better than any map app,' claims Mrs Zhihui. While she never uses her Baidu map for finding her way to a destination, it becomes important every two weeks when her son's whole family drive to visit her. Seeing real-time traffic levels on the Baidu map allows Mrs Zhihui to know when her son's family will arrive, and accordingly what time to cook dinner.

> Once there was a traffic accident on the overhead line and my son's car was stuck in the traffic jam for more than one hour. They called me in the end, saying they will be late, but I said I already knew they would be half an hour late, as I was following the whole overhead line on the Baidu map, and saw that it had turned from the usual green and orange to red.

Here we see another example of Perpetual Opportunism in the ability to exploit real-time information.

All of these involve a process of adoption and adaption. Older people may start out resistant to Google Maps; they may try to use the app like a traditional map, memorising or printing out the details. Many later acquiesce in being directed as drivers, or discover when in Moscow or Lisbon that it can also help them when walking. Later still, they may follow hints that reflect local cultural interests. For example, Google Maps is probably used a good deal more in Ireland for finding the way to funerals than in any other fieldsite. This is simply because people in Ireland expect to attend funerals, even of people they barely know, to show support to the family. Google Maps would come into play following reference to the website RIP.ie. This lists every funeral in Ireland on any given day and provides directions of how to reach that funeral, as well as times of related activities such as the repose (wake) or the mass.

For older people in Ireland, it is hard to exaggerate the importance of travel in their lives. Holidays are one of the most common of all conversation topics, and middle-class people often have a property abroad. Others may regularly travel to the UK for a weekend, perhaps to go to horse racing at Liverpool or to visit their children working in the UK; yet others simply take a short trip to another part of Ireland. Such activities generally involve a particular configuration of apps. Some are mainly used for holidays, especially Tripadvisor, Booking.com or Expedia. Most of the research participants seemed comfortable using smartphones for checking in, both remotely and on arrival at airports. Prior to departure they may have used Duolingo or listened to a local radio station to brush up on their language skills.

Once abroad on holiday, people may use Google Translate and currency converter apps. Social media and webcams become important tools for keeping in touch with family and exchanging photographs; a weather map is also likely to be consulted. If people are on a walking trip, the step-counter may come into play as evidence that they are using the holiday to keep healthy. Older people may also feel more comfortable venturing forth in unfamiliar terrain, knowing that thanks to GPS they are unlikely to become completely lost. The smartphone is also used for envisaged or alternative travel. The most conspicuous example of this was Liam from Ireland (Fig. 5.11). Liam used a virtual reality (VR) app linked to

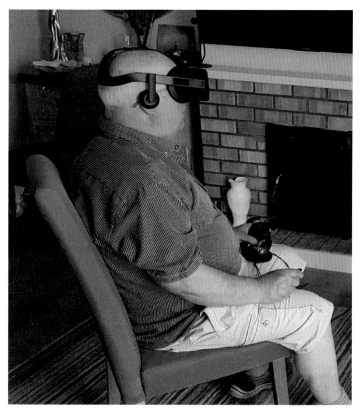

Figure 5.11   Liam 'travelling' to the US using his Oculus goggles. Photo by Daniel Miller.

his Oculus goggles, enabling him to 'travel' to places in the US that he will probably never actually visit.[19] Space was no barrier to Liam's ambition: he enjoyed a VR 'trip' around a space station in the same way. He also browsed different parts of the world using Google Earth, planning a trip to a wedding in Italy or revisiting former holiday destinations.[20]

News and information

This brief survey of how people use smartphones in relation to news and information starts with individuals, progresses to consider circulation within community and ends with an example of news disseminated by the state. In many regions of the world today the verb 'to google' has become synonymous with seeking online information on almost any topic. In practice, this may have various local connotations that extend well beyond its capacity as a search engine. For many older people in Santiago there is no real differentiation between Google as an app, a website and a search engine. Similarly for some people 'the internet' is simply Google.

If Google expanded from being a search engine, YouTube has expanded to become a search engine as well. Sometimes YouTube is viewed as a site for alternative information to that available in the mainstream. For example, a Peruvian brother in Santiago used YouTube to find non-religious arguments against abortion. YouTube can also be used to relocate materials from people's original homeland. In NoLo this was typically music for Egyptian research participants and recipes among Sicilian ones. For example, on the saint day of Santa Lucia, Maria located a recipe online via her smartphone for Cuccìa. This is the day when Cuccìa, a typical Sicilian dish made with boiled wheatberries and sugar, is traditionally eaten. Maria went on to share this recipe and photos of the Cuccìa she had made on Facebook and WhatsApp with family and friends; she also shared batches of the dish itself with her daughters in Milan and close neighbours in her NoLo apartment block.

One effect of Perpetual Opportunism is potential addiction – although, as pointed out in chapter 2, it is often unclear quite what that term means. In Ireland there was a perception of young people as being 'addicted' to smartphones because they were clearly a bit 'twitchy' if an extended period had elapsed since they last checked to see what their friends – or haters – might be saying about them. If older people looked twitchy, the most common reason seemed to be a similar 'addiction' to the news, mostly political news, although for older men it could also be sports news. Many people talked of spending several hours a day looking at news on their smartphones, usually because they had become absorbed in a particular political situation. During the period of our research this usually meant either the US politics of Donald Trump or the UK politics of Brexit. No-one reported a similar fascination with Irish politics.

Anne, for example, spent two or three hours daily on 'Trump' news. She shifted between Google and newspaper apps including *The Washington Post*, *Al-Jazeera* and *The Guardian*, as well as the local Irish papers and various radio and media sources, including US stations such as *Fox News*. These sources were complemented by YouTube, which she used for various satirical programmes or reading satirical tweets from *The Onion*, as well as news shared by others on Facebook. In the morning Anne listened to the news with headphones on so as not to wake her husband. It is possible that the majority of older people in this fieldsite now look at the news on their smartphones before they get out of bed in the morning and before they turn off the light to go to sleep.

People had many ways of deciding what news to trust. On social media this had more to do with their appraisal of the person sending them news. More generally even their online news was often associated with traditional news sources such as newspapers, radio and television stations. In a more detailed investigation of how people judge health information in Dublin people provided many clear criteria, such as dismissing any sources connected to sites from which something could be purchased. In other fieldsites people often showed a preference for news sourced in countries other than their own, depending on their opinions about local politics.

In journalistic coverage of the use of smartphones as a news medium, the main concern is with fake news. However, for participants across several fieldsites what dominates news on the smartphone is ridiculing politics and politicians through sharing satire and other jokes. For example, the meme below was shared across NoLo in WhatsApp groups (Fig. 5.12). It depicts the Italian Prime Minister Giuseppe Conte as whispering to Matteo Salvini, then the far-right Interior Minister:

> Do you have a boat to block to distract the people, since I don't know what the f**k more to say regarding the recession?

The reference is to an episode in June 2018, when Salvini controversially blocked the disembarking of 600 migrants from Libya onto the Italian island of Lampedusa. His action dominated the news at that time. The

Figure 5.12 Satirical political meme shared among NoLo WhatsApp groups.

meme proved particularly popular in NoLo, in a context that regularly sees offline and online protest against Salvini and his hostile policies towards migrants.

The sharing of jokes suggests a more communal consumption of news. In Yaoundé, for example, YouTube has become an important component of socialising within the various sports groups, but the primary content of what is shared will be funny videos. A very common expression that might accompany them is that 'you should not laugh alone', and there is a sense that these shared jokes help develop a generally good and happy ambience. This may start from very early in the morning. As soon as a member of the group wakes up, they may find hundreds of funny videos and images to view and comment on. This activity is said to become one of the favourite occupations for retirees in Yaoundé, sometimes even surpassing watching television at home. One woman said that she spent at least three hours a day looking at the videos shared by relatives in the various WhatsApp groups she belongs to. She finds them very funny. She added that she never shares them herself, but she appreciates what people send her as a kind of care practice. In addition, humour is a key component of the way in which people circulate political news – providing them with a means to participate and become active in political discussion, rather than feeling that they are only passive recipients of news dissemination.

Humour is also a significant component of the WhatsApp messages shared within Dar al-Hawa, usually still images rather than video. Information and news can both be couched within such frames. Often several people will react to the message by adding another image of a joke or by writing 'hahahaha'. A genre that may gain more attention than humour, however, is the sharing of riddles. For example, there are riddles about Islam, about a place, about a specific 'tricky' image or even mathematical issues (Fig. 5.13).

These riddles gain more attention on the WhatsApp group compared to the sharing of jokes. They create a more intensive dialogue,

Figure 5.13   Screengrab of a riddle shared by Laila Abed Rabho and Maya de Vries with research participants in Dar al-Hawa. The text reads: 'How many pencils do you see in the photo? Who is smart and knows the answer?'

accompanied by a competition over who will be the first to give the correct answer. This corresponds to what Laila and Maya could observe in the offline activities of the seniors' club, where games accompanied by competitions were part of their regular activities. Often these were supplied by young students who were doing an internship through the university. The atmosphere was generally positive during these activities, since people were aware that these competitions kept everyone alert. The online WhatsApp versions seemed a clear extension of these goals.

Large WhatsApp groups are also used for sharing information in Uganda. For example, in Gulu the Lusozi chairwoman's brother is part of a council WhatsApp group for which he can access Airtel data, funded by the municipal council at a reduced price. This helps him to find out about council health sensitisation sessions and mobilise the community to attend. As platforms for information sharing, these WhatsApp groups 'bring news', something that people often like to hear of 'from around the world'. This is the case for single grandmother Flossie, who listens to preachers and pastors in the morning and BBC news in the evening. Okida, a biblical counsellor, is particularly interested in UK news:

> I love BBC. I'm following your Brexit so much … we were colonised by you people, so we still have an interest. When there's a mess in your place … when you people are not doing well then we start panicking a bit!

When she visited people's homes in Lusozi, Charlotte would often arrive to find the radio playing from a small mobile phone. It might be playing music, national or international news or, most often, evangelical preaching. Emmanuel, a born-again Christian, is glad of this use of smartphones and social media to preach the gospel. He also listens to preaching on the radio through his phone, specifically 'Voice of America' and news about Israel:

> You know it is better to know about Israel because we are in the end time. If anything happens there, you get it in the prophecy of the Bible.

Atim shares one phone in her household, the phone itself belonging to her and the SIM card being her daughter's. They generally use it to enable relatives in the village to call and let them know about their problems and request financial assistance, for instance for their aunt's recent hospital visit. They also like to listen to preaching, but fear this will use up their battery as they do not have electricity in the house, so have to charge their phone in the church office. They restrict use of the phone to a maximum of 20 or 30 minutes as 'they're always mindful of the power', particularly because, they note, 'if you overcharge it burns the charger'.

Finally, as well as individuals and communities seeking news and information, the state can also employ mobile phones directly as a medium for news dissemination. The clearest example of this was the practice of the government in Japan to issue emergency warnings via the mobile phone.[21] The government started these mobile notifications following the triple disaster of March 2011, commonly referred to as 3/11, which consisted of an earthquake, tsunami and nuclear meltdown. In the years that followed the disaster the government has faced criticism over their handling of the situation, much of which has been mobilised on social media.[22]

In Japan natural disasters such as strong earthquakes, typhoons and heavy rains are an annual occurrence. The smartphone has become the first line of defence for many people, providing a way to prepare and look after themselves in an era where trust in the government is low.[23] A man living in rural Kōchi explained how every morning he checks a prefecture-wide website about natural disasters from his smartphone. It is not an app, but the website is bookmarked so that he can open it in one click from his home screen. It tells him the water level and currents of the ocean. If the tide has fallen, he knows that he should be prepared for an earthquake. In such a scenario he would buy emergency food and not travel into the city.

> I think looking it up yourself is much quicker ... than waiting for the government alarms. I mean this whole tide thing is only a hypothesis, but it's better to know. I think it's [the emergency notifications] good but it's always 5–10 minutes before it happens, so it's too sudden ... I fear that Kōchi prefecture will be the last to be taken care of. There are other prefectures more important than Kōchi. This is why I want to know sooner so I can evacuate more quickly.

The problem faced by the official emergency notifications is that given their high frequency, being used even for minor earthquakes, they are mainly ignored. During the heavy rains and flooding of summer 2018, it was common to be in a restaurant and hear everyone's emergency notification sounding simultaneously, to the amusement of the diners. Some research participants mused that this was because the government does not want to be seen as having failed to alert people to possible disasters.

Figure 5.14 Screenshot of emergency notifications subsequently shared on Instagram by a participant in Kyoto. The accompanying comment remarked on the frequency of such alerts.

The result is a kind of 'crying wolf' syndrome, where the number of false alarms may reduce their efficiency (Fig. 5.14).

On the other hand, the potential for direct warnings about possible infections through smartphones text messaging became a global issue with the rise of contact tracing during the Covid-19 pandemic. This development will be discussed in the final chapter.

Audio entertainment

A recent article in *The Economist*[24] noted that, in retrospect, the main interest that has led people to go online across the globe is probably leisure activities.[25] Before the smartphone, entertainment generally represented a discrete activity, such as watching a television series or listening to a radio programme broadcast via traditional media. The

Perpetual Opportunism represented by the smartphone, however, has transformed this into a potentially constant presence that can be attended to during any interlude during the day. In five minutes, while waiting for a bus, a person can check on a vlogger, scroll through some funny memes circulated by friends, listen to a new music track, see what a friend is up to or read about what the government is doing. If they missed their favourite radio serial, they can stream it at any time.

This is a significant change in people's lives because there is no knowing at what part of what day a person will feel bored, dispirited or simply low. Before this state of Perpetual Opportunism, they may not have had access to television or radio at that particular moment. Smartphones also increase the range of what can be accessed; if one person wants to listen to football scores, another can hum along to their favourite hymns. One person may turn to sports news; another to celebrity news; a third to political news. The stimulus for one may be political memes, for the next memes of kittens and for a third a messy grandchild.

Since entertainment is such a vast topic, the following discussion is confined to a small example: the way people listen to music and other audio genres. In Lusozi, out of 35 research participants, 24 used music players on their phone and 4 had music search apps such as Shazam. More commonly, however, they would source their music from one of the three vendors in the area who sell music, television and film downloads. Customers bring or buy a memory card or their 'flash' and choose which genre they prefer from the vendor's regularly updated selection of downloads. The vendors try to keep up to date with the latest options, playing music from their sound system to attract customers. As one vendor explains, 'people come, especially guys who are current … they get interest from outside and they know what they want'. Younger men can be responsible for choosing the music and placing it on flash disks for older relatives.[26] He notes that older customers are 'quite rare … they come once in a while looking for old songs', such as gospel music, Lingala or traditional Acholi music; the vendor adds that 'it makes them happy'. As in Pype's[27] analysis of older people and popular media in Kinshasa (DRC), in an urban environment where entertainment is often youth-focused, older people can reconnect with contemporary society through their knowledge of music.

Typically people go to buy music at the end of the month once they have been paid. Using a 4GB memory card that costs the equivalent of around £4, they can load about 500 songs onto the device and send them on through Bluetooth. They may pay (the equivalent of) 4p per song or can buy a bundle of five songs for 21p. The vendors have videos as well, which they sell for 6p or give to regular customers as a bonus. 'People want to have videos, especially aged people with families,' one vendor explains. Music vendors typically provide comedy, series, action movies and movies from Hollywood, Nigeria and Ghana.

Launched in 2013, Shanghai-based 'Ximalaya' is one of China's most popular platforms for podcasting and audiobooks, with about 500 million registered accounts.[28] Podcasts, or digital audio programmes available for download or streaming, have grown rapidly in popularity among older people in China.[29] There is a probably a podcast episode for every taste and need in this era of on-demand content. The total number of podcast listeners in China surged to 425 million in 2018, according to data from iMedia Research. The average user spends 150 minutes a day on the 'Ximalaya' app.[30]

For older people who are starting to have difficulties reading, this audio alternative is a considerable boon. They are also a generation that grew up with radio and feel comfortable with it. Mrs Tong, a big fan of 'Ximalaya', explains that 'it is just like the radio on the smartphone, but it has much more content'. Mrs Tong also listens to the children's education programme (she takes care of her grandson during workdays) while doing the housework. She could not remember the last time she sat down to watch a television show when it aired live, because nowadays she watches it only on iQiyi, the streaming video platform on her iPad. Nor could she remember the last time she listened to a radio programme through a terrestrial radio channel. This is because the radio was replaced first by the television and then, two years ago, by 'Ximalaya', which she started listening to at the recommendation of a friend.

Another resource that is commonly used for finding music in some fieldsites is YouTube. Margarita, a retired nurse from Santiago, sold her radio and bought a Bluetooth speaker when YouTube became her primary source of music. YouTube also allows older adults to reconnect to old tunes that are usually no longer available. Now and then, older adults in Santiago use their WhatsApp groups to share links to tunes from the good old days. Similarly, in NoLo, Egyptian women participants saw YouTube as a way of accessing Egyptian and Arabic-language music. They would play this aloud, either on smartphones or connected to speakers, at festivities, parties and gatherings in the neighbourhood, for example during Eid, the festival that marks the end of Ramadan.

Finally, smartphones may also play a role in making or participating in, as well as listening to, music. Brendan founded and promotes a ukulele group in Dublin. The only specific music app he has on his phone is a ukulele tuner. However, discussing the ukulele group now constitutes his primary use of WhatsApp, since this is now a very active group with more than 70 members, resulting in daily posts to the group. Brendan may first download a song from YouTube and then change it into an MP3; this allows him to send it to a Bluetooth speaker, which is what he uses when working with his ukulele group.

Facebook is also the main way that Brendan interacts with other ukulele groups around Ireland and beyond. He uses text and phone to make more detailed arrangements, for example, when the group plays at events or at nursing homes, quite a common occurrence. He also makes use of maps to find the place in question, as well as his calendar to organise these events. His music app has thousands of songs stored on it. Many of these uses are relatively new, but they all derive from his taking up the ukulele. So while Brendan may have only one specific ukulele app, it is clear after some time that he has turned his whole smartphone into a kind of 'ukulele app'. This example reinforces the point made in chapter 4 regarding our focus upon tasks rather than single bespoke apps.

Conclusion

This chapter started by noting how many of the developments in the way smartphones are used derive from a combination of two properties. The first is the internal design that allows the immediate juxtaposition between apps, facilitating their use in combination. The other is the issue of mobility and its inverse. The smartphone may travel around, but in so doing it remains constantly present alongside the individual. Together these two aspects account for the rise of the property that has dominated this chapter, which we have termed 'Perpetual Opportunism'.

The impact on the very nature of photography turns out to be profound. It is so much more than just a matter of tweaking our use of cameras and images. Rather, photography has become in several respects almost the exact opposite of what it once used to be. The camera started as a huge beast of a thing that had to be set up as an apparatus; the photographs themselves took time to process, making photography a rather long, formal and costly procedure. With more portable cameras, it became easier to take pictures. Digitisation had as significant an impact on what people could do with their photographs as it did on how they took them. The photo album, the family collage on the mantelpieces, the shoeboxes and the display portraits[31] are now joined by many new possibilities. The vast majority of images taken today are for immediate sharing through social media, an established part of conversation and general communication. Strip away our nostalgic tendency to think the old ways were better and we have to acknowledge that it was not just the camera that could be clunky. The whole regime of analogue photography was highly constrained.

By contrast, the smartphone camera lends itself perfectly to Perpetual Opportunism. Even schoolchildren are constantly aware that as they walk along the road or across a field there may be entirely unantici-pated possibilities of an Instagram-worthy shot. We never know when that butterfly will spread its wings with just the right background or the grandchild will decide to be utterly adorable. Thanks to Perpetual Opportunism, as long as the activity lasts for the few seconds that it takes to whip out the smartphone and snap the scene, it has been captured and acknowledged. An important component of this opportunism is the lack of cost. Images are easy to take, store, select, delete and replace. Taking a picture as an act of framing contains an element of sacralisation, separating its subjects out from the mundane that remains unframed. This is often a social project, as when the people of Dar al-Hawa use the sharing of images with those unable to join them on their travels as a means of including the entire community.[32]

Each field of smartphone usage discussed in this chapter turns out also to be an example of Perpetual Opportunism. At the very moment when the driver seems lost, GPS is there on their smartphone. The trans-lation app is there for those moments on holiday when someone seems to be frantically telling you something, but you have no idea why or what they are saying. Older people in Bento, who are more likely to use public transport after retirement, no longer have to plan their evening so strictly; now they can simply summon transport back to their homes when they feel ready to leave. A Peruvian migrant in Santiago can check a Peruvian recipe on YouTube while in the middle of cooking the dish. It is all there and then, available in an instant. There is an old expression deriving from music: 'playing it by ear'. Thanks to the smartphone, this has become a literal description of everyday life.

As always with the smartphone, less benign consequences also lie beneath the expansion of its capabilities.[33] The very term 'Opportunism' has negative connotations. The other side to Perpetual Opportunism is perpetual vulnerability. Wherever you are, that person may be stalking you. There may seem to be no plausible excuse for failing to respond when WhatsApp has already informed the sender (perhaps your boss?) that you have received their message. There are also certain relatives who clearly expect you to respond immediately. We have already mentioned the schoolchild who is constantly anxious about what is being said about them. They may lose sleep at night because they can now take their smartphones from beneath their pillow at 3 a.m. and check that they did not miss the insult or revelation that they fear. So we discover that Perpetual Opportunism can also be experienced as perpetual pressure. Demands from an employer may be made at any time and on any day. The smartphone makes possible the 'gig' economy that has so dramatically changed work practices, now based on being available as required.

Perpetual Opportunism has therefore seen the rise of calls for breaks from the smartphone or 'digital detoxing'.[34] If we can use the smartphone at any moment, then it follows that we are tempted to use it all the time. We can always think of another person to phone, another thing to look up. Perpetual Opportunism thereby contributes to the discourse around addiction that was discussed in chapter 2. In turn, all these consequences can be exaggerated. Perpetual Opportunism is not just the property of smartphones: the human voice is also a perpetual opportunity. People have always had to 'bite their tongues', trying to resist the temptation to reply, or to have the last word, even if we know that what we want to say is inappropriate or it is the other person's turn to speak. Human beings have always lived with, and accommodated themselves to, perpetual temptation in various forms. The evidence from this chapter is much more focused upon how people have accommodated the smartphones to their own purposes, rather than blindly surrendering to every possibility that the smartphone presents.

Perpetual Opportunism does not necessarily mean we have become more shallow or short-term in our attitude to life. Smartphones are used just as much for long-term planning as for immediate gratification. In Ireland people love to spend time planning holidays months in advance using Tripadvisor, Google Earth and Booking.com, or by learning a language through Duolingo. Perpetual Opportunism is everywhere an affordance,[35] but most of this chapter has been about the different ways in which people avail themselves of that possibility. People in every fieldsite listen to music through smartphones, but the way people of Kampala and Yaoundé circulate music through vendors is very different from how people access music in Dublin or al-Quds. The way the Japanese government uses the smartphones to alert people to emergencies is distinctive to that region. The opportunities may always be there, but in each region people avail themselves of those opportunities in different ways. This explains the title of the next chapter – 'Crafting'.

Notes

1. Katz and Aakhus 2002.
2. Ling 2004; Ling and Yttri 2003.
3. Sarvas and Frohlich 2011.
4. In the nineteenth century cameras were also deployed in early recording and documenting; see Pinney 2012. However, the clunky apparatus this involved was entirely different from smartphones and could not be used for the kind of functional photography that is described here. See Gómez Cruz and Meyer 2012.
5. Morosanu Firth et al. 2020.
6. A series of essays around the uses and consequences of digital photography can be found in Gómez Cruz and Lehmuskallio 2016. For the relation to memory see Dijck 2007.
7. In the nineteenth century these images were deemed more 'scientific' and 'truthful' than earlier forms of visual representation such as painting and art. See Walton 2016.
8. Miller 2015.
9. Drazin and Frohlich 2007.
10. See also Mirzoeff 2015.
11. For a more general discussion of the impact of framing see Goffman 1972.
12. See Susan Murray's article on 'everyday aesthetics', which she highlights as being central in marking the practice of everyday digital photography. Murray 2008.
13. To spell out this analogy Ernst Gombrich, one of the most renowned art historians, wrote a major work (*The Sense of Order*) that was not about art, but about the frames in which pictures are placed. His point was that the thing that made people stop and contemplate may not be the quality of that which is within the frame, but rather the fact of the framing. In the same way, the argument here is that the photograph is an act of framing nature or other subjects; it is this framing that determines the change in perception more than what lies within that frame. See Gombrich 1984.
14. Hendry 1995.
15. See Favero 2018.
16. Bell and Lyall 2005, 136.
17. For young people also posting on social media, there is the awareness that the family is always watching. See de Vries, under review.
18. As in Trinidad. See Miller 1995.
19. Virtual travel in this manner via VR headsets marks a digital-technological moment in a longer history of imaginative 'room travel'. The concept was first described by French aristocrat Xavier de Maistre (1763–1852) in his text *Voyage Around my Room* (1794), a satirical autobiographical account of a young official imprisoned in his room for six weeks. The work was based on his own experience of being placed under house arrest in Turin as a consequence of a duel. See Maistre and Sartarelli 1994.
20. See the film about Liam at http://bit.ly/VR_Liam.
21. For another example of disaster warnings see Madianou 2015.
22. Of particular interest, perhaps, is the telephone. When it was first invented as a device, its inventors thought that it would be primarily used for the dissemination of information, rather than for social conversation. See Fischer 1992.
23. See Slater et al. 2012:
    'At 2:26 p.m. on March 11, 2011, a magnitude 9.0 earthquake hit Japan. A few minutes later, wave after wave of a massive tsunami struck the entire Pacific coast. As if the natural disaster alone was not enough, at 3:35 p.m., the waters from the tsunami –15 meters high – damaged the Fukushima Daiichi Reactor, spreading rumors and fear of mass nuclear contamination (Ito, 2012, pp. 34-35). Almost everything we know now, and especially what we knew of the quake and tsunami in the hours and even days after the events, was significantly shaped by social media. In fact, the generation of information and images occurred at such a fast pace that social media not only represented, but also directly mediated our experience of the disaster more than in any other event to date. If Vietnam was the first war fully experienced through television (Anderegg, 1991), 3.11 was the first "natural" disaster so fully experienced through social media.¹ This is the result of a number of factors, some a function of the way that

technology use has developed in Japan, especially the fact of mobility of hand-held media, others due to the particular ways that networks of people reacted in the time of crisis. But social media was also much more than a source of information; it was also a tool of social and political action.'

24. See *The Economist* 2019.
25. For papers dealing more generally with mobile music see Gopinath and Stanyek 2014.
26. Pype 2015.
27. Pype 2017.
28. Abacus News 2019. One user may have several accounts.
29. Abacus News 2019.
30. Shuken 2019.
31. See the fascinating ethnography of what people did with analogue photographs by Drazin and Frohlich 2007.
32. See Jurgenson 2019.
33. Jovicic, under review.
34. Sutton 2017.
35. Costa 2018.

6
Crafting

Fieldsites: **Bento** – São Paulo, Brazil. **Dar al-Hawa** – Al-Quds (East Jerusalem). **Dublin** – Ireland. **Lusozi** – Kampala, Uganda. **Kyoto and Kōchi** – Japan. **NoLo** – Milan, Italy. **Santiago** – Chile. **Shanghai** – China. **Yaoundé** – Cameroon.

Crafting: the artisanal transformation of smartphones and lives

As part of the wider monograph series upon which this volume draws, several of the authors have adopted the term 'crafting'. The word refers not only to the way in which people adopt and adapt their smartphones, but also aligns this activity to the larger concern of those monographs with the crafting of life itself. The idea of considering life as craft lends itself to those fieldsites where the focus was on retirement – a time when people find their lives less structured by work routines or family obligations. This freedom may enable them to become more actively involved in constructing the content and rhythms of everyday life. This alignment between smartphones and the experience of ageing will be the subject of chapter 7. But first we need to explore the term 'crafting' further. As the core to this chapter, it will be used to examine the way in which people align their smartphone with individual, social and community life.

Crafting is another example of our 'smart from below' approach, introduced in chapter 1. The purchase of a smartphone handset is just the beginning of several transformative processes that result in the specific smartphones we encounter in fieldwork. It seems appropriate to respect such processes by regarding them as examples of artisanal craftsmanship. After all, crafting does not mean people are totally free to do what they like with smartphones, or with their lives. Artisans too are constrained by the material properties of the substance with which they work. They have to chip away certain elements carefully and add or mould others, according to the malleability and nature of those materials. But, unlike an artwork, the crafting of smartphones is always relative to context and usage. It is not a question of creating something autonomous. The aim of this crafting is rather to create alignment with everyday life.

The chapter is therefore organised as a sequence. It starts with reflections on the relationship between an individual and their smartphone(s). How is a smartphone crafted to fit the specific person? It then shifts to consider the way in which smartphones align not with individuals, but within relationships. How is the smartphone moulded to fit in the spaces created by relationships between people? Finally, it examines the wider field within which smartphones reflect or facilitate more general cultural values. While the presentation is as a linear sequence, what has been revealed by the end of the chapter is clearly a circle. Every individual is, after all, crafted in turn by the norms, values and expectations of the society in which they were parented and educated, and is subject to religion or other moral forces. It will therefore become apparent that crafting encompasses a much wider range of influences than merely the individual personality. In crafting the smartphone, people also craft their relationship *to* the wider world in which they live; they are also crafted *by* it. The smartphone then has to become a habitual part of our everyday habits, something that anthropologists call habitus.[1]

Crafting the individual

The iPhone belonging to Eleanor from our Dublin fieldsite is a marvel; she has effectively turned it into a kind of life manual. No individual app icons are displayed on her screen: all are organised into a nested hierarchy, grouped in fields such as finance, sports, news and utilities, or based around work-related functions. This vertical order is complemented by a horizontal order where she makes full use of the capacity of different apps to be linked. For example, Eleanor's calendar would have a task, such as paying a particular utility bill. She explained that she would link an event to her notepad. On this notepad she has created a step-by-step description of the process involved in paying that utility bill, including the password and relevant website address, which may extend to several pages. She then describes how on her smartphone an app related to a job function might be linked to a collection of all the PowerPoint slides from the presentation she had attended about how most effectively to carry out that particular work. In addition Eleanor has also provided herself with visual aids to help locate relevant information. These include a whole series of emoticons such as pins for medical information, a car for transport and a flash sign for any payment that has to be made that day. As a result, at any time and faced with any particular task, she claims she can locate within three or four taps of her fingers the instructions for the most effective sequence of actions to accomplish it.

An example would be Eleanor's use of the app provided by Laya, her health insurance company. She reports that she immediately takes photographs of any receipts from her doctor and sends these in through the app; she is then guaranteed to be paid back within 10 days. All her photographs are date-stamped so that she can more easily order and share them. Her camera has also become central to her organisational work because she sees it as the main device for collecting and storing evidence. This might relate to a car repair, but it might equally be her timetable for water aerobics. She uses her phone alarm not just for getting up in the morning, but also to tell her when she needs an injection, or when she is supposed to be leaving the house for some task. A whole section of Eleanor's smartphone is devoted to money management. Although she does not have much money, she likes to move her funds around frequently to keep them active.

Eleanor talks about her phone in terms of cleaning, organising and housekeeping. The mass of PowerPoint slides and other photographs on her device can easily become disorganised. They therefore require constant editing, deletion and re-ordering to remain useful and close to hand, which is what she requires from her smartphone. She is also constantly updating her calendar. Everything is securely backed up in the cloud so that, although she sees her whole life as present on the smartphone, she was able to neutralise its functions immediately when her smartphone was recently stolen in Spain. Having backed up her data securely, she found herself able to repopulate a new smartphone with the entire contents very quickly. The one thing she has not really taken to is the phone assistant Siri.[2] She tried to customise it by using both male and female voices, but is not comfortable with either. Indeed, she resents not only the interference by Siri, but also the whole trend towards using artificial intelligence to pre-empt her with suggestions. She dislikes the way that Netflix tries to give her ideas of what she might like to watch based on previous viewings, commenting 'it's trying to be helpful but it's clumsy'.

Eleanor's transformations seem to fit a pattern. At this point in her life she has less control over her work or her health than she would like, something for which maintaining this rigorous order partly compensates. Her iPhone is appreciated as one of the few things over which she does have control. It is then not surprising that she might resent any competition to that control, represented by suggestions emanating from the smartphone itself.

The reason for starting this chapter with Eleanor is not just to provide a rather impressive life instruction manual. It is because she has so clearly crafted her phone to express her sense of herself as a consummate professional, with the organisational aptitude that this implies. She spent her whole working life trying to secure a professional role that would fully utilise these abilities. Sadly, this never came to pass: her employers mostly failed to appreciate these qualities and she never found the opportunity to become the person she thought she could be through her employment. In the end, it has become mainly through her smartphone that she has been able to demonstrate, at least to herself, both who she wants to be and who she believes she really is. Eleanor and her smartphone may appear highly individual, even unique. There is no one else quite like her. But it is also a reflection of a wider sense of order that we associate with terms such as 'professional' or a 'well-organised' approach to life. This is a cultural order as well as a personal one, and it is as apparent from a detailed inspection of her smartphone as it would be from getting to know Eleanor herself. What has been crafted is the alignment between the two.

The same point applies to another Irish individual, Eamon, whose family have been fishermen for 150 years. Displaying a rugged and practical self-sufficiency, he comes across as the icon of a particular version of masculinity. He is adamant, for instance, that he does not need other people or television because he is never bored and always active, whether in sporting activities or practical tasks. Every single use of his smartphone has to be legitimated under the strict criterion of necessity. Eamon was entitled to use Skype for the two years that his daughter was in Australia, but he was very firm in maintaining that he has never used Skype either before or since. One advantage of smartphone development is that he no longer needs to use the voice function, which he dislikes; he can replace a conversation with terse text messages about, for example, when his train will arrive at the station. In Eamon's case, the aesthetic comprises a particular form of social minimalism that derives from one of the established stereotypes associated with masculinity. Again it is a style that applies equally to the man and to his smartphone. Such an approach to life is probably mainly relegated today to older men in Ireland, having been superseded by more modern ideals of gender.

There are other examples where the individual's sense of order is less evidently tied to a cultural norm – people who might be considered somewhat eccentric. Gertrude, a sports physiotherapist, has two phones. This is because she is obsessed with the idea that at any given time she might see some image or composition that she needs to capture with her phone and then immediately send out on Instagram, or sometimes also Facebook and Twitter. It might be a landscape, a selfie of herself in a particular situation or just a constellation of colours. The sentiment is common enough today, but it is not usually taken to this extreme. Gertrude admits she is terrified that if her phone ran of battery, she would not have the means to capture this image at the precise moment she encounters it. As a result she not only carries with her the two iPhones, but also a dongle (in case there is no wireless network) and a spare battery charger.

Gertrude always carries two phones – but when Melvin, another participant in the Irish fieldsite, emptied the pockets of his jacket, he revealed no less than four Nokia mobiles, plus a Huawei smartphone (Fig. 6.1). In his case the primary concern is music rather than photography. Melvin is constantly recording 'sessions' and live performances,

Figure 6.1   The five phones from Melvin's jacket pockets. Photo by Daniel Miller.

usually in pubs, of traditional Irish music; when one phone runs out of storage, he turns to the next.

Melvin also travels a good deal to places such as the UK and Corsica. He uses different phones to communicate with people in each place using local phone plans. He also duplicates information, since he worries that a phone might be lost or stolen, and may also carry a spare battery. Melvin is used to people seeing him as eccentric in various ways. He clearly enjoys playing to the expectations that this reputation creates in other people that know him, and is fond of extravagant, often generous gestures. He was perhaps the least surprising person in the two Dublin fieldsites to be carrying five phones in his pockets.

In each of the cases so far presented, the characterisation of the person preceded their possession of the smartphone. Smartphones have neither accentuated nor repressed their personality. Melvin is generally thought to be eccentric, while Eleanor was equally comfortable with being considered hyper-organised. None of this makes their use of smartphones any less extraordinary – it is hard to imagine anything in the world that would have aligned with their character quite so effectively as the smartphone. Their stories show just how intimate the device has become and reveal the ways in which each smartphone can be individually reconfigured to develop this holistic, seamless relationship between user and device.

By teaching classes on how to use smartphones, Alfonso was in a good position to see how the way in which people use their smartphone is also expressive of their attitude to the devices themselves. For example Francisco, a Chilean student, was very serious by nature; he could be a bit grumpy towards the more flippant students, but also showed a tendency towards nostalgia, along with a very dry sense of humour. He quite likes gadgets and devices, preferring ones that he can take apart, repair and put back together again; this has the added bonus of demonstrating that he is also a person still in decent functioning order. As a student, Francisco is alert and tends to have already installed the relevant apps. However, he has a sceptical pragmatism that means he will gravitate towards older analogue methods, unless entirely convinced that the digital one is better. He does not feel this is the case, for example, with the smartphone calendar.

Francisco is extremely wary of the smartphone's ability to follow him, for instance through GPS, to the extent that he refused to perform any GPS-based tasks during classes. He had recently been mugged and had his smartphone stolen, after which he started to worry about anything that would allow people to follow him. He turned off the GPS after Google Maps had itself pointed out its powers of surveillance by reminding him that he had been in the park for three hours. The point here is that Francisco's fears derive from his sense of himself as a rational actor who bases his conclusions on observations. He has seen – and experienced – the fact that people are being mugged and also the way in which Google Maps operates, from which he deduces that gangs of thieves must be using Google Maps to mug people. The fact that other people regard his behaviour as bizarre is of much less concern than his continued faith in his own reasoning.

Peter is an Irish engineer who had grown inordinately fond of his Nokia phone: it was reliable, lasted for ages, was well designed and convenient. He felt he had betrayed it on taking up the smartphone, abandoning a true friend for the bells and whistles gimmickry of the smartphone. Peter is a very skilled engineer, however, so he decided to re-engineer his smartphone. By disabling many of its core elements, he crafted his Samsung Galaxy to last 120 hours without a recharge. It was also programmed so that once he was at home all calls were redirected to his landline, while the smartphone could be left behind in a drawer. In effect, he had turned the new smartphone device into a replica of his old Nokia mobile phone.

A final example from Ireland is important because the smartphone may not reflect a whole personality, but simply their dominant interests. When Matis came from Lithuania to live in Ireland, he certainly landed on his feet. His current employment is perfect, given his lifelong passion for cars. Once he has finished taking out the rubbish, fixing tables and other required tasks in his role at the Mexican-style restaurant where he works, he is free to concentrate on restoring classic cars for his employer, who shares this passion.

Matis has lived in Ireland with his wife, children and grandchildren since 2008. It is thus fairly unsurprising that the apps dominating his smartphone's screen are those associated with car parts, such as Donedeal (a digital Irish marketplace for buyers and sellers of cars) or Mister Auto (a shop for car parts). Equally important are the YouTube videos he watches of other car aficionados, which explain how to deal with particular problematic tasks. The camera is also essential, since specialist buyers will expect a complete photographic record of every stage in the restoration, which may take a whole year. Matis notes with pride how his employer drove the last one, a classic racing car, for 1,500 km to Italy and back with no problems. Attached to his phone is a wire leading to a micro-torch/camera. He uses this as an endoscope which he can push down cracks inside the car to investigate the state of that particular section before starting work on it. What the torch does not reveal can be viewed thanks to the pictures, taken with a flash attached to the endoscope and sent in turn to his smartphone. Matis sourced the gadget in China. In showing us how it works, he seems to be beaming as much as the torch.

The inverse to engineering the smartphone is using the smartphone for self-engineering. This has been the approach of Fernanda, from Bento, who has noticed a decline in her ability to remember things. Together with many older people, she says that she does not fear death, but has a horror of dementia. Her response to sensing some memory loss, a common experience for older people, is to embark upon brain training exercises. These are often seen as analogous to keeping physically fit and extending life expectancy. She therefore exercises her mind daily, using games such as Freecell, Lumosity, Wood Block Puzzle and Codecross. Learning English also helps Fernanda to stimulate her cognitive functions through apps such as English Conversation and Google Classroom, which allow her to submit and correct her exercises through the app. She recently downloaded and installed Duolingo, an app that she uses to learn Italian, as her son is now going out with an Italian woman.

Having retired from being a worker in a food company, Fernanda's plan is now to become an entrepreneur. She has recently accepted a friend's invitation to start a telemedicine company for the care of the elderly. Prior to this, she had recognised the benefits and the dangers of sharing cognitive processes with her smartphone:

> If you lose your smartphone, you lose your life. Awesome, it's all there. Your life is there. I am scared to lose the smartphone. Without this, I would have a problem remembering. I'm talking to you and tomorrow I already think … 'what did she say?' So for me the smartphone is fundamental.

Having observed the way Fernanda is ceding cognitive abilities such as memory to her smartphone, she now seeks ways to make her smartphone a device for repairing, or at least maintaining, memory.

Memory is also central to the reason why Toriyama san from Kyoto, who now uses a smartphone, also still keeps her old feature phone (*garakei*) and charges it up every night, even though she is no longer using it. 'I don't want it to die,' she explains. She is scared that the photos she took, and therefore the memories she has made, will be lost forever if she lets her old phone 'die'. Several others in the Japanese fieldsites also said they would not dream of throwing their old mobile phones away for this same reason, preferring to keep them together safely in a drawer. The feeling that the phone becomes a repository for memories is complicated by the way in which the smartphone assumes other cognitive functions and can come to feel like an integral part of a person.

All the examples so far have been quite individualistic in orientation. A contrast is provided by Mario, who is retired. Mario has a passion for horticulture and is a proud environmental activist. As he explains:

> I have always been this way, ever since I was a kid, I remember. Always I was curious about the community, the environments in which we live, about people, about their collective experiences of work and life.

Mario has turned his smartphone into a device for running one of the local community allotments (Fig. 6.2). He also co-organises social events, such as a tour of the bees they keep in the allotments and the honey collection (a popular event with local children), or a shared community dinner where people from different cultures bring traditional dishes from their country to eat outside together in the gardens. He also enjoys using

Figure 6.2   A community allotment in NoLo. Photo by Shireen Walton.

the plant-identifier app PictureThis. Over time, the apps that were not well integrated into Mario's daily routines have been deleted.

Mario also exemplifies the contradictions of discourse discussed in chapter 2. As an environmentalist, he does not feel he is particularly reliant on smartphones. At the same time, it is through the use of his smartphone that Mario has been able to facilitate the community and environmental values that he cherishes.

Some further examples of how smartphones fit into the lives of individuals in NoLo can be seen in the short film below (Fig. 6.3).

Finally, we can return to the starting point of this discussion, describing how smartphones can also become an instrument in the crafting of life in retirement. Marília was examining the smartphone of Eduardo from Bento, who had retired just two weeks before their conversation. When he was working, the clock and alarm were crucial since he had to get up at 4 a.m. every day. After retirement, however, he found that he was struggling to make his body adjust to his new routines, so he abandoned the alarm and started avoiding the clock. Now he uses his smartphone to follow recipes and tutorials so that he can please his wife at dinner. Using Google on his phone, Eduardo can now also research decorative ironwork, something that he plans to turn into a second hobby. Furthermore, his smartphone allows him to extend his work with a group of new Christians who plan to open their own church in São Paulo's inner city where his daughter and grandson live. The smartphone includes

Figure 6.3   Film: *My smartphone*. Available at http://bit.ly/italymysmartphone.

apps for hymns, an app for Bible study and others that help him organise church donations – including how to calculate these with his Calculadora as well as bank them using a credit card app called Payeven Chip. In all of this, Eduardo is deliberately adjusting his smartphone to facilitate the new life he envisages.

The implications of these cases will be explored further in chapter 9. In this final, more theoretical chapter we consider the smartphone as something that goes 'Beyond Anthropomorphism'. What has been observed here is not just a property of resemblance. The smartphone appears more like a prosthetic[3] to which certain bodily or cognitive functions have been ceded, such that the loss of the smartphone would constitute the loss of an integral part of the person. This has been justified by the many ways in which the smartphone has extended the capacity of that individual, but clearly it is a strategy of investment that carries with it a risk of subsequent loss. When reading about these individual cases, it is again hard to think of any prior device that has reached quite this potential for intimacy.

Relationships

Chapter 3 has already provided a critical perspective on the temptation to focus on the individual and their smartphone. The term 'Social Ecology' was used to discuss the way in which the phone could equally well express relationships to others – for example in Kampala, where the phone can be shared among family and neighbours. Chapter 4 used an example from Shanghai to show how the apps on a phone are expressive of the relationship between couples. Smartphones are also becoming increasingly important in the formation of relationships thanks to dating apps.

The smartphone can then express relationships, not just an individual. Rachel from Dublin works as a PA to a high-flying boss who is herself deeply concerned with the problem of maintaining her family relationships alongside her demanding work schedule. After decades of working together, they have none of the usual work boundaries. Rachel may be in touch at 10 p.m. doing something on her boss's behalf, such as finding the answer to a question from her boss's children. But she discusses her assiduous support in terms of loyalty and friendship, is proud of her work and has no desire to retire. Her smartphone is geared to this way of working. If her smartphone were not always on, she would feel she was betraying the trust of her boss. The device also allows her to deal with her boss's anxieties about travel and her family far more effectively.

Although it is Rachel's personal smartphone, it is mainly used for her work. Yet she is extremely comfortable with the way the smartphone fits snugly into this relationship and makes a point of how she feels she has achieved just the work-life balance she wanted. In fact she uses the smartphone for whatever her work requires and hardly at all otherwise. Outside of this working relationship, she still happily uses a pen and paper diary for herself. The smartphone delineates as it facilitates.

Smartphones may also be involved in the transformation of family relations over time. This observation may emerge more clearly in chapter 8 because, in terms of group identity and formation, the key app seems to be LINE/WeChat/WhatsApp. In the case of Japan, the 'family' as an institution can be in some senses remodelled by the smartphone; many people have 'family' LINE groups for their immediate family relations. However, connections between people are always subject to change. Yamashita san, a woman in her sixties from Kyoto, explained how she lived with her daughter, who was in her forties; after her husband passed away they became distanced from his family. On her LINE she created a group called 'family' (ファミリ) which included her and her daughter, followed by a small number of close friends.

The patriarchal family unit (*ie*) was historically central to Japanese domesticity and society. The nuclear family gained prominence from the end of the nineteenth century in accordance with the Meiji government's plans to modernise the country.[4] Yet now, with families shrinking and many people not having children or having only one child, 'chosen' families, made up of close friends, have become increasingly significant for some people. Friendship groups on LINE, especially for female participants, were important sites where one could draw on the support of friends, something that became increasingly valued as people aged. Such groups can often be formed of old high school friends or former colleagues, with people commonly being members of several groups simultaneously. Wada san, another woman from Kyoto, started living alone after her husband passed away, while her daughter and grandchildren live in Tokyo. She responded by creating a visual code for categorising her LINE contacts on her smartphone. By using emojis that represented how she knew those people, Wada san was able visually to map the various 'families' that were contained within her phone. For example, having worked for an airline, all of her old colleagues were listed with an aeroplane symbol next to their names; her immediate family members were indicated by a house emoji. Crafting her smartphone in this way enabled Wada san quickly to see exactly how she was related to each contact. In a sense she was constructing a visual family tree, with various branches connecting her to everyone in her smartphone.

Religion

In the introduction, it was suggested that the linear narrative structuring this chapter is actually better seen as a circle. Take the example of Eamon, the gruff Irish descendant of fishermen. This characterised him as an individual but he, in turn, manifested a traditional understanding of how proper men should behave. Similarly, in the last example, the way in which friends are being incorporated into pseudo-family relationships through the smartphone reflects wider changes in Japanese society. If individuals craft smartphones, they are in turn crafted by these wider values. The most clear-cut example of the imposition and maintenance of cultural values comes in the consideration of the role of religion because it is usually the most overt.

For example, Rosalba comes originally from a village in rural Southern Italy. For her the smartphone, even more than a good book, is her material companion. One of her most frequent smartphone practices is googling, especially recipes – often traditional ones from her region. Rosalba is an avid searcher, however, and she also frequently googles information. 'Let's ask him,' she will say frequently, referring to Google. The availability of information at the touch of a button is something Rosalba finds endlessly marvellous. Growing up in a large rural family homestead, her father a fruit seller, she describes her childhood as 'another world, another time and another place – a time when talking, school and the church were one's core sources of information'. For many years the family did not have a television and she grew up playing outdoors, on the land, with her brother and cousins her age.

By contrast, information today comes to her within seconds from the comfort of her own home, thanks to Google. She googles health information and, if she hears something on television that she would like to know more about, she 'asks him'. She has recently discovered an app for measuring her pulse, which she uses on her tablet. For Rosalba, a devout Catholic, the technology is an omnipresent force that both accompanies and informs, enlightens and instructs. At the same time, actual prayer and weekly (Sunday) churchgoing remain a staple of her life.

The way individual crafting of smartphones helps to bring people within the compass of religious faith is evident among the Catholic

Peruvian migrants living in Santiago. This group has carefully maintained their devotion to Peruvian sacred virgins and patron saints. Marcelo, for example, uses his smartphone to listen to the Rosary prayer when walking or taking the Underground from home to work. He follows the webpage of the Knights of the Virgin (Caballeros de la Virgen), which includes audio recordings of people praying. In silence, with his headphones on, he prays back. In Catholic Christianity, the Rosary is prayed as a dialogue. The Catholicism of another research participant, Tomás, is not so immediately apparent – the wallpaper on his smartphone is not the usual Purple Christ (the Lord of Miracles) or Saint Martin of Porres, but Gohan, a Dragon Ball Z character. During his commute, Tomás likes to play a game similar to Space Invaders. His religious devotion comes more to the fore through watching Christian television series and films. 'I like series or movies that deliver a message,' he explains.

In Dar al-Hawa smartphones are commonly used as a 'handy muezzin', as the majority of the population have downloaded an app that reminds its user of the time for prayer. In Islam there are five prayers a day. The app is programmed to create an alert based on the local time zone, using a voice recording of a muezzin call to prayer (Fig. 6.4).

Figure 6.4 The Salatuk app as shown in the Google Play Store. This app acts as a 'handy muezzin', reminding the user of the time of prayer.

God is the greatest (*Allahu akbar*); I testify that there is no God but Allah. I testify that Mohammed is God's Prophet. Come to prayer. Come to security/salvation. God is the greatest. There is no God but Allah.

If an individual is out of earshot from any actual mosque, this digital version comes into play. Most people just hear the first few seconds before turning it off. The call to prayer crafts the daily routine for many in Dar al-Hawa, especially older people who stay at home for much of the time. The app is simple and requires no code or password. It is also popular with younger people who are working and afraid that they may miss the time of prayer.

These are examples of relatively private crafting of the smart-phone influenced by religion. Once again, however, this can also apply to relationships, not just to individuals. Many of the Irish participants take part in a traditional pilgrimage trail along northern Spain called the Camino de Santiago.[5] This has been the custom for centuries, but numbers have grown considerably in recent times. Traditionally, walking the Camino might have been associated with an exercise in spiritual contemplation, even for those who are no longer formally religious. This may have included a feeling of retreat from, or distance from, the mundane world, although the journey always had a strong social component as people walked together or socialised in the dormitory inns they generally stayed in overnight. The anthropologist Nancy Frey has argued[6] that the fundamental values of traditional pilgrimage have been undermined by smartphones, as these create a constant mode of retained contact with the wider world. Her observation was not confirmed in the attitude of pilgrims who went on the Camino from one of the Dublin fieldsites, although admittedly most of them are not especially religious. In any given year, when only one or two can afford the time or money to go, they delight in sharing the experience with their friends and relatives back in Dublin, sending daily messages and updates as to where exactly they have reached. Those on the Camino view this as an ethical expression of fellowship, rather than as an assault upon individualised spirituality.

Cultural norms

It is relatively straightforward to suggest that a deeply religious population is likely to allow their religious faith to dominate the meaning and use of smartphones. Religions tend to be explicit about their role in determining cultural values. But at the heart of anthropology is an emphasis upon the wider impact of normativity. The term itself is clearly related to the word normal – what gets taken for granted as simply the natural order and so casts other behaviour as abnormal. Mostly normativity is not achieved through religious control, nor even educational teaching. It emerges as part of everyday life where people make often quite subtle hints to one another about what they find to be appropriate or inappropriate in another person's actions. They may look a bit shocked or surprised that someone is standing a bit too close, for example, or at someone's choice of clothing. The degree of moral pressure will vary, but research participants in the Japanese fieldsites often talked about how awareness of proper social behaviour is integral to being 'Japanese'.

Bullying and being ostracised were talked about by many participants in Japan as commonplace if someone is not skilled at 'reading the air' (*kuuki ga yomenai*) and behaving in a socially appropriate way. For schoolchildren, learning the importance of reading between the lines in social interactions starts young. However, texting through smartphones has complicated the way that social interaction relies on these multiple layers of communication. Yumi, a 17-year-old girl in Kōchi, explained:

> Because it's so difficult to read between the lines on text messaging, the teachers are always raising awareness for bullying on SMS. Misunderstanding can spread and that can be the beginning of bullying. Because words spread quickly and create bad vibes sometimes among teenagers. You know, it can create fights and rumours. I think you just get left out. If they have groups (and) someone makes a mistake they might get left out of the group. It's not like it hurts physically, but it hurts mentally.

Yumi used to use Facebook and Twitter. She gave them up a year and a half ago because they had become too stressful and because of the social scrutiny she was feeling. These issues are not confined to teenagers as a Kyoto woman in her sixties revealed.

I think people in Japan are in a hierarchical society, so people feel obligated to comment or like someone's post. Here, you have to be loyal and respond to their post. You feel duty (*giri*) to compliment them. If it's someone you know and you look up to, or you want them to have a good impression of you, then you have to like their posts. Sometimes I feel that people's likes are just 'duty-likes' (*giri-iine*). I'm not pleased with 'duty-likes'. If someone genuinely likes the post I'm more happy. I think some people get sick of it and they quit Facebook. Some of my friends said it's just too much trouble to put 'like, like, like'. They don't like the pressure.

Most research participants from this fieldsite would never post anything controversial, for example about politics, because 'you don't want to make enemies, or contradict people. Don't create conflict – even face to face. People like small talk about the weather and food and their health'. This is the reason why many participants created multiple anonymous Twitter and Instagram accounts, enabling them to engage with topics they were interested in without the social scrutiny. The smartphone in Japan thereby becomes a space for negotiating the difference between a person's true feelings and what one says in public. This distinction is expressed in the cultural notions of *honne* (本音, 'true sound') and *tatemae* (建前, 'facade'), which generations of scholars have seen as lying at the heart of Japanese normative society.[7]

As in the previous discussion of individuals or relationships, it is possible to study the capacity of smartphones to express these cultural norms. Equally, however, smartphones can be examined as factors in social change and the development of new social norms. In the case of Yaoundé, the primary focus of Patrick's research was the emergent middle class. 'Emergent' suggests a coming into being. One of the best-established academic approaches to the formation of the middle class, initially in reference to European examples, was the idea of a 'public sphere' promoted by the sociologist Jurgen Habermas.[8] He discussed how, at a certain point in European history, there arose new sites, such as coffee shops, which encouraged the development of public debate. These changes, in turn, led to an unprecedented form of politics based on the discussions among members of this nascent middle class. Considerable subsequent research examines the role of the media as another space in which this public sphere could subsequently develop.[9] Patrick makes a similar argument, this time considering the role of the online world as a place where the new middle class in Cameroon create their public sphere, once again characterised by intense political discussion.

In the case of Cameroon, several topics dominate such discussions. These include the long-term struggle between the English-speaking sections of Cameroon and the dominant French-speaking areas. In addition, since 2014 there have been many attacks in the northern regions of the country by the jihadist terrorist organisation Boko Haram. Both topics continually circulate on social networks present on smartphones. The inhabitants of Yaoundé carry with them violent images on a daily basis, along with journalists' comments on these images, videos of individuals and the everyday life of conflicts and tensions. These are spread through WhatsApp groups. In turn, discussions or information that began within Cameroon then circulate through its diasporas in France, Germany, the United States and the UK, including what can turn out to be faked images (Fig. 6.5).

Examining the emergence of a smartphone-based public sphere suggests both continuities and discontinuities.[10] Smartphones have taken on a central role in these incessant discussions of the key topics of the day.

Figure 6.5   Images of war circulating in Cameroon through a WhatsApp group. Photo by Patrick Awondo.

These naturally gravitate towards the smartphone, where there is relatively free access and free debate. The middle class see this as a form of 'informational citizenship', implying that these topics are subjects about which it is their duty to have opinions.[11] News and politics seem to burst into their smartphones daily, at times even without them even wanting it. At the same time, the shift to a new medium has other consequences. For example, the constant circulation of images has led the Cameroonian government to ban smartphones from the military, which is involved in suppressing Boko Haram and also separatist movements. At the same time, leaks circulating on social media create thousands of reactions from the population, which then require a response from the state.

This process has gone still further with the response to the Covid-19 pandemic, a situation in which it became increasingly important to combat misinformation and the circulation of fake news. As a result the Minister of Health started to tweet details such as the number of deaths and measures being taken by the government. Such direct contact between a minister and the public was quite unprecedented for a government that tends to work in secret. It created new possibilities of accountability for government and the further democratisation of information, as details are further circulated on WhatsApp. The emergent public sphere came into its own, with online circulation creating transparency where previously opaqueness was the norm. Often this created new tensions in its wake.

These can also be very long-term trends. At the heart of Xinyuan's monograph is a study of a generation who lived through China's Cultural Revolution (1966–76), which remains a key influence on how they respond to the smartphone decades later. The idea that every individual must sacrifice themselves to the revolutionary cause through constant self-reform[12] has become aligned with a growing 'techno-nationalism'.[13] Older people view it as the citizen's duty to help China achieve its stated goal, a 'digital leapfrog' of economic development, which explains the more positive attitude to smartphones in China discussed in chapter 2. The values of the Cultural Revolution have become imprinted upon people as a sense of duty to their craft. The ethnography in Shanghai showed how older people naturally view the smartphone as the chance to craft their own lives in the tradition of 'self-reform' and 'self-perfection', continuing the struggle to become the 'New Man' (*xin ren*) with the help of technology. In short, the Communist Party is constantly encouraging people actively to craft their own lives according to its ideals. Crafting the smartphone then fits into this process snugly.

Conclusion

This chapter opened with the suggestion that the concept of crafting should be applied equally to the way in which people craft their smartphones and the way they seek to craft their own lives. But this is really a triangle, as underlying both these processes are the influences of deeper cultural values, manifested in the subsequent appearance of both people and their smartphones. The gruff, pragmatic Irish descendant of fishermen and his minimalist smartphone are equally expressions of older cultural traditions about what makes Irish men masculine. The final example, from Shanghai, is unusual because the cultural pressure towards crafting smartphones as a contribution to an established idea of crafting life is here quite explicit. The same would be true when people overtly present their smartphones as engineered to express their religious principles.

In most cases, however, the links are not overt. These are subtle and implicit processes. It is only through long-term ethnography that we have been able to bring the pieces of this jigsaw together. Initially, the main emphasis is on what is most apparent, which is the intimate alignment between an individual and their smartphone. A series of portraits of people from Ireland, such as Eleanor, were used to illustrate this process. To give more dynamism to this observation, they were supplemented by cases such as that of Francisco, where Alfonso could observe how the smartphone is being crafted to fit the man, or Eduardo, where Marília could watch a smartphone being employed in anticipation of further crafting of its owner's life on retirement. The chapter also considered how people subsequently reflect upon what it means when a device has become so much a part of themselves – an intimacy that can manifest itself in their reliance upon the smartphone as externalised memory – and the vulnerability they may feel about a part of their own memory which potentially could be lost.

From these observations about individuals, the chapter proceeded to examine the way in which smartphones are integrated into relationships, such as those between an employee and her boss or within the wider Japanese family. It then turned to explore wider social norms, with examples that reflect more general principles of the normative and the consensual, both especially apparent in Japan. Smartphones are not necessarily just a static representation of values. Other examples consider the smartphone as an active participant in the formation of a new set of values. This was evident from the case of smartphones being used to circulate political discussion, a key factor in the formation of the middle class of Yaoundé as a new public sphere.

How the balance between the individual and social normativity ends up will vary considerably. On the one hand, there were examples where even the individuals concerned would expect others to regard them as somewhat eccentric, such as Melvin's decision to carry five phones in his jacket. The opposite situation may be found in a religious community or where many people may also be devoted to consensus. Every individual may be different, but there is a moral imperative towards conformity.

We should also acknowledge the role of the smartphone designer, whose greatest contribution has in some ways been their modesty. Why modesty? The craftsmanship found in this chapter is made possible because the most advanced technological possibilities found in the use of algorithms and artificial intelligence are not being employed to complete the smartphone's design. Algorithms and artificial intelligence have created some capacity for autonomous learning. But what really matters is the way that smartphones are created with an unprecedented facility for subsequent change. Designers have provided a sufficiently open architecture to allow smartphones to become myriad artisanal forms that they would never have themselves envisaged. In 2008 Chris Kelty showed in his book *Two Bits*[14] how giving freedom to successive software designers through Open Source allowed for more imaginative democratised development on a broader scale for digital technologies. In addition, in the same year, Clay Shirky's book *Here Comes Everybody*[15] imagined people coming together in collaborative crowdsourcing to take advantage of these new possibilities. Neither of these visions came quite to fruition as the authors imagined. Instead the smartphone has in effect become the means for a more modest revolution that did succeed. It has shifted Open Source from a concept within software design into the myriad possibilities of crafting through the subsequent consumption of smartphones. This has in turn been seized upon, not by new forms of popular collaboration, but through the already established sociality and norms of culture and relationships, as well as by the manifestation of cultural values through individuals.[16]

In most of the fieldsites, very few people make much use of the pre-installed apps on a phone, mainly preferring to use apps that they have chosen to download. The result is that the smartphone can be truly personalised, not just with respect to content, but also in terms of how it is organised and what it is constructed to be able to do. In the case of Eleanor, what really makes her smartphone special is content – all the instructions she has patiently written out and linked to various functions. It is Eleanor's intelligence and craftsmanship that make her phone unique. Designers have allowed for this possibility, but it is then the 'smart from below' crafting that makes a smartphone into what it ultimately becomes.

Notes

1. One of the most influential books in the history of anthropology, *Outline of a Theory of Practice*, was written by the French anthropologist Pierre Bourdieu (1986). In this book he discusses a concept that he calls 'habitus', a term now commonly used throughout the social sciences. The term conveys an obvious link to the idea of habit – a pattern of things we do without even reflecting upon them. Bourdieu argued that 'habitus' derives from the way that habits in many different domains are linked as an underlying order. Among the Berber population in North Africa, where Bourdieu studied, he could discern this order across a range of activities, from the way in which people organised their kinship to their agricultural systems to the way their calendar is arranged. In such a society, a single individual was likely closely to reflect the normative order that we also can call culture. In a place such as contemporary London, there is a greater diversity of people. However, in a volume called *Anthropology and the Individual*, Miller (2009) argued that, even at the level of the single person, we can see a similar underlying sense of order that may relate to many of their different engagements and activities; this often appears to other people as what we commonly think of as their personality. So the concept of habitus can be applied to an individual, not just to the wider society.
2. Siri is the voice assistant on Apple devices.
3. See Lury 1996.
4. Daniels 2015.
5. Frey 1998.
6. Frey 2017.
7. Benedict 1946; Doi 1985; Hendry 1995.
8. Habermas 1989.
9. Garnham 1986; Couldry et al. 2007.
10. Schafer 2015.
11. To access a wider discussion of informational citizenship in the digital age, see Bernal 2014. Bernal discusses what she calls 'infopolitics' – that is, the way that 'Power is exercised and expressed through communication and control media, traffic, censorship and authorization' (p. 8). This is 'particularly crucial in the state-diaspora relationship' (p. 54) and changes the vision of citizen interactions and commitments.
12. A deep-rooted idea that can be traced back to Confucius. See Cheng 2009.
13. Wang 2014.
14. Kelty 2008.
15. Shirky 2008.
16. For a rather esoteric argument that envisaged this kind of consumption as the realisation of the possibilities of culture as envisaged in Hegelian philosophy see Miller 1987.

7
Age and smartphones

Fieldsites: **Bento** – São Paulo, Brazil. **Dublin** – Ireland. **Dar al-Hawa** – Al-Quds (East Jerusalem). **Lusozi** – Kampala, Uganda. **Kyoto and Kōchi** – Japan. **NoLo** – Milan, Italy. **Santiago** – Chile. **Shanghai** – China. **Yaoundé** – Cameroon.

This book takes what is probably a rather unexpected approach by focusing upon older people rather than youth. The expression 'older people' here stands mainly for those in middle age, and our focus was on people who saw themselves as neither young nor elderly. The actual age of individuals varied considerably across the fieldsites. This approach makes adults the ordinary population to which youth would be the exception, rather than the other way around. In most sections of the book this demographic is simply assumed. However, there are many facets of smartphone use which are age related, so this chapter is dedicated to these.

The discussion starts from a consideration of a youthful population and moves on to intergenerational relations. It then examines the problems older people encounter in learning to use smartphones, followed by other more general problems they face. Finally, it considers the development of apps that are targeted specifically at this demographic. This chapter also stands for something larger: the study of smartphones from the perspective of a social parameter. It might then have been a chapter about gender or about class. Each parameter would bring forth analogous observations to those discussed here.

Youth and intergenerational relationships

The use of smartphones as a medium for expressing the experience of ageing may occur for all age groups, whether for young people trying to establish their identity or for older people starting retirement. Young people in NoLo have a strong attachment to the online worlds facilitated by smartphones and other devices. There are social media groups specific to constituencies, for example the 'second generation' of young people with foreign-born parents. These have become spaces in which young people collectively explore issues of group identity that may become the basis for growing social and political awareness and foster forms of activism.[1] This generation, referred to in Italy as 'second generation'[2] or 'new Italians',[3] is growing up with a sensitivity to any exclusionary gestures from other Italians.

This generation is also concerned to retain commitments to the places their parents came from. As an example, children of Egyptians in their teens and early twenties took to rap music to express their relationships to both Egyptian and Italian pop culture. They listened to, rapped in and recorded their voices in Arabic and Italian on their smartphones, while gathering in public spaces such as the local park. In Milan, poetry has become the medium of choice for reflecting upon identity and themes of inclusion, exclusion and otherness in Italy for Hazara refugees and migrants from Afghanistan who are in their twenties and early thirties and who speak Hazāragi.[4] The smartphone notebook function offered a convenient way to pen thoughts digitally, in Persian or Italian interchangeably, on breaks at work, at a restaurant or travelling on the bus or metro. Smartphones and social media have also played a crucial role in Hazara online activism and awareness campaigns concerning the ongoing persecution of the Hazara in their homeland of Afghanistan.

To pre-empt the discussion of the 'Transportal Home' in chapter 9, when groups of young people have this profound experience of ambivalence about where they reside, the smartphone can become of huge significance. It is the one place where friends, relatives and strangers, sometimes in different cities or other parts of the world, can congregate, irrespective of where they physically reside. All of this may make the smartphone a more comfortable place within which to see oneself 'living', rather than any alternative potential candidates for a sense of home. This is particularly true when set against a backdrop of economic crisis that makes renting and/or buying houses challenging for many young people. Having a place to identify with matters a good deal for migrants and their children – especially amid a backdrop of exclusionary practices concerning legal citizenship rights[5] and where the state formally and informally determines those who are 'welcome' in the 'home' of the nation.[6]

For older people, by contrast, the smartphone can become an instrument of destruction and rupture. In Yaoundé, seniority and respect for older people was paramount within historical social orders. In former times, knowledge itself was configured in relation to age. In agriculture, for example, experience mattered: the young learned from the old, an order that was disrupted by the rise of formal education. When it comes to smartphones, however, older people are no longer necessarily respected as the repositories of knowledge. Instead they find themselves constantly obliged to learn from the young (Fig. 7.1). Retirees in Yaoundé now rely on young people, whether their grandchildren or just young people they see in the street, and commonly admit that they are 'embarrassed by the dexterity of the youngest'. One research participant, a 59-year-old high school teacher, feels this technological generational gap quite deeply; he explains how his generation found they had to master first the computer, then the internet and now the smartphone. At each stage of the process, the young were already there. When a teacher has to learn from

Figure 7.1 Grandfather Tom learning how to use his new smartphone in Yaoundé, assisted by his grandson. Photo by Patrick Awondo.

the youngest in the society, there is bound to be a commensurate sense of humiliation for older people, many of whom find this re-direction of education, at least initially, quite unnatural. By contrast, older people in Lusozi may feel that the time young people take to show them how to use their smartphones confirms traditional forms of respect.

Some older people may have spent decades developing skills that are no longer needed precisely *because* of smartphones. A female research participant in Dublin had developed considerable expertise in finding her way around the countryside after years of delivering flowers for a flower shop. But thanks to Google Maps, her ability to locate places without a map has become redundant. Older people in Japan deplored the way smartphones have devalued key skills, such as mental arithmetic, for which they once received respect. They expressed concern about younger generations forgetting how to write now that predictive messaging suggests the right *kanji* (character) to use based on their phonetic input. Older people remember how much time they spent at school learning to write *kanji* by memorising the correct stroke order and practising calligraphy for hours on end – hard-won knowledge now in danger of being lost.

In other fieldsites, older adults complained that their younger relatives lack the patience to teach them how to use their smartphones. A 63-year-old woman in Chile has a typical complaint:

> My daughter bought this phone for me and taught me [how to use it] on the first day. After that, if I ask something, she says 'I already taught you'!

A 67-year-old man in Santiago describes another problem that is frequently encountered:

> When you ask them [younger people] how to do something, they do it very fast on your phone, 'pa, pa, pa, it's done!', but they don't show you *how* to do it.

Children and grandchildren often fail to understand the difficulties that their parents and grandparents have. In Bento, for example, they claim that their parents and grandparents have worked with technology in their employment before, exclaiming 'You worked with that, how can you not know how to use it?' In reality, many of these older people may have retired from their working life decades before. Many older people also fear becoming a burden to their families and so decide not to seek help. They stoically accept the fact that their children work hard and have many other commitments. As a 71-year-old woman in Bento put it: 'considering all of that [her children's busy lives], do you think I would bother them?'[7]

There are numerous ways in which young people can be unhelpful. Abu Zaki, a research participant in al-Quds, was facing some technical problems with his smartphone, an old Samsung Galaxy given to him by one of his sons after the latter purchased a newer handset. It is quite common in Dar al-Hawa for older people to end up with an old version of these smartphones, no longer required by a younger member of the family. Abu Zaki complained that his grandchildren had downloaded too many games to his phone; he has no idea how to delete them and the young people simply will not help him with this. Some older people may not be confident of even remembering their own number (Fig. 7.2).

In Dublin, young people in turn often expressed their frustration in trying to teach older people how to use smartphones; they found them frequently slow to learn and in need of constant repetition. Younger people claimed these struggles were surprising since smartphones were 'intuitive'. When several members of our team subsequently decided to

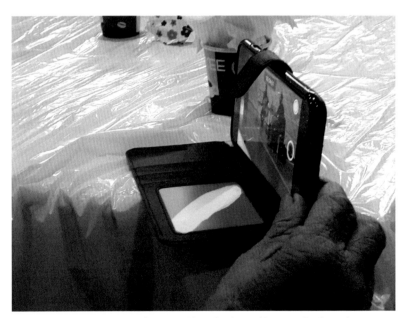

Figure 7.2   A woman taking a video during a live music show in al-Quds. Her own phone number is tucked inside her case. Photo by Maya de Vries.

teach older people smartphone use, however, it soon became clear how untrue this claim is. Smartphones are not intuitive devices for those who are not familiar with them. For example, an older person is told to download an app. They look at their phone and see an icon called 'Downloads'; they duly press this, which gets them nowhere at all. How would they ever guess that the appropriate icon is called 'Play Store'? Why should they not logically assume that this would be a place specifically for games, rather than for downloading a banking app? How should they know the new meanings of packages, the cloud or when something crashes? All of these bear little resemblance to the previous use of these terms, and are all the more misleading for being apparently intelligible.

In another example, older students are told to 'go on the internet'. When they look on their phones, they see one icon called 'Internet' – but then realise that the young people teaching them may be accessing the internet through something called Chrome or Google or Firefox. No one bothers to explain the difference between these various potential routes to the internet. What is the difference between a pre-installed icon called 'Gallery' and another called 'Google Photos'? Smartphones are many things, but intuitive is not one of them.

How smartphones make people younger

Despite the initial struggles, the potential inversion of traditional roles, the lack of patience of younger relatives and the counterintuitive design of smartphones, older adults generally do manage to use their smartphone. So why do they make such an effort to adopt this 'young people's thing'? At the start of this project, many people across these fieldsites still assumed that smartphones would be more 'natural' for younger people, the alleged 'digital natives'. At that stage, smartphones represented a boundary that reinforced the distinctions of age – a kind of age-based digital divide. The reason why older people persevere is that using a smartphone provides them with more than new capabilities. The adoption of a device once so deeply associated with young people can actually make older people feel younger. Once they have mastered the smartphone, it shifts from being a barrier between younger and older people to something that signifies the collapse of that barrier. In the era of 'successful ageing', adopting this new device and all the possibilities it entails amounts to staying active and constantly reinventing oneself.

Every author of this book is also writing a monograph, each under the title of *Ageing with Smartphones*. Ageing means very different things across these very disparate fieldsites, with chronological categories of age often less significant than how a person feels or is perceived. As already noted, someone in Kampala may be considered an elder at 40, whereas others may not feel old at 80 in Japan. The Palestinian population tended to shift to different clothes and other signs of belonging to a more senior segment of the population when relatively young, at 40 or 50. Usually the women will wear a long dress in dark colours and a veil such as a hijab.

For most of the other fieldsites, however, these monographs suggest a radical change in the experience of ageing. The traditional category of an elderly person, imagined by The Beatles as sitting on a rocking chair surrounded by grandchildren at 64, has largely disappeared. Several of our research participants say they expected to feel old on their 60th, 70th or 80th birthdays, but that just did not happen. Instead, the new separation is between the experience of frailty, at whatever age this strikes, as against sufficiently good health which has led to the experience of continuity over the decades. Older people in some fieldsites just continue listening to the Rolling Stones, though now through Spotify, and perhaps consider dating, though now through online sites such as Plenty of Fish. As such, this mastery of the smartphone as an opportunity to feel younger fits neatly into a much more general pattern of change in the overall experience of ageing. This factor was particularly important in Shanghai, where older people felt they never had a proper youth because of the disruption of the Cultural Revolution. Only now on retirement, aided by smartphones, can they embark on the project of being youthful.

The impact of the smartphone thus depends a good deal on the wider context of how the experience of ageing is changing. In Ireland smartphones helped people to feel younger partly because of many other ways in which these affluent retired people were apparently able to reverse the ageing process. Another example was the way in which they had more time to be involved in cultivating wellness or being active in green and environmental pursuits, to the extent that sustainability applied as much to older people themselves as to the planet. By contrast the Palestinians tended to be comfortable in retaining traditional ideals of seniority and age, changing their behaviour accordingly.

A final version of these changes can be found when a traditional category of ageing, such as being a grandparent, remains but is given a new twist or increased vitality through its manifestation in digital form. An example of this is the concept of Nonna, or grandmother, in Italy. In popular culture, particularly outside of Italy, 'Nonna' has become a kind of idiom signifying various idealisations associated with local traditions, homeliness, cooking and caregiving. It is widely used in advertising as a

Figure 7.3 Film: *Nonnas*. Available at http://bit.ly/_nonnas.

marker of 'authentic Italy'. Today a major role played by grandmothers in Italy is their active involvement in childcare; they provide practical, economic and social support to families. A number of women in NoLo had either moved to be near their adult children or already lived close to them, and were actively involved in grandparenting in their sixties and seventies. The smartphone is a main instrument in this, and in other aspects of everyday life. WhatsApp, for example, is used widely for arranging schedules and practical matters, for sharing photographs and videos among families and friends, and for pursuing individual interests and activities. Examples can be seen in the film featured here (Fig. 7.3).

Smartphones in this instance are not so much making people feel younger as making an existing traditional category of 'older person' more appropriate to contemporary life.

Teaching and learning smartphone skills

What is involved in learning to use a smartphone? Dijk and Deursen[8] suggest there are six levels of skills that are required to develop digital literacy, outlined below.

1) Operational skills, such as how to use a particular button
2) Formal skills, such as the ability to understand and use aspects of the interface, such as menu structures or hyperlinks
3) Information skills, such as how to look things up
4) Communication skills, such as the use of social media

5) Content creation, such as one's own playlists for music streaming
6) Strategic skills, such as how to use smartphones for personal or professional goals

Each one of these can be the cause of a new digital divide between the competent and the unskilled user. It is a point considered in some detail by Donner in his book *Beyond Access*[9] because it is often a feature of the developing world: simply because a population now has access to smartphones and the internet does not mean the end of inequalities. Mastery of one level may simply accentuate other constraints and divisions based on what people know and do not know about the subsequent deployment of smartphones.

These points emerged with particular clarity because our research also included some more elderly people who are starting to experience age-related frailties – reduction of cognitive skills, arthritis, trembling fingers, impaired vision – that also affect their usage of the device. So, on the one hand, the smartphone stretches the issue of skill because it increases complexity. Yet – literally 'on the other hand' – older populations are confronted by their loss of dexterity. The result could be observed ethnographically as part of everyday life, but was most easily discerned through the method of teaching smartphones to this population. The wider context includes considering older people's motivation for wanting to learn smartphones, the everyday situations in which they employ them and – especially – the wider intergenerational relations and tensions that are often involved in their obtaining, learning and then deploying their smartphones. Finally here, as in every chapter since the topic was broached in chapter 2, we find a subsequent issue of ambivalence towards the device.

Several of the team committed to teaching courses either about smartphone use in general or the use of WhatsApp specifically, in one instance for over a year. The students taking part had a very wide range of concerns and expectations. In Santiago one woman wanted to take HDR (high-dynamic-range imaging) pictures to post later on Instagram, while a man wanted to download an app to scan the QR codes he came across on flyers. Some participants found it difficult to grasp the difference between (paid) mobile data and (free) Wi-Fi, or to understand the notion of 'the cloud'; others had trouble with the touchscreen interface. Factors that affected their difficulties and expectations include familiarity based on previous use, the degree of family support they enjoy, their general level of education, their motor skills and the length of time since they began retirement.[10]

According to a recent survey of older Americans by the Pew Research Center,[11] one-third of seniors feel only a little, or not at all, confident when using electronic devices (including smartphones). As a result of that feeling, three-quarters of them say they need help to set up and start using a new device. According to one of Marília's students, a man aged 72, fear of making mistakes is the key difference between older people and their younger counterparts. He explained:

> If youngsters see something wrong, they laugh at themselves, because they are allowed to make mistakes. People, however, are not so tolerant with older adults.[12]

Because of this, he says, many of his friends feel so embarrassed when they fail that they become too scared to even try. In addition, they fear 'being overcharged', 'erasing important information' or 'pushing the wrong button' and thus ruining the device itself. Since they regard smartphones as machines, they assume that when it does not work it must be broken. It is hard to understand what young people mean when they tell them that they cannot break a phone and that they just need to go back some steps and take a different path.

Often this fear is associated with the more general sense of being stigmatised by virtue of being elderly, such that new technologies are not 'natural' for them. By contrast, older people see their difficulties as 'natural', commenting that 'I don't understand technology' or 'my head is not good for this' and insisting they could not do something unless the teacher was standing beside them. Others were more adventurous from the start, however, or became so after a while. The same Pew survey mentioned above also notes that once older people do go online, they proceed to engage 'at high levels with digital devices and content'. Among older adults who own a smartphone, for example, 76 per cent use the internet several times a day.

Older students commonly reported feeling overwhelmed by the vast array of menus, gestures and *different ways* of doing the *same things* on the smartphone.[13] Mostly, there is no clear or logical hierarchy in the arrangement of apps and functions. Many students with Android phones do not recognise the difference between the home screens and the app drawer, especially when they have the same wallpaper. One of the most noticeable obstacles encountered by students was being faced with too many choices. For example, when trying to share a picture from their gallery app after having selected the picture, they are confronted by various possibilities, including a heart, three vertical dots, three circles

Figure 7.4 Which one of all these icons is 'share'? Photo by Alfonso Otaegui.

intersecting, a square with an arrow, a square with a smiling face and a T, a paint pallet, three dots forming a V and a 'trash' icon (Fig. 7.4). Which one is 'share'?

Many older students also found it difficult to distinguish between a 'tap' and a 'long press'. The lack of self-confidence may contribute to a desire to press the button long enough to be sure it has been pressed, the analogue equivalent of pressing a doorbell. The effect is often completely different from a tap. They may also have a problem hitting the exact spot, which again may result in a completely different outcome. The film about Valeria from Santiago provides an example of these concerns (Fig. 7.5).

The students came to these classes with different motivations, but the one that dominated was often the desire to use WhatsApp. In describing her decision to get a smartphone, 70-year-old Maria Teresa from Santiago observes 'It's due to WhatsApp, everybody had it and I was feeling left out so I had to buy this'. Indeed some older adults in Bento buy what they see as a 'WhatsApp device', rather than a smartphone. This

it meant starting from zero

Figure 7.5   Film: *Valeria*. Available at: http://bit.ly/valeriasmartphone.

devotion to WhatsApp is why the whole of Marília's class, around three-quarters of Alfonso's and also some of Maya's workshops ended up being dedicated entirely to the use of WhatsApp.[14]

When Maya offered a course in smartphone use in Dar al-Hawa, Nura immediately announced that she wanted to sign up, even though she already knew how to use a smartphone and was a frequent user. In practice, given this and also her good knowledge of Hebrew and some English, she ended up helping other pupils to translate relevant materials into Arabic. Nura was especially helpful in teaching WhatsApp, which the pupils learned to use for everything from voice recording to sending their location to sharing contacts and images and backing up. On one occasion, while explaining how to back up files and folders, Nura accidentally deleted her WhatsApp. After a few minutes of failing to find it again, she called for help with great emotion in her voice. Nura clearly felt a deep despondence at having to be parted from WhatsApp for even a few minutes. She decided she was not learning anything and did not want to stay in the class, both expressions of her frustration and anger about the 'lost' app. At first Maya was not able to help her because the smartphone's memory was full. Eventually, after clearing out many of the WhatsApp backup folders, enough room was created to download WhatsApp again. Everything worked. Nura relaxed and a sigh of relief spread through the room.

In Nura's case, the stress consisted of two elements. The first was the prospect of being cut off from her constant connection with relatives, friends and activities. The second was the fact that she had not been able to deal with this problem herself. This exposed the degree to which she is still not entirely clear about the workings of something that is now the infra-structure of her life. This is an experience that exacerbates the worry that as people age, they become increasingly dependent upon other people. This is yet one more contradictory aspect of smartphones: learning to use them simultaneously produces greater autonomy, but also often creates new forms of dependency.

The discussion around learning to use smartphones ends with an example that shows how almost everyone can find some sort of engage-ment with the device. Mary from Kampala is in her fifties and never learned to read or write. She owns a Samsung smartphone, indicative of her relatively privileged position as a beneficiary of her ex-husband's allowance. Her children, both in their late teens, claim to have taught her how to use her smartphone, as well as apps and social media. Mary explained that she did not use the internet much except for WhatsApp, for which she buys daily data bundles. Most of her apps go unused, but she listens to music on her phone, which her friends send her through Bluetooth, and now has just over 100 songs. Mary is often home alone when her children are at boarding school. When she feels bored, she also uses her phone to play a game featuring a cat avatar that requires looking after, feeding and cleaning. Her 16-year-old daughter downloaded this game for her. As previously noted during the *Why We Post* project,[15] illit-eracy is less of a barrier to the use of smartphones and social media than might have been expected.

In conclusion, this section has largely followed the initial pres-entation of six levels of skills set out by Dijk and Deursen. We start by considering the struggles that some people have just to master basic motor skills. By the end, what is striking is the way each person, at their own level of competence, is resourceful in adapting the smartphone to their own interests and tasks. The smartphone does not necessarily reduce inequalities, but it may be accommodated to them.

Specialist apps and devices

Given the ageing population of some of the most affluent countries, there have naturally been attempts to create technology specifically designed for their usage. One such is the Doro 8040 smartphone (Fig. 7.6). The home screen of this phone includes four circles which can be filled with the images of selected contacts. A frail 91-year-old who is extremely technophobic is unable to negotiate her smartphone sufficiently to make a phone call that would require her to dial a number. Thanks to the

Figure 7.6 Example of a Doro phone, showing its quick buttons for accessing important contacts. Photo by Daniel Miller.

design, however, she has regular contact with her four closest relatives whose images appear on her screen.

As well as specialist smartphones, there are also apps intended for older people. One example is Meipian, which is used in Shanghai. 'Meipian' literally means 'beautiful piece' in Chinese, referencing the apps' use in editing photos and templates. The app boasts 150 million active users mostly aged between 40 and 60; they are economically well-off, with plenty of leisure time and a desire to express themselves.[16] In Xinyuan's survey of her informants, Meipian is one of the most popular apps. Every week one of the Shanghai research participants, Mrs Shen, shares at least two Meipian posts on her WeChat. She learned how to use Meipian at the computer class held at a community school for elder citizens, where handy apps considered suitable for older people were introduced by tutors.

Unlike WeChat, which only allows nine photos for each post, Meipian allows up to 100 photos. This appeals to Mrs Shen, who explains:

I want to record great moments in my life properly and completely … usually, I have hundreds of photos from a two-day trip.

This desire to 'record life properly and completely' is widely held among older people in China. They also commonly regard posting online as having a form of status analogous to publication. Their attitude is very different from that of people, who do not take their daily 'stream of consciousness'-style postings particularly seriously. Another research participant, Mr Shi, highlighted the difference:

Once my younger son said nobody would make a fuss about posting online, but I told him that I believe every word that I share should be good enough to stand the test of time. I take it seriously.

The older generation grew up in a time that lacked the communication tools for this type of continual expression of everyday life. A person could only be present to an audience through mass media such as radio, television or newspapers, all of which were highly restricted. They therefore regarded any public presentation of the self as something worthy of careful consideration. Furthermore, the main source of Meipian's revenue is its printing service, which offers users the option of turning their digital items into printed brochures. Printing has become increasingly popular among older people. 'I never thought I could write a book through my smartphone!' Ms Zhu exclaimed when she received a nicely printed 'book' of her Meipian posts with a photo of her grandson as its cover.

The use of Meipian returns us to a previously discussed contrast between China and most other fieldsites. In Brazil and Chile, older people often see their difficulties in using smartphones as 'natural' results of their age. They may assume that any physical problems, such as poorer vision and unsteady hands, are part of a wider inability that extends to cognitive functions. They thus assume that young people are the natural users of smartphones rather than themselves, and so it is natural to be technophobic.

By contrast for Shanghai, the film featured here shows that on retirement a woman like Dan can develop whole new dimensions to her life, including creating a following for her singing (Fig. 7.7). The smartphone has become so significant to Dan that she wants to ensure it is burned after her death as a way of conveying its contents to her afterlife.

These positive associations reflect the way that China has tended historically towards political gerontocracy. As noted in chapter 2, skill in

Figure 7.7   Film: *It carries all my love*. Available at http://bit.ly/
carriesallmylove.

using digital media is seen as the duty of a good and productive citizen;
it expresses their contribution to building modern China. Shanghai is
a place where older people may surpass young people in their affinity
with smartphones. But the conclusion is this. If there exists a population
that sees their usage of smartphones as natural, then it follows that for
people in Brazil and Chile it may actually be the stigma around age that
is the real issue. Certainly this stigma may be just as debilitating when it
comes to mastering smartphones as anything to do with the device's
technology.

Problems and benefits associated with smartphones

There are other, more general ways in which ageing impacts upon people's
relationship to the smartphone. One is the greater fear of being mugged. As
a research participant in Bento remarked, 'If you have white hair, you are
already a target'. Helen, aged 67, was frustrated when she could not show
Marília all the pictures of her grandchildren that she keeps on her smartphone. 'I came here with nothing. It is not safe,' she said. These women
were talking in a large square where people come to walk and exercise
every day; it is also one of the 200 points with free Wi-Fi provided by the
City Hall in São Paulo. Yet most of Marília's research participants agreed it
was not safe to make and receive calls or to text on the streets.

They have good reason to be wary. In the first quarter of 2019
an average of 13 mobile phones were stolen every hour in São Paulo,
representing 63 per cent of all thefts registered in the city.[17] More than

half of the 60 research participants to whom Marília spoke in April 2018 had had a smartphone stolen at least once, or knew someone in their family who had experienced such a theft.

People have consequently developed different strategies to protect themselves and their smartphones in public spaces. For example, Lucy, aged 65, said that she would never answer a call on the street: 'I just let it ring'. Lilly, aged 67, explains why she makes some exceptions: 'I take a quick look inside my bag. If it is one of my children who is calling I just go inside one of the stores on the street, so I can answer the call'. Other users even leave their smartphones at home when they go out. José, aged 72, is one of them: he never takes his iPhone outside his house, preferring to travel with his less expensive Android smartphone. Another strategy is to retain one's smartphone for as long as possible, hoping that an old device will be less attractive to thieves. Bia, 59, uses this strategy. 'This one is old, nobody wants it,' she says. The last, least common option is to have a second device in one's bag, known as 'the thief's mobile phone'. This is an old ruse, especially popular with women who drive. If they were robbed in traffic, they would hand over the fake bag, specially prepared for this eventuality. Nor is Brazil alone in having high rates of street crime. A Chilean research participant describes his reluctance to check his smartphone in public, and the tactics he has adopted as a result: 'I allow it to buzz in my pocket. When I get somewhere safe – a gallery – I might pull it out and take a look.'

The opposite was noted in the Japanese fieldsite of Kyoto, where smartphones are often highly visible in public as crime rates are low. Here research participants did not fear petty theft and phones were often carried in outer pockets of backpacks and back pockets of trousers. In Kyoto, where kimonos are often rented and worn around town by both international and Japanese tourists, women can be seen at scenic sites tucking their smartphones into the 'obi' belts for easy access for selfies.

More generally people fear the loss of privacy or the possibilities of surveillance or intrusion by technologies they know they only partially understand. During the smartphone course in al-Quds, teaching about connecting to open Wi-Fi networks revealed that only 3 out of 15 participants were aware that Wi-Fi spots were free. They had only used their mobile company's network. When people learned about this additional resource, they worried about whether it was safe for them to use it. Having connected to the Wi-Fi of the community centre with Maya's help, Amina, aged 74, asked if this same connection will remain when she gets home. As noted above, many people confuse Wi-Fi and data when it comes to payments.

Another extension of age-based digital divides that was addressed in earlier chapters appears in government's increasing propensity to favour online-only access. For example, the Chilean Senate has recently approved a bill called the 'Digital Transformation in the State',[18] which aims to make most state services digital. Those who wish to communicate via paper or hard copies must specifically request this service and justify their request. In 2017 Israel also embarked upon a large-scale reform intended to create a digital nation; it was based on digitising governmental services, both on local and state levels. Those in favour of the reform proclaimed that it would lead to benefits ranging from 'accelerating economic growth' to 'reducing socio-economic gaps and making government smarter, faster and more accessible to citizens, making Israel a global leader in the digital domain'.[19] However, Laila and Maya sensed that neither older nor younger Palestinian adults in Dar al-Hawa were part of this vision, since so many lacked either digital literacy or Hebrew language literacy or both (Fig. 7.8).

Figure 7.8 An emergency alert app for older people. It is only available in Hebrew, not Arabic. Photo by Maya de Vries.

The problem is not just governments. In Shanghai a major state-owned hospital in the city announced that one of its most popular departments would no longer provide an 'on-site' hospital appointment service. Instead the service would be limited to an online system. The hospital saw this as a solution to problems of overcrowding and long queues. However, such policies can cause consternation among those people, often older or economically disadvantaged, who lack the necessary skills to obtain an appointment digitally.

Such developments serve to create increasing dependence upon smartphones which some older people may resent, experiencing it as a form of age-based exclusion. Sarah, a research participant from the Dublin fieldsite, noted with dismay that she had had no difficulties working with computers before she had children. Unfortunately during the long period since then technology had progressed enormously, leaving her stranded. She seems to live in a different time zone, barred from participating fully in the present. Some cousins invited her to join Facebook, to which she suggested they call in for coffee instead.

Sarah knows that family members and friends modulate their behaviour to accommodate her and she finds this humiliating. When she wants to travel she relies on her husband, a bank executive, to book their flights and communicate with family members while away. She knows this gap might get worse and is conscious that she cannot bridge it on her own. However, she has a friend, Aoife, who strives to help her rejoin the digital world. Aoife keeps an eye out for messages on their book club's WhatsApp which Sarah might otherwise miss. Sarah now wants to attend classes to learn these skills, but does not want to go alone. Fortunately Aoife has offered to attend them with her.

The final item in this litany of problems is also crime-related. Older people are often the primary target for scams, cyber attacks and frauds. In Brazil, for example, there were almost 26,000 attempts at online scams per day in the first half of 2018.[20] These scams were very creative and diverse, ranging from links to freebie redemptions to apparent job vacancies or offers of benefits.[21] False promises often result in extortion or the capture of bank details that will be used for future scams. Apart from being the main target, older people tend to share stories about friends, neighbours or family members who have suffered at the hands of scammers. All of this contributes to a general fear regarding new technology and promotes the negative discourses discussed in chapter 2. They worry that while they might learn enough to be able to use the smartphone and to be contacted through it, they may not be able to learn it well enough to realise that they are being scammed.[22]

The reason for listing this series of issues is that they are problems that impact particularly upon older people. But at the same time there are an equal number of benefits arising from smartphone use which pertain particularly to that same population. This is most obvious in the realm of health: consider the degree to which it is elderly people who are the most likely be stricken by frailties, disabilities or various problems of (im)mobility. An example from Ireland is Chris, a disabled man of 67 who was brought up on a social housing estate. He was sent out to work at the age of 12 and has mainly worked on building sites. Since 2005 he has been disabled and can only go out with a mobility scooter.

Chris regards his smartphone as a lifeline and has more than 40 active apps. Of particular importance to him is online shopping, since there are no cheap clothing outlets in Cuan, one of the Dublin fieldsites. Chris has apps that allow clothing to be delivered from the US and he also makes use of a Chinese app called Wish (an eCommerce platform). His real passion is Radio Caroline – an old pirate radio ship that he used to listen to when working on building sites. Radio Caroline still has three channels: one for contemporary music and two more for music from the 1960s and 1970s respectively. In addition, Chris will also use Facebook and listen to local Dublin community radio. He follows sports including, through the Manchester United app, the local darts team, and will also 'visit' places using Google Street View. Chris spends a considerable amount of time in hospital and will often google all sorts of health information, looking into both biomedical and complementary treatments. His smartphone also connects Chris with the taxi that takes him into the hospital.

Kamila, aged 79, has been widowed twice. She does not have children and lives on her own in a two-bedroom, ground-floor flat in Dar al-Hawa. Kamila is acutely aware of the fact that she is vulnerable and does not let anybody enter her home. Only recently did she buy a smartphone and it was a slow process for her to start using it. However, Kamila now appreciates that the device helps her to connect easily with her sister, who lives an hour's drive away. The women can now chat much more often thanks to WhatsApp, Kamila's favourite app, followed by YouTube. She uses YouTube to watch the *Burda Magazine* channel to learn new designs. Sewing, knitting and making pastry are her principal hobbies, and she is always looking for new ideas and inspiration. Previously Kamila had been unaware that a YouTube app existed; she would just open a browser and search for YouTube. It was only after attending the smartphone workshop that she started to use the bespoke app.

Similar stories about the positive benefits of smartphones specifically for older people occur throughout this volume. In a way the experience of lockdown that followed the Covid-19 pandemic gave much of the world a taste of why online communication has become so important for older people who lack mobility. While for months people missed face-to-face contact and hugs, the mere idea of lockdown without online communication felt like a nightmare they had been spared. Given this litany of negative and positive effects it seems once again clear that ambivalence is not an inconsistent reaction; it is about the only rational response to the smartphone's impact on our lives.

Conclusion

In the introduction we referred to ethnography as 'holistic contextualisation'. We try to understand the smartphone as embedded in cultural values and social relations. It would have been possible then to write chapters on the relationship between smartphones and any more general social parameter, for instance class or gender. Age was selected as the example because it was the parameter upon which this research project was based. When combined with the evidence from the previous chapter on the smartphone in the context of individuals, relationships and society, the evidence from these two chapters reveals the complexity that comes with these considerations of context. We often tend to use terms such as 'expressing', 'embodying' or 'representing', but we also need to delve deeper into the characterisation of context.

In some cases, what has been presented could best be described as co-evolution. A group such as the Italian 'second generation' youth are simultaneously developing their relationship to their smartphones and to their wider identity. Co-evolution might also serve as a description of how smartphones work in alignment with other life changes, such as when people pass from full-time employment to retirement, which is often a major shift. Smartphones may then become a hub for organising this new way of life. This use of smartphones on retirement is described in detail in several of the monographs summarised in this volume. They frequently use the same term, 'crafting', that headlined the previous chapter.

Quite a different role for the smartphone emerges when the device is approached not as a technology, but as an idiom. From this perspective it becomes a device that stands for a certain relationship to age itself. This became clear when considering why older people elect to master smartphones, however daunting the task. Quite apart from technological capacity, this reverses the meaning of the smartphone. Before taking our classes the smartphone had represented a digital divide, with older people excluded and separated from the young, the 'digital natives'. Once older people become proficient in their usage, however, smartphones are transformed into an idiom for their own youthfulness. These new users may feel younger because they are also now associated with a youthful technology. Potentially, they may use Spotify to find their rock music favourites – but more importantly the smartphone may facilitate for them a more immediate connection with the contemporary world.

A third characterisation of context becomes apparent when smartphones are considered within intergenerational relations. These may incorporate a wider field of power. A major historical shift has developed from a time when most societies simply assumed wisdom and seniority were integrally linked, based on experience and knowledge accumulated over decades. Today the smartphone consolidates a shift towards knowledge that may require older people to be educated by younger people – a dependence that may be resented, though not necessarily. An association with power is suggested by our evidence of how young people often seem dismissive, impatient and unhelpful in teaching smartphone use to older people; it is as though they are not willing to cede their advantage in this relationship. Power is also involved in what was described in this chapter as a series of hurdles involved in mastering the smartphone. Even when populations gain access to smartphones and knowledge of how to use this technology, they may then encounter subsequent hurdles and barriers posed by the contexts within which smartphones are used, such as knowing where to find information or how to be strategic in smartphone deployment. These then create new digital divides.

A fourth, again quite different, way of conceiving of the smartphone in context can be understood as a form of struggle. An appreciation of how people struggle emerged most clearly from time spent teaching smartphone use to older people. This exercise revealed the many features of the smartphone that are difficult for some older people. A person with arthritic fingers may struggle to find the exact position of an icon and be unable to switch easily from a long press to a short press. The issue is not one of power, but of frailty and the lack of dexterity. It can also be a problem of declining memory or having to learn skills about unprecedented devices, which often fail to work without being broken. For many older people these factors seem inextricably linked with issues of stigma and lack of confidence. Our researches revealed by evidence that this is less true of the fieldsite in China than the others, so we cannot see this technological dissociation by older people as somehow 'natural'.

Finally, this chapter discussed some of the wider contexts that are equally relevant to the relationship between smartphones and age. An example of these occurs when older people in Bento regard the smartphone as dangerous because it attracts muggers. Then there is the intervention by commercial forces that create apps or handsets specifically designed for older people. There is also the proliferation of a digital divide created when older people, less skilled in smartphone use, are differentially excluded from online state services.

In summary, this chapter focused upon one major parameter, that of age, to explore the way in which we situate smartphones in their social, economic and cultural context. These conclusions would mostly likely apply irrespective of which social parameter had been chosen. They would also be evident from many other previous academic studies, to which we are beholden, such as the excellent work on the digital lives of young people discussed in chapter 2.[23]

The five perspectives outlined here each highlight a different facet of this task of contextualisation. To make things still more complex, there is one further factor which applies to them all: the rapidity of change. Each year the way we interact with smartphones is evolving in its complexity and depth. Whether it is the co-evolution of youth identity, the example with which this chapter started, or the new state regulations for digital infrastructure, an example given near the end of the chapter, the processes by which smartphones relate to social relations and cultural values are extraordinarily dynamic.

Notes

1. Examples include the 'Yalla Italia' Twitter page, which can be accessed at: https://twitter.com/yallaitalia, or the Young Italian Muslims Facebook group of the NGO of the same title: https://www.facebook.com/GiovaniMusulmanidItaliaGMI/.
2. See Clough Marinaro and Walston 2010.
3. See the EU report on New Italians by Antonsich et al. http://newitalians.eu/en/.
4. Hazāragi is the language spoken by the Hazara people in Afghanistan and within the Hazara global diaspora. It is a dialect of the Persian language closely related to Dari, one of the main languages of Afghanistan. The linguistic boundary between Hazāragi and Dari is not so clear cut. See *Encyclopaedia Iranica Online* 2020.
5. The children of migrants are not awarded Italian citizenship before the age of 18. Many young people in Italy today continue to feel ostracised by these laws, which they question on online forums and through other channels such as NGO and community groups. See Andall 2002.
6. See also Giordano 2014.
7. See the survey 'RG033 – Resultados – POnline 2017' from Accessa 2018. This is an initiative for digital inclusion in São Paulo that provides free internet access and many free courses to help users to improve their digital skills. The research found that over 70 per cent of respondents learned to use the internet by themselves or attending to courses; only 4 per cent could count on their relatives' help.

8. Dijk and Deursen 2014, 6–7.
9. Donner 2015.
10. Behind this lies the variety in the experience of ageing, for which see Thumala 2017 and Villalobos 2017.
11. Anderson and Perrin 2017, 3.
12. See Leung et al. 2012.
13. Kurniawan 2006.
14. Duque and Lima 2019.
15. Miller et al. 2016, 170, 207.
16. Zhao 2018.
17. Henrique 2019.
18. This bill, called 'Transformación Digital en el Estado', was defended on the grounds of saving paper and time. The president argued that this law aimed to 'modernise the functioning of the State. We are in 2018 and we still handle most of our bureaucratic procedures on paper'. See *Mensaje Presidencial de S.E. el Presidente de la República, Sebastián Piñera Echenique, en su Cuenta Pública ante el Congreso Nacional*, available at: https://prensa.presidencia.cl/lfi-content/uploads/2018/06/jun012018arm-cuenta-publica-presidencial_3.pdf.
19. See Israel's Ministry of Social Equality 2020.
20. Travezuk 2018.
21. O Globo 2018.
22. For the scammers' perspective see Burrell 2012.
23. For references to the work of Sonia Livingstone and others who have created this exemplary series of studies see endnotes to chapter 2.

8
The heart of the smartphone: LINE, WeChat and WhatsApp

Fieldsites: Bento – São Paulo, Brazil. **Dar al-Hawa** – Al-Quds (East Jerusalem). **Dublin** – Ireland. **Lusozi** – Kampala, Uganda. **Kyoto and Kōchi** – Japan. **NoLo** – Milan, Italy. **Santiago** – Chile. **Shanghai** – China. **Yaoundé** – Cameroon.

Why is the penultimate chapter devoted to LINE, WeChat and WhatsApp? What does it mean to describe these as the 'heart of the smartphone'? The justification comes from our evidence that, for many users across most regions, a single app now represents the most important thing that the smartphone does for them. These may dominate day-to-day usage to such an extent that they become almost synonymous with the smartphone itself. As noted in chapter 7, for some Brazilians a smartphone is simply a device for using WhatsApp, while other users in Japan refer to their smartphone as 'my LINE'.

The second reason for calling these apps the 'heart of the smartphone' is that they have often become instruments for expressing people's devotion to those they most care about. For many of us, what matters above all else in life are our core relationships: to children, parents, partners and best friends. These apps are the platforms where siblings come together to take care of elderly parents, proud parents send out endless photographs of their babies and migrants reconnect with families; they are the means by which you can still be a grandparent even if living in another country. All these uses come with concomitant issues of surveillance, dependency and stress. The first section of this chapter will be concerned with this affective dimension. The word 'affective' refers here to moods, feelings, emotions and attitudes, each a facet of the way in which these platforms have enabled the capacity for affection and care.[1]

The second section of this chapter provides another reason why it has become the penultimate chapter of this book. It is the chapter that makes perhaps the most forceful argument for an ethnographic perspective. The three apps discussed here tend to dominate more intimate and private communication, often within families. This makes them simultaneously not only the most important apps but also the most difficult to gain access to and to research. To teach about the use and consequences of such important modes of communication, we need direct participation and observation. It takes months to build up the trust, guarantees of anonymity and friendship that this required. It is thus difficult to see what other research approach could provide the scholarship required.

In the second section of this chapter we will examine how such apps may also spell the end of a discrete category that we have become used to calling 'social media'. One of the conceptual advances made during the *Why We Post* project was a perspective called 'scalable sociality'. Before social media, people generally had two main ways of communicating: privately, for example on the telephone, or publicly, for instance broadcasting to the general public. The earliest social media scaled down broadcasting so that people could post on Facebook or tweet to just a few hundred or to 20 other people. Subsequent social media, including the channels discussed in this chapter, grew out of private messaging services to reach an audience first of just a few individuals and then many others. The overall result was scalable sociality: a range of platforms that could scale from small to large groups, from the highly private to completely public. By 2021 there have been further developments. The three apps discussed here blend equally with other texting, messaging, voice calling and webcamming elements of the smartphone. The result is that today there is little left of any discrete arena which could be called social media, separated off from the more general capacity of the smartphone for communication.

The other major development has been in the way in which platforms such as WeChat now incorporate a wide range of services that might once have been independent apps. WeChat and LINE can feel more like a kind of smartphone in their own right, with WeChat replacing apps with its own mini-programmes, as discussed in chapter 3, and LINE becoming a kind of super app.[2] These changes form the basis of the third and final section of this chapter, which acknowledges the multifunctional nature of these dominant apps and their consequences. Once again this has to include a consideration of the commercial forces that have led to these developments.

A brief history

LINE was initially launched in 2011 by the South Korean company NHN in Japan. It was designed to be a messaging service for their employees to communicate following the Tōhoku earthquake and tsunami, which severely affected Japan's telecommunications network. During the disaster phone lines were down but data channels remained open, and so became the most effective way to stay in touch.[3] The application was subsequently released to the public in June 2011; by 2013 it had become Japan's most popular social networking service.[4] By 2018 LINE had gained 78 million Japanese users,[5] complemented by 165 million monthly active users worldwide;[6] major markets had been established in Thailand, Taiwan and Indonesia. The penetration rate for LINE is higher than for smartphones in Japan because of access via tablets. None of Laura's research participants were without LINE and many were using the app for a wide range of services, from making payments and keeping up with the news to reading manga.

WeChat is the most popular and most frequently used app in contemporary China and in Xinyuan's fieldwork. It is a smartphone-based, multi-purpose social media app launched in 2011 by Tencent, the company that also owns QQ – the social media app that dominated Xinyuan's previous research on migrant factory workers and remained the fourth most popular app on research participants' phones.[7] WeChat provides text and audio messaging, audio and video calls, location sharing, multimedia sharing and a payment service (Fig. 8.1), as well as a wide range of functions from taxi hailing to online shopping and many more. The growth of WeChat is impressive: by 2014 it had become the most popular messaging app in the Asia-Pacific region.[8] The total number of monthly active WeChat users passed one billion in April 2018.

The app also has a feature called 'Public Accounts', part of WeChat's 'Official Accounts' offering. This is where organisations, businesses and other entities can create a page that broadcasts news, sends people to ecommerce sites and more.[9] Users can subscribe to information from more than 10 million accounts, ranging from media outlets to personal blogs. Information on WeChat is storable and searchable. In 2015 a WeChat user read an average of 5.86 articles per day, meaning that it has also become a 'reading app'. Its expansion through mini-programmes was discussed in chapter 3.

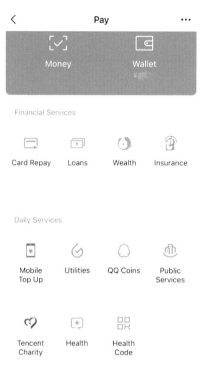

Figure 8.1 WeChat's Pay function. Screengrab by Xinyuan Wang.

WhatsApp was founded in 2009 by two ex-Yahoo employees. By 2011 one billion messages per day were already being sent;[10] by 2013 the app had acquired 250 million users.[11] It was purchased by Facebook in 2014 for $19 billion dollars.[12] By 2016 it had ceased to charge an annual fee[13] and by the end of 2017 it had 1.5 billion active monthly users.[14] Since 2016 WhatsApp has claimed to be entirely encrypted from end to end and, at least so far, it includes no advertising. It has also grown from text messaging to phone calls and video calls. A key function is the app's ability to show that a message has been received, which has changed the etiquette of messaging. While these are its technical possibilities, much of the development in how it is deployed comes from the creativity of users themselves. This is very clear in Marília's publication on the use of WhatsApp for health purposes in Brazil.[15] Research participants almost never mention that WhatsApp is owned by Facebook. This may be to protect what seems to be a largely positive view of WhatsApp from the increasingly negative connotations that have accrued to Facebook and its corporation.

The visual expression of emotions and care

Social media has transformed the basis of human conversation. Although there has always been visual communication, what we regard as conversation has tended to be oral, recently supplemented by conversational-style texting. Social media added a visual component, as summarised by the name of an app popular with young people called 'Snapchat', literally talking through photographs. These may be simply images of the face, to express how the user is feeling. For some time Japan has been at the forefront of global visual digital communication, as seen with the development of emojis[16] in the late 1990s and more recently with the advent of 'stickers' (large format emojis), which are available to download in themed sets on LINE. As of April 2019, approximately 4.7 million sticker sets are available from the LINE Sticker Store.[17] According to figures released by the company in 2015, up to 2.4 billion stickers and emojis were sent by its users each day. Stickers convey a vast range of meanings: 48 per cent express happiness, but the rest reflect emotions ranging from sadness (10 per cent) to anger (6 per cent) and surprise (5 per cent).[18]

Both urban and rural research participants in Japan stressed the role of stickers in the demonstration of everyday care via LINE (Fig. 8.2). They gave several reasons for their heavy use of stickers. Participants explained how stickers are a more relaxed form of communication; you do not have to check them carefully for embarrassing mistakes or typos which could lead to miscommunication. This was particularly important for older app users who were less familiar with touch-screen keyboards. Where stickers truly shine is in their ability to express feelings difficult to put into words through characters displaying extreme emotions, ranging from the most profound sadness to passive-aggressiveness to bursts of enraptured happiness.

Many participants said that stickers allowed them to maintain daily closeness between family and friends living remotely. Stickers enable people to make their textual messaging 'warmer' and more fully expressive of their personalities and feelings than if they had to rely on digital text alone.[19] This 'warming up' of communication became an important part of 'care at a distance', for example through the maintenance of regular contact with elderly parents who might be living in a different part of Japan. The daily flow of stickers and photos allowed people to send 'little bits of nothing', enabling care to be simultaneously more immediate and less burdensome. One research participant in her eighties communicated her daily activities to her daughter via a sticker set which

Figure 8.2 Example of LINE stickers expressing 'good night' wishes.
Screengrab by Laura Haapio-Kirk.

she had downloaded. It featured a humorous grandmother character
with whom she identified (Fig. 8.3, p.187). She would send her daughter
a sticker to let her know she was awake and to communicate with her
throughout the day, in addition to texting short messages and occasional
phone calls. Her daughter explained that stickers enabled the women to
keep their communication flowing throughout the day and formed part
of how she communicated care at a distance.

Middle-aged women also benefited from having their female friends
in close contact while they managed care responsibilities. This peer
support was especially welcomed in the form of stickers which would
express how they were feeling quickly and efficiently, in the moment
when they especially needed support. As Sato san, a woman in her early
sixties in Kyoto, noted:

> If I have a particularly hard day with my mother, smartphones can
> give me a quick window to reach out to my friends and get sympathy.

That aspect of being able to reach out to someone right at that second when you need them is a great thing about smartphones and receiving stickers that tell me 'It's okay!' is great.

In NoLo, research participants from Sicily were fond of using WhatsApp to leave audio messages. These were mainly family communications, often laden with emotion and affectionate greetings and salutations. For example, a Sicilian mother greeted her children, starting the message with 'Joy of my life, how are you?', followed by a detailed rundown of the day's events, specific goings-on in their lives, reminders about birthdays and an account of what happened at a grandchild's nursery that day. This reflects the person's style of offline communication – which often involved multimedia and was tactile as well as vocal, aiming to create colourful and vibrant forms of expression.

Into this came the new visual dimension as we can see in memes in NoLo (Fig 8.4 and Fig 8.5). Elena particularly enjoys communicating socially via memes circulated on WhatsApp. It is through memes

Figure 8.3   Screenshot from the LINE sticker store (Ushiromae). Screengrab by Laura Haapio-Kirk.

circulated on WhatsApp that she particularly enjoys being sociable and communicative.

> I send up to seven or eight memes a day, mostly to friends but also to particular family members, one of my sisters, a cousin who lives abroad…

These memes express a mixture of humour, irony and satire, love and friendship, and sometimes spiritual content. Elena is not necessarily expecting a response; merely being able to express herself in this way makes her happy.

Whether at her desk at work, on the metro or at home in the evenings curled up on her sofa with her cat, Elena has developed this light-hearted but meaningful communication.

Figure 8.4 A greetings meme in NoLo. The text reads: 'Hello/good morning full of hugs'. Screengrab by Shireen Walton.

Figure 8.5 A meme sent in NoLo. The text reads: 'Tell the truth, you were waiting for my good morning!!!'. Screengrab by Shireen Walton.

The visual and the audio are not at the expense of textual or oral. Now, suddenly, all phone calls are 'free' thanks to WhatsApp, meaning that they too can be expanded. People feel no qualms about calling on WhatsApp, or indeed using the webcam facility, especially when it comes to friends and relatives abroad. This represents a significant change across many fieldsites, which has greatly contributed to care at a distance.

We have rarely evoked the global within a book called *The Global Smartphone*, but in this example from China there is a sense of the world, and also what transcends it.

It is 11 a.m. UK time, 13 September 2019. In China, it is already early evening. Mrs Jinwei in Shanghai sends Xinyuan in London a WeChat animated sticker. It features a bright full moon surrounded by three joyful bouncing bunnies saying 'Happy Mid-Autumn Day!' (Fig. 8.6).

This is just one of the hundreds of stickers, emojis, short videos or animated albums related to the full moon or moon cakes that circulated

Figure 8.6 'Happy Mid-Autumn Day!' Animated sticker on WeChat sent to Xinyuan Wang in 2019.

among friends and family members on WeChat on the day of the Chinese Mid-Autumn Festival (Fig. 8.7).

The Mid-Autumn Festival is said to be the second most important national festival after the Chinese New Year. Traditionally family members gathered to offer sacrifices (for example, moon cakes) to the moon, and to express affection and their feelings of missing family members and friends who live far away. Yet 'afar' is precisely what is being transformed. As Mr Liguo notes:

> Even living in the same city, friends meet on WeChat. Live near or afar, it matters much less once you are on WeChat. So see you on WeChat.

This is discussed in the final chapter which explores care transcending distance. In 2019, wherever they lived, people in China could reflect upon the melancholy words written 900 years ago by the great poet Su Shi. They share the knowledge that WeChat was bringing them together in this contemplation as a global Chinese population.

Figure 8.7 Xinyuan Wang with friends and research participants, as seen in a picture circulated on WeChat. Screengrab by Xinyuan Wang.

> How long will the full moon appear? Wine cup in hand, I ask the sky… Why then when people part, is the moon often full and bright? People have sorrow and joy; they part and meet again. The moon is bright or dim; it waxes and wanes. Nothing in history has ever been perfect.

In Dar al-Hawa memes are also a major component of WhatsApp traffic. In this religious community, care is often expressed through prayer. The most popular images are a mixture of morning blessings and Suras from the Qur'an. These commence in the early morning, around 4 a.m., when the muezzin starts calling people to the dawn prayer. The WhatsApp group then starts to fill with messages, including images of flowers and tea or coffee combined with all sorts of positive texts about the glory and the beauty of the morning, always accompanied by the blessing 'good morning' (Figs 8.8a to 8.8e). These are generally shared images, rather than ones created by the community.

Here it is the norm both to greet people as appropriate to the specific time of day and for people to respond similarly. As a result, these postings initiated a torrent of messages in the two WhatsApp groups that Laila and Maya were following. The response may be a similar image,

Figures 8.8a to 8.8e Early morning memes circulated through the WhatsApp group of the Golden Age Club of Dar al-Hawa.

but it can also be a text written by a recipient, although some of the older people find it difficult to write texts on smartphones. Audio messages might have been expected to solve this problem, but these in fact caused other problems as some of these people are hard of hearing. One of the advantages of WhatsApp is the range of media that can be employed there, as well as the fact that it is free. As Lama, aged 77, puts it:

> Smartphones really help, it brings all the people together. I communicate with my brother in Amman, I communicate with my sister and her daughters in Ramallah.

Sending religious memes through WhatsApp fits within the intense religiosity of this community. They usually keep up with the five prayers a day and more generally try to act throughout the day in ways intended to please God and to follow the commandments. In so doing they believe that their acts are recorded in each person's 'book' so that after death God will be able to decide whether this person will go to heaven or to hell.

Eman, a research participant aged 42, suggested that:

> As long as people get older the countdown for facing God starts, so they care about doing things that please God, such as worship and treating people properly. Maybe, because people don't know when

their lives will end, so they must be prepared for anything. We hear about sudden death, maybe if a person fears God he protects himself and avoid many things, such as punishments in life that affects him, his children or his family. There are many things, but they depend on how people think, if they are religious, educated, how do they think about life, about hereafter, what do they know about punishments, about intimidation, heaven and so on. All these things are related to how people think.

In alignment with these sentiments, many of the memes shared through WhatsApp are intended to encourage others in their religious practice. They are carefully tailored both to the specific time of day and through the crafting of their religious iconography, such as the drawing of hands at the bottom of the first illustration which symbolises prayer. People within this community regard such exhortation to religion as the best way of expressing care for each other, thereby looking after one another in this life and beyond.

In Yaoundé, as in other fieldsites, a whole culture of memes and stickers has developed around platforms such as WhatsApp. These stickers are 'Africanised' to reflect local sensibilities better. They became common for young people, but are now conquering the networks of older people as well. In the WhatsApp family group of Marie, a 69-year-old former schoolteacher, her children's intensive use of stickers has led her to adopt them too. Marie explains:

> In the beginning, as soon as the discussions got heated one way or the other, often while trying to keep the atmosphere good, the children would start throwing around all sorts of funny little pictures to relax the atmosphere. The funny pictures of the children were the most amusing; often they made the whole group laugh. So that if there was any tension, you have to be more relaxed.

These stickers seem to be created by users, but are difficult to trace. Some characters have appeared in Nigerian films (Fig. 8.9a). Others are derived from private individuals whose stickers have gone viral (Figs 8.9b to 8.9f).

Still other stickers are more locally identifiable because they are drawn from political and public life, and so become part of daily circulation and satire within private networks. They may originate in transnational politics and popular cultures, as in the case of Barack Obama, or international sports figures such as the French footballer Paul Pogba (Figs 8.10a and 8.10b).

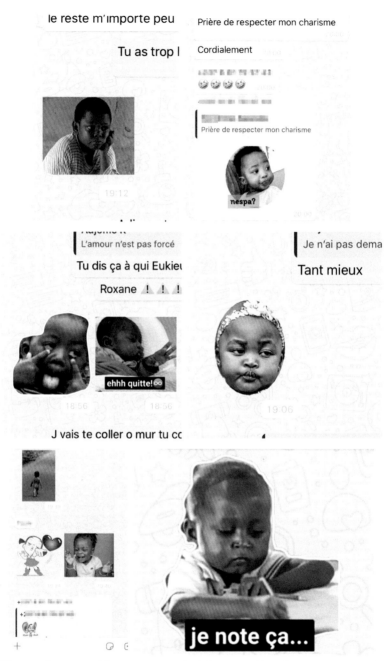

Figures 8.9a to 8.9f   User-created stickers used in WhatsApp groups in Cameroon. Screengrabs by Patrick Awondo.

Figures 8.10a and 8.10b   User-created stickers used in WhatsApp groups in Cameroon, depicting Barack Obama and Paul Pogba. Screengrabs by Patrick Awondo.

This example brings out a final point in the way that visual materials are used to express care. This is the emphasis upon humour and sharing meme-based jokes.[20] Across most fieldsites, personal communication includes the additional affective dimension provided by humour. Pauline, for example, found that the endless circulation of jokes around her Dublin fieldsite often incorporated a kind of life-affirming element. Often it is not even content that matters, rather the fact that there exists this ongoing communication as well as the sheer frequency of the communication. This point is also made by Tanja Ahlin with respect to transnational care.[21] The mere sending of a message can ensure that people feel less lonely or isolated – a consideration for their being included within a WhatsApp group in the first place. Although people complain at length about the constant interruptions of their phone, it becomes apparent that it would be worse if they had been excluded.

This is often felt most intensely in family situations. Sinead from Dublin brought her smartphone with her on Mother's Day on the off-chance that her adult daughter might make contact. The phone was turned on, fully charged all day, and carried in her coat pocket. No message came. 'I knew she wouldn't ring,' Sinead said later. 'After I threw her father out for being constantly on the drink – years and years ago – this daughter will not speak to me.' But still her smartphone had given her hope.

These examples show just how effective the visual image and these apps can be when trying to convey care and affection, sometimes surpassing the capacity of speech or writing. Yet these media should not be viewed in opposition to each other. As Gunther Kress argued,[22] a world of conversation through speech, text within books and images on screens is a multimodal world; there are important genres such as the meme which comprise combinations of these. What mostly matters to people is not *how* they convey care and affection, but the feeling that, one way or another, the person they care about has been supported. Care is not without its contradictions and more difficult aspects, however, which will be discussed in chapter 9.

The transformation of family

The initial discussion has demonstrated the capacity of these apps to communicate depth, as in care and affection, changing the mood and impacting the underlying atmosphere for social relations, including being the primary evidence for their maintenance. These apps also have a capacity for breadth, the range from private to public communication that was earlier described as 'scalable sociality'. Scalable sociality is evident within these apps where the groups range from quite large associations, for example those playing a particular sport, to the close communication between three or four close friends. They also vary from long-term commitments, such as one comprising all the siblings, to quite ad-hoc groups, for example those involved in arranging a birthday party. The organisation of this section will follow this principle of scalable sociality by starting with the family and then moving outwards to consider community and larger groups.

Most of the 18 people asked in the Dublin fieldsite have a range of WhatsApp groups, as illustrated below (Fig. 8.11). Almost everyone has at least one family group and some may have four or five. Typical examples are groups interested in seeing their grandchildren, or looking after elderly parents, or simply everyone within a given household. Sometimes these are further afield, such as all the cousins living in California. The next most common category for these Dubliners would be friends' groups, such as some women who meet periodically around the town or reciprocally celebrate their respective birthdays. Also common are sports groups, especially golf, but also swimming and triathlon. Several of these

Figure 8.11 Breakdown of the number of WhatsApp groups on each Dublin participant's phone. Taken from fieldwork in the Dublin region conducted by Daniel Miller.

participants also have walking groups on their WhatsApp. Then there are committee groups, such as the Men's Shed, or the Town Twinning group, or those taking the same course. In one case (but reassuringly only one) there is a puppy with its own WhatsApp group.

These groups reflect the dominant demography of the research, mainly retired people who have grandchildren and have joined in various leisure activities. They generally regard WhatsApp groups as a positive increase in people's capacity for communication and a time saver in that messages do not have to be repeated. But there are also negative aspects. Research participants often complain about the sheer number of WhatsApp messages that flow into their phone daily. They may see this as a reflection of the intense sociality of this vibrant community. On the other hand, they note that whenever the parent or grandparent of a child has texted that their child/grandchild has scored a goal in a football match, everyone feels obliged to make an appreciative comment. The aggregate number of notifications or accumulated messages can then become extremely tiresome. Similarly, someone on a town committee complains when certain members hijack the group, deviating from its main purpose and instead using the group to tell everyone about their latest holidays.

An impact of very considerable importance lies in the realm of family relations, which in most fieldsites still represents the fundamental unit of sociality. In places such as Bento WhatsApp has promoted a return to the more extended family of aunts and uncles, first and second cousins and nephews, a situation that reverses the prior historical retreat to the more nuclear family. Previously, encounters with extended family tended to coincide with the rituals of special dates such as Easter, Christmas and birthdays, making them more formal in tone. With WhatsApp contact is more constant and can therefore be permitted to be more trivial. WhatsApp turns the 'good morning' messages, jokes, photos of meals and reports from holidays into casual encounters, changing the very nature of the extended family.

An example is the establishment of Bete's WhatsApp family group. When it was created, the group's goal was to organise the next Christmas dinner, distributing roles and allocating responsibilities for cooking each dish. Yet when Christmas had passed, the group remained active while retaining the name 'Christmas supper'. Its function had transformed into a means of addressing the distance and disruptions of living in different neighbourhoods within a big city where travel by road is becoming a nightmare. Today, members can keep in touch and share daily occurrences without the endless traffic jams of contemporary São Paulo. In chapter 3 a similar point was made during the discussion of 'Screen Ecology' by the presentation of Mr and Mrs Huang. The chapter explored the way they now constantly interact with family outside of their household, through their use of multiple devices, so that the distinction between a household and those living elsewhere is much reduced.

Bringing together such a diverse group as family, including several generations, will obviously create its own problems. People complain about both the quantity and genre of the messages circulating in family groups. Older people who traditionally acted as hubs for the re-sending of information about the family find that this latest technology seems to be precisely designed for this most ancient of roles. Simple, often visual messages are easy to handle even for a novice and can be forwarded at will. Younger members of the family may be somewhat disconcerted by this barrage of family small talk. Roger in Bento complains that the older generations are sharing jokes he first saw on email 10 years ago, but he does not want to be the spoilsport who criticises his great-aunt's joke. He recognises that at a certain age people have an awful lot of spare time to commit to the role of being the family hub.

There are some subjects that are taboo in these family WhatsApp groups, however. Generally politics, football and religion are not discussed; it is not uncommon for children to drift into other Christian denominations than their parents, which can be a sore point. Politics in Brazil has also become toxic. The impeachment of former President Dilma Rousseff (2016) and the imprisonment of her popular predecessor Lula (2017) have been followed by the election of the extremely div-isive Jair Bolsonaro (2019). All of this could easily tear apart the fragile weaving together of family sociality. Many people can give examples of family members who have had to leave a WhatsApp group or who have caused significant offence because of political comments.

From family to community

The capacity of WhatsApp to enable older people to develop roles as family hubs may extend to wider layers of sociality. People who have had influence in their professional careers often have the skills needed to filter information and re-direct it to those who they know would benefit. In the Dublin fieldsite almost everyone seemed to have some relative in the medical services that they could contact with anxieties about symptoms. Individuals may also become curators of events, letting people know about what is going on or what is coming on television as a kind of public service. A good example in Bento was the way that people came to hear about various educational courses. Every class will create a WhatsApp group to support teaching, but in practice the students quickly appropriate this space to share other information, for instance details of new courses. This proliferates yet more WhatsApp groups. Another potential expansion of sociality occurs in retirement, when people have time to recontact people they knew at school or college. WhatsApp is therefore now viewed as a kind of time machine, allowing people to reconnect with their distant pasts. Last but not least, WhatsApp is also used to retain links to former work colleagues, which is often very important to retired people.[23]

This expansion of contact from the family to beyond flows easily into the wider community, as was found in NoLo. The fact that Shireen's research participants saw WhatsApp as their most used app reflects the intense sociality of the people she came to know there across a range of social and cultural contexts. In NoLo it is impossible to separate out family and community usage. While the community announces events on Facebook, WhatsApp bears the brunt of organising them. As well as their various WhatsApp family groups there are those associated with hobbies, the children's school, work, recreation and volunteering. By now there seems to be a WhatsApp group for almost every activity at every scale, from the more public activities of a gym class, Italian language lessons, knitting groups and cultural groups such as 'Sicilians' or 'Egyptians in Milan' through to community allotments and semi-private residential apartment building groups.

For example, Giovanna in NoLo had found retirement quite difficult. Very different from her active life as a public secondary school teacher, the retreat to her home and domestic routines seemed at first a bit like a prison, albeit a relatively comfortable one. She felt herself to be in a space associated with loneliness and isolation, compared to the wider, buzzing, multigenerational sphere of her work environment. This changed when a colleague recommended that she join a women's choir. Hesitant at first, Giovanna soon found the group vibrant and friendly. While they met only weekly, she could also enjoy the constant buzz of over 40 participants on the WhatsApp group. Members of this group shared photos, videos and song lyrics, as well as emojis full of hearts, flowers, shooting stars, laughs, cries and hugs.

In addition, the group's administrator made this quite an active political space for expressing liberal views and singing/marching against racism and the plight of refugees in Italy. As one of the choir members stated, 'The piazza is our natural environment!' This women's group and its activity has given Giovanna a new lease of life, as she became increasingly committed to this expressive dimension, both in relation to singing and politics. Retirement now feels like something she can participate in and shape, carving out spaces for herself and her need for sociality as well as broadening her social and political horizons. It helps Giovanna to deal with her feelings of anxiety and melancholia as she approaches her seventies.

If Giovanna represents the first stage of retirement, then Pietro, in his late seventies, is an example of WhatsApp usage among older participants. Pietro has severe walking difficulties and rarely leaves the apartment. He often wears his smartphone tied around his neck so that he is never without it. WhatsApp has become one of Pietro's main portals to the wider world. He checks it regularly for messages from family, friends and his family doctor, engages in a few chats and then has another cigarette as he dives into the *Corriere della Sera* newspaper or continues with a novel until lunchtime. Then Pietro's wife Maria, who is also retired and volunteers in the local neighbourhood, will prepare lunch for the two of them. In the afternoon he will take a nap, return to his reading, browse the internet on the home PC in the study and watch television in the evening after dinner.

When Pietro was added to a new WhatsApp group representing the apartment building where they have lived for 30 years, the couple had different reactions. Maria welcomed the wider sociality and its usefulness for communicating on practical matters, such as concerns about communal spaces and corridors or other issues that needed to be shared and discussed. Pietro was initially more uncomfortable with this unfamiliar mode of sociality, especially as it quickly shifted from the supposed function of information exchange to the wider postings of emojis, memes and even poems. At the same time the pings he receives on his phone, including news notifications, bring him much pleasure throughout the day. They make him feel connected to a certain buzz of being-in-the-world, a place from which his physical condition had otherwise gradually removed him.

Not every region is dominated by the use of these apps. Patrick's survey in Yaoundé revealed that 78.2 per cent of people use simple voice calls to contact their loved ones, but only 18.6 per cent make use of WhatsApp to call. Mostly this reflects a generational divide: phone calls have been more or less replaced by WhatsApp for the 16–35 age group in this fieldsite, but this has not really spread to older people as yet.[24] Currently it is mainly middle-aged people who are in that transition phase, in which people are increasingly asking them 'are you on WhatsApp?' rather than the previous 'do you want my MTN or my Orange number?' question about different phone networks. In Yaoundé there is, in addition, a class dimension to WhatsApp use. As younger people, or those who cannot afford to make phone calls, have quickly colonised WhatsApp, more middle-class users have started to see WhatsApp as associated with people unable to pay for 'normal' voice calls. This connotation of lower status means that, at least for the moment, it is not just older people but also more affluent ones who may use WhatsApp less. This is likely to be a short-term development, however, as WhatsApp proves too useful to avoid. For example, the app is becoming increasingly important for internal communication within communities living in Yaoundé, each of which originated in a particular region of Cameroon. For now, however, WhatsApp is not only less dominant here, but also enjoys a lower status than in other fieldsites.

Despite this, there has been a growth in the use of WhatsApp for communities. In Yaoundé community groups are often interconnected. One group, for example, known as the 'Bafout veterans', is a sports and leisure group, initially created by people originating from the region of Bafout in the western part of Cameroon. Such groups gradually accrue structures of mutual support and solidarity, mostly through the formation of *tontines* – rotary savings and credit associations. In many parts of Africa these organisations are one of the most common ways to finance projects, as access to credit is otherwise limited.[25] In effect, they combine ethnic association and leisure such as sports with financial support to become the basis of cultural identity.

Figure 8.12   Film: *Community uses of smartphones*. Available at http://bit.ly/communityusesphones.

Despite initial problems, the growing importance of WhatsApp within Yaoundé was made clear by the president of the veterans' group mentioned above. The group has 83 members and has been in existence for 33 years. Mr Sing, a retired engineer, noted how WhatsApp accelerated the possibilities of sharing news and other items around the group, thereby enhancing members' sense of belonging. Previously they had a tradition of going for a drink after a sports activity, but now 72 members also share a WhatsApp group which circulates information and amusing videos daily.

Some illustrations of the use of WhatsApp and smartphones within communities are given in this short film (Fig. 8.12, p. 202).

WhatsApp and religion

The trajectory from family through to community and wider groups culminates in what is often the largest sphere where these apps tend to flourish: that of religion. One of WhatsApp's growing organisational functions in Yaoundé is for prayer groups. Didi is a retired schoolteacher and a mother of four children. She separated from her ex-husband, a colonel in the Cameroonian army, four years ago and now has time on her hands, much of which she devotes to the church. Although she had a smartphone, Didi resisted WhatsApp until 2018, worried that social media contributes to the 'loss of benchmarks' and 'restrains moral and Christian values'. For the last five years she has been a member of 'ekoan Maria', a network devoted to the Virgin Mary, gradually assuming responsibility for a whole district within the diocese of Yaoundé. As a result she has had to become involved with WhatsApp, the primary means of coordinating activities for this devotional group. Almost every day she publishes an agenda for these networks based on the Catholic Diocese of Yaoundé. She also uses WhatsApp for a variety of messages, including messages for Sunday mass – this can be the reading and the homilies (a sermon or speech that talks about a specific part of the scripture), or even the entirety of the mass. Didi will also share images (Figs 8.13a and 8.13b) – mainly representations of the Virgin Mary, but also more general information for the diocese, such as press releases of a religious nature.

Peruvians are second only to Venezuelans in number among the diverse migrant communities of Santiago.[26] Many Peruvians are devoted to their religion and the Latin American Church in Santiago, famous for being welcoming and supportive of migrants, is a common gathering point. Among all the Peruvian Christian brotherhoods, the Hermandad

Bonne rentrée sous la protection de
notre Seigneur à nos enfants,
les enseignants, le personnel scolaire et
à Tous les parents!
Force, intelligence, sagesse et surtout la
faveur pour réussir cette année scolaire!
Edenespoir.org

Figures 8.13a and 8.13b Examples of photos shared by Didi in Yaoundé via WhatsApp groups. The texts say: 'Happy anniversary to all mothers!' (Fig. 8.13a) and 'Happy return under the protection of our Lord to our children, teachers, school staff, and to all parents! I wish you strength, intelligence, wisdom, and above all, the good fortune to make this school year a success' (Fig. 8.13b). The meme on the left is a special message sent on Mother's Day.

del Señor de los Milagros ('Confraternity of the Bearers of The Lord of Miracles') is the most diverse in terms of regional origin; it also includes non-Peruvians. The brotherhood consists of three battalions of men and three groups of women, each with its own WhatsApp group.

Within the battalion joined by Alfonso, WhatsApp group messages are sent daily. The most common use of the group is for sending daily Bible readings, but WhatsApp groups are also used for the organisation of fundraising events, processions and meetings. Every now and then chains of prayers are passed through these WhatsApp groups on behalf of a member or their relatives, to which everybody responds with a blessing and messages of hope. The battalion's chief is Enrique. At the time of the fieldwork he had been living in Chile for almost 10 years. His work hours change from month to month, so sometimes he has to wake up at 4 a.m. Enrique would read the Bible at his desk and then prepare a message to send to the brothers in his battalion. He would google 'New Testament reading of the day' and then copy and paste the result (Figs 8.14a and 8.14b). He would also google an image of Christ or the Virgin Mary to attach to the message, then capture a screenshot and paste it into the

Figures 8.14a and 8.14b Examples of the types of messages Enrique would send via WhatsApp. The image on the left (Fig. 8.14a) is a 'good afternoon' message followed by a passage from the Bible and accompanied by an image of Jesus on the Cross. The image on the right (Fig. 8.14b) was sent on Peru's National Day (28 July). The text reads: 'I didn't ask to be born in Peru. God just blessed me.'

WhatsApp message. Most brothers would respond with 'Amen' once the message was received.

A scroll through the WhatsApp groups of Santiago's Peruvian migrants is in effect a tracing of their life stories and place in the diaspora. Many have WhatsApp groups of their high school and college friends from Peru – people with whom they may meet up every now and then when they return to Lima. Secondly, they also have numerous family groups: the cousins' local nuclear family, the siblings' group (through which care for an ageing parent is coordinated), as well as various extended family groups. Thirdly, they have the 'professional groups', including all their workplace colleagues; these last groups expand to include people other than Peruvians. Finally, there are social groups such as Christian brotherhood or Peruvian city clubs. WhatsApp, along with Skype and Facebook Messenger, has made the entire experience of living in a diaspora much more seamless; migrants can now remain connected not just to others in Peru, but also to Peruvians who have migrated to Japan or the United States. Those who can remember often reflect on the early 1990s – a time when they had only expensive and infrequent communications for which they had to go to the phone booths in the Snail Gallery next to the Plaza de Armas (the main square in Santiago).

For people living separately from those they love, technology has been a profound transformation in their lives.

The trajectory of this section has echoed the more theoretical argument of scalable sociality. As in this final example in Santiago, WhatsApp is important to not only to the individual family, community and religious groups, but also for enabling each of these to flow into the others. Yet there are problems associated with this breadth of usage, one of which has become known as 'context collapse'.[27] An example outside the field of religion comes from China where again the very scale of usage, in this case of WeChat, invites leakage between different domains of sociality. In the Shanghai fieldsite, for example, Mrs Ruyun complained about how bad she felt when she found she had been 'blocked' by her daughter Qing on WeChat. In fact Qing had not really blocked her at all, but had applied the strictest WeChat privacy setting, such that all her WeChat contacts can only view posts from the last three days. Unlike 'blocking' certain contacts, this setting is indiscriminate; even if users want to exclude only a few contacts from viewing their previous posts, there is no way to apply the 'three days only' rule just to those unwelcome contacts. The reason for Qing's action was that she had recently attended a large industry conference and had just added several conference attendees to her WeChat contacts. She went on to explain:

> On my WeChat there are too many posts about my private life, including photos of my honeymoon a few years ago and my son's photos, and I really don't want to give these new professional acquaintances the chance to know so much of me. However, it will be considered rude if I really block them from viewing my WeChat posts as they will able to tell that they have been blocked by me specifically.

The challenge Qing faced was how to remain friendly with new contacts in the workplace while protecting her privacy. The 'three days only' privacy setting[28] offered by WeChat came to her rescue. This is something that most people will not take personally (or will take much less personally), precisely because the setting is applied to *all of* someone's contacts. Her mother, of course, did take this personally as discussed above. Here the issue arose precisely because it is becoming more difficult to separate family usage of these apps from more general usage.

The useful app

The discussion so far has yet to emphasise one of the most significant aspects of these apps, especially in the case of LINE and WeChat. This is the way they have expanded through the incorporation of as many different functions as possible, often replacing the app-based organisation of the smartphone. While the previous two sections have emphasised the role of these apps in communication, this final section acknowledges an equally important development in these core apps: the way they have become central to the underlying utility of the smartphone. They are increasingly the way people do 'stuff'. To illustrate this point two examples are provided from the field of health, because this was the area where our project focused upon the potential use of smartphones for more applied purposes.

In January 2019 LINE launched a remote medical consulting service for Japan, in partnership with the largest Japanese medical platform called M3.[29] The potential of LINE as a medium for health interventions is the subject of Laura Haapio-Kirk's applied research in a collaboration with Dr Kimura, a social nutrition researcher from Osaka University. Potential benefits of LINE might include the increased privacy of messages when compared to voice calls, especially when living in close proximity to family members, or the reduction of stigma associated with visiting a psychiatric clinic. The same point suggests the suitability of LINE for other stigmatised topics, as suggested by a recent local newspaper advert in Kanagawa, an urban area close to Tokyo (Fig. 8.15). It announced that the town council will start to provide LINE consultations on the topics of parenting, single parents, domestic violence and social withdrawal (*hikikomori*).

This advert is particularly directed at young people (those under the age of 39) who are having difficulty finding a job or who are struggling to socialise. It is hard to put a precise figure on the number of socially withdrawn people who do not leave their homes since their families tend to hide the issue. However, a survey in March 2019 by the Japanese government estimated that the country contains over one million *hikikomori*, 613,000 of whom are aged between 40 and 64.[30]

Similarly, the utilitarian importance of WhatsApp is evident in Alfonso's applied health project. This is being carried out as part of his ethnography at an oncological centre in a public hospital in Santiago – the

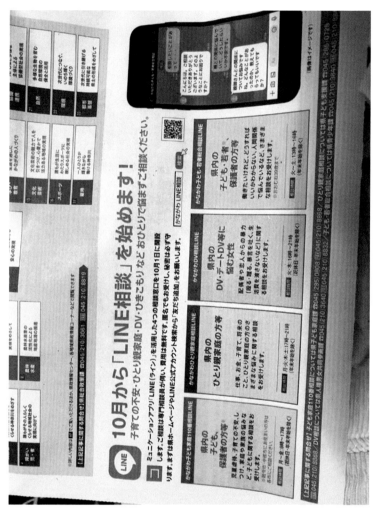

Figure 8.15 Newspaper advert announcing the availability of LINE consultations for topics such as domestic violence and social withdrawal in Japan. Photo by Laura Haapio-Kirk.

only public hospital in Santiago to have implemented a 'nurse navigator' model of healthcare. The nurse navigators work as mediators between oncological patients, helping them to navigate the medical and bureaucratic systems often found in a public hospital.

Cancer treatments present two complex systems for the patient to navigate. The first represents the complexity of medicine. Different cancer treatments can each have several effects on different systems of the body, so managing the treatment implies handling a lot of information. The treatment is based on a series of procedures (image exams, chemotherapy sessions, blood tests, etc.) which require prescriptions and appointments, and have to be carried out in a specific order within a certain amount of time. Failures in this regard can reduce the probabilities of their success. Nurse navigators have the expertise to deal with both the complexities of the treatment itself and those of the bureaucracy and convey both to the patient. For these purposes, the nurses need to make appointments for exams, blood tests and the like, which requires a good deal of paperwork. They also need to stay in touch with patients in case the latter have doubts or questions. At one level, these dedicated nurses constitute a human factor in healthcare that no smartphone app can replace. They were also the group making by far the most extensive and creative use of WhatsApp.

According to the nurses, WhatsApp was ideal for developing a variety of modes of communication that correspond to the particularities and necessities of individual patients. Some patients prefer a phone call and others need to see the information written in a text message; yet others will be reassured if they see a picture of the prescription or a request for a medical test. Some need an audio message, which they can then listen to several times, in order to understand the meaning, because most of these patients have low incomes and very limited education. WhatsApp has also become integral to the way that nurse navigators make themselves available to the patients for any doubts or questions they might have, as well as providing comfort and maintaining the essential social relationship that sustains them through treatments and care at a distance. Such a relationship may last for several years.

A third example illustrates our project's general shift in orientation. As noted in previous chapters, the main finding of our research in regard to health has been evidence that suggests we might refocus the field of mHealth from an emphasis upon bespoke apps to concentrate instead on the potential of the apps that people mainly use. In this short film (Fig. 8.16) Marília Duque discusses what she has learned about the

Figure 8.16   Film: *What I learned from using WhatsApp*. Available at http://bit.ly/learnedfromwhatsapp.

potential of WhatsApp for improving health communication and care in São Paulo.

Commerce and corporations

Another area that has not yet been discussed in any detail but which is increasingly important to the way these apps are becoming the core to overall usage is commerce. This is the field where WeChat, in particular, has been in the vanguard. In the *Why We Post* series we saw that WeChat had already started to become important for payments and ecommerce.[31] Today the most frequently used mobile payment apps in China are WeChat Pay and Alipay. The majority (72 per cent) of research participants[32] in Shanghai aged between 45 and 70 regard mobile pay as their first choice in daily payment; more than 90 per cent had made a payment via smartphone. It is common for people no longer to take along any cash or bank card when they are out as long as they have smartphones with them. The staff of convenience shops in the neighbourhood where Xinyuan lived reported that less than 10 per cent of the shop's daily income was in cash.

The widely recognised starting point of the monetisation of WeChat was on 28 January 2014, when it launched 'WeChat red envelope – a scheme that allowed users to send 'digital red envelopes' of money to WeChat contacts online. The WeChat red envelope extended

Figure 8.17 Photo showing the variety of payment QR codes provided by a street food vendor. The green one is WeChat Pay. Photo by Xinyuan Wang.

the long-standing Chinese tradition of handing out red envelopes of cash as festival or ceremony gifts to an online activity, making it more fun in the process. Today through WeChat Pay, service accounts can provide a direct in-app payment service to users. Customers are allowed either to pay for items or services on webpages inside the app or to pay in-store by scanning WeChat QR codes (Fig. 8.17).

In 2015 WeChat launched the 'City Services' project, further expanding the scope of services by allowing users to pay utility bills, book a doctor's appointment, send money to friends and obtain geo-targeted coupons, etc. WeChat has also become important in the wider development of ecommerce in China. Aside from business institutions, all the WeChat public accounts can sell products or services on WeChat. Many of the mini-programmes discussed earlier are used for ecommerce as a rival to Alibaba, the dominant company in Chinese ecommerce.[33]

This is one side of the equation. But it at least implies another, which is that while we did not study corporations ourselves, it is impossible to consider these developments without acknowledging the forces that lie behind their creation and which facilitated their adoption and adaptation by users. Most of the uses described in this chapter, especially the highly monetised ones discussed in this section, are likely to have developed from the corporations on the basis that they, not only anthropologists, conduct research on how users adopt and adapt their products. It is obviously of interest to the companies to keep their apps aligned with any developments in the way they are used.

A good example of this is the way that Tencent (the Chinese company that created and owns WeChat) now tries to develop functions that reflect the company's understanding of how social relations operate in China. This became clear when, in June 2018, WeChat Pay launched a new function called a 'kinship card' (*qinshu ka*). This allows people to combine their WeChat Pay function with a maximum of four relatives, including parents and two children.[34] The 'kinship card' payment further includes older people and children who, for various reasons, have not got involved in mobile pay yet – they may not yet have a bank account or they are anxious about using mobile payment systems. Kinship card beneficiaries do not need to provide the bank card information to WeChat because the card operators will acknowledge their payment via WeChat. The way this operates encapsulates several principles of Chinese kinship. By including parents, it may be regarded as an expression of filial piety. It also builds on their prior incorporation of the red envelope (Fig. 8.18).

One research participant, Mrs Zhong, was attracted to this kinship card partly for reasons of financial security, as the card operator can always check every expense. It has also helped her with her mother, who has always been worried about online fraud. Before the kinship card was

Figure 8.18   The digital red envelope on WeChat replicates the physical one in which people traditionally placed gifts of money. Screengrabs by Xinyuan Wang.

developed, Mrs Zhong would transfer £250 every month into her mother's WeChat pay account. Now she simply sets up the kinship card link based on this budget. The upper limit allowed by the kinship card is around £450, which is enough for daily expenses while preventing any major fraud. On the other hand Mr Guo, who has some major financial commitments such as a new house, began to worry when his mother asked him for a kinship card; he was concerned that it could impact upon his own expenses. Mr Guo was somewhat relieved to find that his mother only spent a symbolic £2 in the first month through the kinship card. He explains:

> I then realised that it was not about money at all. She just didn't want to 'lose face' (*diu mianzi*) in front of her close friends because some of them have got kinship cards from their children.

A recent book describes the relationship between users and WeChat[35] as 'super-sticky', a label that references the difficulty users would have in leaving this mega-platform once they are on it.

> Super-sticky WeChat responds to users' needs and established ways of life in China, and in so doing, it also reshapes Chinese lifestyles on its mobile interface.[36]

The fact that 'super-sticky' WeChat effectively 'glues' people to the app is at least in part because the design and strategy applied by the app developer follows a similar concern to our own study; both seek to understand how the smartphone can align itself more fully with customary Chinese forms of social relations. So it is not just that commerce creates a device that is subsequently transformed by users. We would expect companies in turn to learn from that usage and develop new possibilities which facilitate or try to commercialise those same patterns of usage.

Conclusion

What has been learned from our consideration of these apps, which at the start of this chapter we described as the 'heart' of the contemporary smartphone? There are three primary conclusions. The first is that they may well indicate the direction of travel for the smartphone itself. The second is that they have accrued this position of centrality through the breadth of their employment. The third is evidence that depth of employment is just as important as breadth.

First, the trajectory of travel. LINE, WeChat and WhatsApp could all be called super apps or platforms. Mostly, however, what these apps emulate is really the handset itself – the single device which serves every purpose. WeChat in particular, through its mini-programmes, has demonstrated a potential for superseding all apps through their incorporation within itself. As has been mentioned in several chapters, plenty of people also see smartphones primarily as a WhatsApp or a LINE device. The direction of travel looks similar to the hegemony established by Microsoft Windows or its Office Suite. There will be rivals, as in that case there is Apple. But, if the analogy holds, then these apps presage the increasing dominance of one particular interface. This would be in striking contrast to the extraordinary proliferations of apps and app developers that populate most smartphones today.

Obviously such a development is precisely the intention of the incredibly powerful corporations, such as Facebook and Tencent, that lie behind these apps. But the evidence is that they are succeeding at least as much because most people do not particularly want a culture of apps per se. As was argued in chapter 4, users are focused simply on ease of use; if that means working through a single hegemonic company, most seem prepared to acquiesce to this dominance by their behaviour, despite verbal protestations about the power of corporations. The culture of apps may well be only a temporary stage in the evolution of communication.

The trajectory is one of expanding breadth, which consists of two main elements. The first is the ability to incorporate as many different functions as possible. These apps have become hugely important in fields such as health and welfare, the main example of utility discussed in this chapter, although discussion of their usage within the sphere of religion serves to make just the same point. The second element of breadth was described in this chapter as scalable sociality – the ability of these apps to span from the smallest and most intimate communication to the most public. This discussion led from a focus upon the family through to the use of apps by groups, then by communities. In this case religion stood for a group of believers that extends beyond, a sort of mega-community.

Yet if we reflect upon what has created the quality described as this 'super-sticky' sense of dependency, it is perhaps the ability of these apps to become the vehicle of depth, as much as any sense of breadth. The core to all three apps is social communication. These apps have exploited or piggybacked on our most basic human dependency, which is upon each other. As long as the smartphone is seen as facilitating communication, its faults and foibles recede; it becomes first and foremost the medium for the relationships that make life meaningful. Much of the early part of this chapter was really about love. The term 'Perpetual Opportunism' introduced in chapter 5 here takes on an associated connotation: the sense of perpetual contact as evidence for care.[37] This potential for being always in contact in turn carries the desire for constant support, both emotional and financial. An example within this chapter were the rotating credit schemes (termed *tontine* in Cameroon), which were central to the ethnographic research in both Yaoundé and Lusozi in Uganda. In Yaoundé these schemes seamlessly combine social and financial support, but then extend to encompass much more. Starting with communication, each app has thereby evolved to become one of the 'technologies of life'.[38]

The main example discussed in the first section was the use of visual imagery. We are naturally conservative in our thinking, and if face to face comes first, we are bound to see it as more 'natural'. For anthropologists, however, there is no such thing as a natural or unmediated relationship. As the sociologist Goffman[39] showed, face-to-face conversation has always involved a thicket of cultural rules that dictate what is or is not appropriate. Face-to-face communication is often so hedged in by etiquette, performance, fear of embarrassment and other cultural frames that we find it hard to say anything at all. Visit a pub in London and much of the conversation seems to be either platitude or banter.[40] Viewed in this light, there is no reason at all why visual images cannot be the more 'natural' medium; they may at times feel warmer than direct speech. Visual images on social media can convey what people cannot utter face to face; alternatively, as in Japan, they become the complement to engaging face to face. In turn, people now have to develop new etiquettes and norms to accommodate their use of the visual within smartphones.[41] Across these fieldsites, the visual did not seem peripheral or superficial; it too has the capacity to speak from heart to heart.

This centrality of depth also builds on the conclusion of chapter 7. One might think that the way in which these apps have ingratiated themselves into the most intimate aspects of family communication would be quite sufficient to make this point: depth is the key to subsequent dependency upon these three apps. But in fact the section on the family argued for something closer to chapter 7's conclusions about co-evolution. These apps do not simply reflect family relationships and communication: here we argue that they fundamentally change what we understand by, and experience as, family. There is then a further trajectory to these chapter conclusions – one that leads naturally to the wider and more theoretical discussion provided by the next, and final, chapter.

Notes

1. See Ahmed 2004.
2. Steinberg 2020.
3. Bushey 2014.
4. Akimoto 2013.
5. Smith 2020.
6. Russell 2019.
7. Wang 2016, 28–37.
8. Cecilia 2014.
9. Graziani 2019.
10. Iqbal 2019.
11. Fiegerman 2013.
12. BBC News 2014.
13. Drozdiak 2016.
14. Iqbal 2019.
15. Duque 2020.
16. Emojis were invented in 1998 by an employee of the Japanese phone company NTT Docomo.
17. See Linecorp 2019.
18. Shu 2015.
19. For other ways in which the emoji can smooth communication in the very different context of business see Stark and Crawford 2015.
20. Shifman 2013, 78–81, 156–70.
21. Ahlin 2018a.
22. Kress 2003.
23. Danny Miller can personally attest to this. People that he had completely lost touch with from both school and university are now, on their retirement, suggesting meeting up again.
24. Based on Patrick Awondo's survey in the field. The sample for this was 65 people.
25. These are important in many other regions of the world. See Ardener 1964.
26. Instituto Nacional De Estadisticas (INE) 2019.
27. See Marwick and boyd 2010.
28. Aside from the 'last 3 days only' setting, one can also set the privacy level to 'the last 6 months only', meaning that contacts can only see their posts going back six months.
29. Pulse News KR 2019.
30. Kyodo News Agency 2019.
31. McDonald 2016, 169–70 and Wang 2016, 37–50.
32. Figures are taken from a survey of the use of mobile pay among 220 people (aged above 45). The survey was conducted in Shanghai (April 2018–June 2018) by Xinyuan Wang.
33. Sheng 2020.
34. Just to be clear, there is no reason to think this feature was influenced by our research. However, it was clearly influenced by an understanding and interpretation of Chinese kinship practices.
35. Chen et al. 2018.
36. Chen et al. 2018, 107.
37. See Singh's article in Prendergast and Garattini 2015. Here she explains that just having a dot that says someone is also on their phone at the same time can be a kind of minimal social contact for an elderly person. Singh 2015.
38. Cruz and Harindranath 2020.
39. Goffman 1971.
40. Fox 2014, 88–108.
41. Horst and Miller 2012, 28–30.

9
General and theoretical reflections

Fieldsites: **Bento** – São Paulo, Brazil. **Dar al-Hawa** – Al-Quds (East Jerusalem). **Dublin** – Ireland. **Lusozi** – Kampala, Uganda. **Kyoto and Kōchi** – Japan. **NoLo** – Milan, Italy. **Santiago** – Chile. **Shanghai** – China. **Yaoundé** – Cameroon.

Introduction

This volume has aimed to weave some more general and theoretical conclusions into every chapter. In several of these chapters the emphasis has been more on organisational aspects of the smartphone as a technology, including its structure, the apps it contains and its relationship to other devices. In this concluding chapter, however, we place greater emphasis on the consequences of smartphones for people. Because ultimately, as anthropologists, we are less interested in technology per se – the question of what a smartphone is – than in using studies of such devices to throw light on individuals, society and culture with the goal of furthering our understanding of humanity.

First, to reprise some of these earlier claims. Chapter 1 began by expounding the approach of 'smart from below' and noting that in most respects, despite being called a smartphone, this device bears little resemblance to the traditional phone. Nor is it dominated by the ambitions of S.M.A.R.T – the ability of a device to learn from its usage. Chapter 2 explored popular conversations about the smartphone, showing that, rather than providing evidence for smartphone use and its consequences these discourses often exploited the device as a means to engage in a variety of moral debates about contemporary society.

Chapter 3 examined the smartphone as a material object and situated it in a series of contexts. These included the notions of 'Screen Ecology', defined as the relationship between the smartphone and other screen-based devices, as well as 'Social Ecology', defined as the social relationships that are involved in the sharing of smartphones and apps. Both of these concepts are important in understanding the smartphone itself. This chapter also linked the smartphone to other networks, forming a kind of remote-control hub both of people and potentially the 'Internet of Things'.

Chapter 4 considered the way in which usage of the smartphone was task-orientated, and the importance of this for understanding the culture of apps and the way these are deployed. It introduced 'Scalable Solutionism'– the spectrum that ranges from an app with a single function ('there is an app for that') to the Swiss Army penknife aspirations of WeChat and LINE, which strive to be comprehensive and able to perform whatever task is required. But we also recognise that users might take an app with much broader potential as simply a single purpose device. The observation of 'Perpetual Opportunism' formed the basis of chapter 5, which drew on a number of genres of usage such as photography, transport, news and entertainment.

Given the diversity of both what the icons on a smartphone screen enable and the ways in which these are deployed, this book avoids using terms such as 'apps' or 'platforms' as its foundation. Instead, it starts from everyday usage, allowing a shift from a more technological focus to one that concentrates on the lives of smartphone users. This shift to an emphasis on people rather than the device itself is largely complete from the start of chapter 6, which examined the way in which the smartphone is crafted to reflect individuals, relationships and wider cultural values. Chapter 7 considered how, as a result of our capacity to transform it, the smartphone can closely align with social parameters – in this case, age. Chapter 8 argued that three apps/platforms should be considered as the heart of the smartphone because the most important apps are those most fully engaged with social relations, as in the expression of care and affection, the family and community. Chapters 6 to 8 also provide the main evidence for the conclusions now developed in this final chapter.

Our conclusions will begin with an attempt to re-direct our understanding of how people experience the smartphone based on the term the 'Transportal Home'. We then explore further some of the issues of intimacy and correspondence between people and their phones, introducing the concept 'Beyond Anthropomorphism'. There follows a summary of the way smartphones enter into – and sometimes transform – social relations under the heading of 'The Relational Smartphone'. The chapter's fourth section starts by considering the more general problem of contradiction and ambivalence. It argues that these reflections have surfaced with particular clarity as a result of the response to the Covid-19 pandemic, including debates around the fine line that exists between the smartphone as an instrument of care and the smartphone as an instrument of surveillance. The conclusion of this discussion leads back to the initial premise of 'smart from below'.

The Transportal Home

There are many precedents to the consideration of the internet and online worlds as some kind of home. A work by the sociologist Heike Mónika Greschke, for instance, has the title *Is There a Home in Cyberspace?*.[1] However, the concept introduced here under the title of the 'Transportal Home' extends far beyond any previous analogy or argument that references the online home. The starting point is an assertion that the smartphone is best understood not just as a device through which we communicate, but also as a place within which we now live. We are always 'at home' in our smartphone.[2] We have become human snails carrying our home in our pockets. The smartphone is perhaps the first object to challenge the house itself (and possibly also the workplace) in terms of the amount of time we dwell in it while awake. As a term, the Transportal Home comprises several elements. In addition to referencing the home, it acknowledges the smartphone as a portal from which we can shift from one zone to another. Finally, there is also an analogy with transport, as a vehicle for mobility.

Consider a common accusation made against the smartphone which was noted in chapter 2. Most people become annoyed when they are sitting with someone in a restaurant and their companion in effect disappears from their company, becoming instead absorbed in their smartphone. What has happened is that the individual has, in effect, gone home. They can use this portal to zone out from the place where they are sitting, to return to a home in which they can carry out many familiar activities, from finding entertainment to organising their schedule or messaging friends or relatives through text and visual media. Previously we entirely respected the right of somebody to take their leave and go back to their own private house. However, it is disturbing when someone who appears to be sitting next to us has, to all intents and purposes, abruptly retreated to some other place from which we are excluded without saying goodbye. They may remain in our physical company, but they have disengaged. We have become used to the idea of the internet as the 'Death of Distance',[3] but now the smartphone seems to implicate a parallel 'Death of Proximity'. Wherever a person appears to be, they can actually be back at their Transportal Home. The effect is radically to disrupt previous conventional notions of public and private, in turn leading to protests at this flagrant rupture of conventional etiquette.

The significance of the Transportal Home is as much to do with the growing fragility of the traditional sense of home as with the smartphone's capacity to compensate for that loss.[4] The world has grown more restless, with movement resulting from migration, work patterns, better transport and multiple other factors.[5] Our fieldsite in NoLo includes a high proportion of migrants from other parts of Italy as well as from abroad. These people have already seen the limitations of the traditional concept of home as a single physical location, which would separate them from much of their family and their sociocultural upbringing. For Sicilians living in Milan, the smartphone helps them to accept that Milan is the place where they reside, because they can simultaneously also remain in 'their land' (*mia terra*) of Sicily, the site of their memories and dreams.

During the latter half of the twentieth century in Japan, the migration from rural areas to the city has led to significant depopulation of the former. However, Laura's research documents a contemporary counter movement from the city to rural towns and villages which has grown in strength since the triple disasters of earthquake, tsunami and nuclear meltdown that occurred on 11 March 2011. The aftermath of these disasters saw diminished trust in state infrastructures[6] and the return to rural regions, motivated by a sense of alienation and rootlessness in the city. Many people in both the urban and rural Japanese fieldsites agreed that the smartphone is now the centre of their lives, connecting them not only to their family and friends, but also to several of their everyday activities.

In contrast to many migrants in the Italy or Chile research, our Japanese participants do not necessarily feel that this is a good thing. Many are ambivalent about spending so much time on their smartphones, while simultaneously saying that it has transformed their lives by allowing them to maintain close relationships with friends, children and grandchildren who often live far away. While they may have adopted the smartphones only recently, especially if they are older, they experience an ever-growing dependence upon it. This is due in part to its centrality to many of their daily activities, but also perhaps to an increasing sense of disconnection outside of the smartphone, caused in turn by depopulated neighbourhoods.

In chapter 1 we made a reference to an argument by Bogost that suggested that, thanks to the smartphone, we now live in a placeless world – a version of what Augé[7] regarded as an increasingly common experience of placelessness. But the concept of the Transportal Home turns this argument upside down. We find that we are by no means placeless. We can always know where we live and how the various components of the home may converge, as long as we are prepared to regard the smartphone as that stable location. What matters is the way the mobile phone is immobile, constantly in our presence.

This argument has a temporal dimension as well as a spatial one. In the Shanghai fieldsite it is common to find older people who have moved to the city to take care of their grandchildren. They may find it difficult to fit in with the new life there, having been uprooted from their previous social networks and from the social support back in their home town. They now cling on to something that which gives them an experience of a home in which they feel secure, and which they hope will be a home thereafter.

This deployment of the smartphone as an alternative home becomes even more important when we turn to the situation of young people in Europe. In places such as Milan or Dublin, a major source of anxiety is that, while the previous generation was able to afford ownership of their own home, at least when they wanted to start a family, this is no longer the case. The issue here is that there has been an expansion of life expectancy, combined with a failure to build sufficient additional homes to cope with the consequential lack of houses, and the selling off of state housing. As a result many young people have little prospect of being able to buy a property prior to raising a family; they are left to wonder whether they will ever be able to make this move. It is therefore not surprising that they too develop a commensurate attachment to the one home that they can afford: one that at least gives them a place where they can always be, with an address that is fixed and belongs to them. When young people are criticised for their attachment to their smartphone screens by older people, it would be reasonable for them to point out in turn that the people making these criticisms generally possess or rent a house of their own, while, on the other hand, they are condemned for paying attention to the one home they actually do possess, their smartphone.

When people do move out from their parental home, the instability of context makes it all the more important that the smartphone collects together the address where we live, our phone number and our email address. It makes life easier for everyone if we are always at 'home' in our smartphone, and therefore always contactable. With WhatsApp, there is typically a tick that acknowledges that we are at home and have received that communication. At other times the smartphone, as with a traditional home, may become a place where people feel relatively private – not just to think our own thoughts, but also to do things unobserved.

Homes in many places are often divided into rooms. We are likely to have a bedroom for sleeping, a kitchen for cooking and a living room for socialising or watching television. The whole works as a kind of organism with circulations of energy, routinised time and dedicated space. The Transportal Home has many similar qualities to the physical home, as this infographic shows (Fig. 9.1). In a similar way to such houses, it is

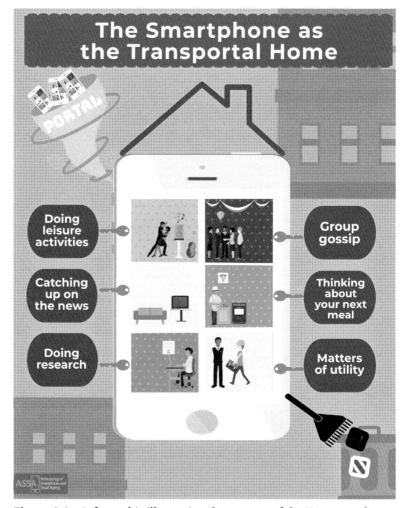

Figure 9.1 Infographic illustrating the concept of the Transportal Home. Created by Georgiana Murariu.

also divided into many domains, each used for different kinds of experience. We can enter through an icon into a place where we can play games or watch television. There is another icon that transports us to a site for research and study, another for listening to music and yet others for dealing with daily chores, such as shopping or banking.

As within larger homes, we can spend time in these various apps or spaces in relative quiet. This may be one of the reasons we are so concerned to protect the feeling of privacy traditionally associated with being inside the place where we live; many people worry that the smartphone is betraying them and breaching that privacy through commercial data collection. The smartphone can provide many private spaces, including those where we can have conversations through LINE/WeChat/WhatsApp that no one around us should be aware of or even store pornography. But we can also invite people to come around and visit, have a group gossip, bore others with our latest enthusiasm for a healthy diet and share fun stuff.

We may similarly think of the smartphone as a domestic space where cleaning and tidying may become a regular task. Susana in Santiago, for example, tells us with enthusiasm how she keeps her phone tidy (*mantener de teléfono*):

> Once a month I download the photos from the phone. I erase photos, I erase videos. Every day I clean it. E-ve-ry-day!

Another research participant, Ernestina, likes the Outlook app since she can easily erase emails. Having something called 'the recycle bin' helps to sustain the domestic analogy. When she explains, 'I like to keep the phone tidy', Alfonso cannot help but notice how tidy the living room of her apartment is. Smartphones can get very messy. We may live with this messiness or we may take action: delete all that surplus rubbish, re-order things that are in the wrong place, replace an app with a better one, agree to have the infrastructure updated. People in Shanghai now use a term meaning 'refusal-disposal-separation' (*duan-she-li*). To keep a home tidy, one needs to refuse to buy more stuff and give up what is not needed. Yet in this fieldwork Xinyuan found the phrase was most commonly used in reference to tidying up and sorting out smartphones. Guanghua, for example, reports that:

> Since 2016, I would do '*duan-she-li*' regularly on my WeChat. I usually feel good after deleting a couple of unnecessary contacts.

In Yaoundé the appropriate expression is *faire le ménage* – a term that means to do housework, but which has now been extended to apply to smartphones: 'cleaning my contacts' (*faire le ménage dans mes contacts*) and 'cleaning my screen' (*faire le ménage sur mon écran*). Even more tellingly, people use the phrase 'there are people that I will not allow to enter my home again' (*il y'a des gens que je ne veux plus laisser entrer chez moi*) in reference to their smartphone, rather than the home where they physically live and sleep.

We can also see many other qualities of the conventional home in the Transportal Home. As Mary Douglas argues,[8] the home is where we organise space over time; what gives the home solidity is not 'the stoutness of the enclosing walls but the complexity of coordination'. The home is where we distribute our attention to other people when making subtle negotiations about whom we should be in contact with and when. It is where we set our daily routines and know what the time is, and what we should be doing at that moment. It is also where we are often most subject to surveillance and control from others. The smartphone is where we edit our world of experience, what we choose to see and what we choose to keep. As the next section will show, this makes it a key site for crafting ourselves in the world.

To a surprising degree, the Transportal Home can also colonise the traditional home. It is often not something separate from that home, but an extension.[9] An example of this appears in the discussion of Screen Ecology in Shanghai found in chapter 3. Increasingly, people today live with a spectrum of screen sizes, which plays off mobility against readability. The smartphone is the most mobile element, but then we also have the tablet, laptop and smart television, all increasingly integrated; some people, when at home, may now read their emails on the smart television. The house considered in Shanghai is populated both with smartphones and their larger cousins, such that in each of its rooms the family can take advantage of the portal qualities of some screen or other. Here screens become in effect the windows of the house; through them the inhabitants can see a world that goes way beyond the view through the physical windows. Smartphones also change the relationship between people within the home. In Bento, for instance, couples can now live together more easily because they do not have to clash over who gets access to the entertainment system or who can speak to another person, thanks to their separate smartphones.

Screen Ecology is also behind one of the most significant changes in our relationship to homes – one that became clear in the lockdown response to the Covid-19 pandemic. This was the realisation of how much many forms not only of work, but also of socialising, shopping, entertainment and life in general can continue from the confines of the house, thanks to digital communication. The experience of lockdown exposed all the limitations of online-only sociality, including the lack of casual encounters at work and hugs from those one loves. But equally there seems to have been an almost universal appreciation of how much worse things would have been without online contact. In some regions the smartphone was an auxiliary to the primary player, which was the computer with Zoom and other communication platforms. But this project also included fieldsites where hardly anyone has a computer and where all of this communication depended upon the smartphone.

Not everything just described is positive – for example, there is the concern that the very object used to maintain or even create a sense of privacy is actually a spy within a regime of surveillance capitalism[10] that transmits data about one's most intimate world to strangers. This Transportal Home does not have the inviolability enjoyed by the traditional home as a place of privacy. In other ways, the smartphone may reduce the prior experience of home as a refuge. Employees may now be expected to remain in contact with their work, for instance, even after leaving the workplace. A child bullied by other pupils at school now finds little or no respite through coming back to her or his home.

Under lockdown during the Covid-19 pandemic, people soon realised that not only did they miss hugging people outside the household, but also that a Zoom party or celebration was a poor substitute for the real thing. Sexual and intimate encounters online are clearly not the same as offline and may be regarded as very imperfect substitutes. A smartphone does not come with a garden to cultivate or the facilities to bake one's own bread. There are probably hundreds of other examples that expose the paucity of a life that is only online. Trawling through the limitations and dangers of the smartphone should not, however, be based on a whitewash of the traditional home, with its own myriad problems ranging from family surveillance and claustrophobia to domestic abuse. All homes come with their contradictions.

The other major difference is with respect to the portal aspect of the Transportal Home. The physical house is immobile and limited in this capacity to interact with that wider world, since it lacks the crucial mobility of the body. By contrast, the Transportal Home[11] provides an easy and instant connection to another world; we can Skype to other countries, shop in a virtual mall or game in an alternative universe, all without leaving our smartphones.[12] The smartphone has its own relation to feelings and affection, from the tactile experience of the touchscreen and scrolling to the secure feeling that it lies snug in a pocket. We also experience the sense of loss, for example when the smartphone breaks or cannot be found. Suddenly it seems as though one is cut off from even the possibility of social encounters or temporarily locked out from part of one's own memory.

Several features of the Transportal Home may be particularly relevant to older people. As individuals become more immobile, the smartphone becomes even more important as a home they can portal *from*, as opposed to the one they are merely constrained to stay *within*. This is clearly demonstrated in the Japanese fieldsites, where older people find increasing value in the support of friends through the messaging app LINE as they age and become more restricted physically. As Komatsu san, a Kyoto woman in her sixties, explained:

> I think when we are elderly, it doesn't mean that we have friends right next to us. So the smartphone might feel more precious to us (as we age) because it allows us to stay sociable.

The same point applies to a further element of the Transportal Home, which is transport. Chapters 6 and 7 presented several cases of older people no longer allowed to drive. The smartphone then becomes the control hub for their relationship to transport, with apps including local bus timetables, Uber and maps. For people in Ireland, this usage extends to their wider mobility; smartphones are employed for many aspects of holidaying or maintaining their properties abroad, including learning a language or using Tripadvisor or Airbnb. But even more important is the sense of transport that comes with the smartphones piggybacking on the mobility of our bodies: they have a unique ability to be with us at all times, giving us access to the smartphone's Perpetual Opportunism. These qualities also have their downsides, of course, such as the fear felt by people in Bento who believe that answering a call or checking their smartphones in a public place may make them a target for crime.

Finally, it should be clear from reading this book that references to general and theoretical arguments are always nuanced, depending on the fieldsite. The Transportal Home, for example, will inevitably mean something different in each fieldsite, because of the different ways in which people understand the home and the varying ways that they use their smartphone. The short film featured here exemplifies this point with regards to Japan (Fig. 9.2).

To conclude, for anthropology the significance of considering the smartphone as the Transportal Home lies as much in acknowledging

Figure 9.2 Film: *The smartphone as a Transportal Home in Japan*.
Available at http://bit.ly/transportalhomeinjapan.

the increasingly problematic relationship that migrants, young people and others have to traditional homes as it does in appreciating the smartphone's capacity to compensate for this loss. The smartphone brings together many capabilities, ranging from the multiplicity of activities that take place in its separate 'rooms' to its ability to care at a distance or act as a remote-control hub, linked in turn to other systems such as transport. From the Death of Proximity to providing a security contact for the frail, the impact of the smartphone as a Transportal Home is profound. For all these reasons we welcome further research that explores the constant change we can observe in the relationship between the smartphone home and other places, for example the domestic home and the office.[13]

Beyond the anthropomorphic machine

For more than a century[14] humanity has been fascinated by the development of the robot and its potential to realise our imagination of the anthropomorphic machine – one that closely resembles a human being. This has been an exercise in alterity. The robot was traditionally portrayed as a machine that becomes increasingly similar to us while remaining other than human. We are therefore fascinated by the robot's potential to turn against us or to acquire rights as a 'sort of' human, a common topic in science fiction. Yet this fascination with robots may have led us to neglect a more profound and more advanced trajectory towards and beyond the anthropomorphic machine – one that proceeds through ever greater intimacy with people, rather than similarity or alterity. This development is most fully realised in the smartphone.

The popular conceptions of the robot as physically anthropomorphic reflects a more superficial encounter.[15] By contrast, a smartphone does not look one iota like a human being. It has, for example, no arms or legs. The smartphone has no need of limbs; it achieves its physical mobility through its placement in our trouser pockets or handbags. Anthropomorphism is advanced through processes such as complementarity (for example, taking over some of the work of memory) or prosthetics (for instance extending our ability to know what is around us). In addition, there is the smartphone's ability to transform the individual to whom it belongs. We may change our daily habits and practices once we have become owners of a smartphone.

The primary evidence for 'Beyond Anthropomorphism' comes from chapter 6. The example of Eleanor from Ireland revealed how completely a smartphone may also express the personality of its owner, in her case the desire to be seen as the consummate professional, or, in another example from the same fieldsite, the traditional masculinity of a gruff descendant of fishermen. The smartphone seems to become an extension of the person, rather like the daemon in the novels of Philip Pullman: something that is somewhat separate from us, but whose absence would increasingly feel like the loss of part of ourselves. The smartphone is a device that extends our capabilities not only to know things better than we can alone, but also to be more organised. This may be particularly the case for those with limited resources or barriers to using smartphones. One of this volume's authors, Laila from al-Quds, is completely blind; she has found both compensation and frustration in her new iPhone, but would not now be without it. There has probably never been a device that could achieve this level of intimacy or continuity with a human being (Fig. 9.3).

Creating a smartphone was treated in chapter 6 as an act of artisanal craftsmanship. In turn, the parallel between crafting a smartphone and crafting life became particularly clear within this research project, since most of the ethnographers were simultaneously studying retirement.[16] Retirement may involve taking up a whole series of new activities and behaviours, the continuation of some activities and the loss of others. Today retirement has often become a joint enterprise with the smartphone. Having more time for sailing in Dublin could also mean using seven different sailing apps. In Yaoundé, retirement might mean

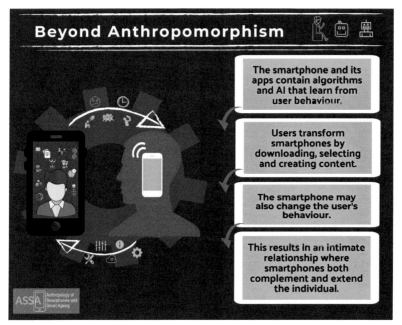

Figure 9.3 Infographic illustrating the concept of Beyond Anthropomorphism. Created by Georgiana Murariu.

having more time to devote to the church and downloading the Bible and other religious apps. In Japan, male research participants of retirement age had often built their entire identities around their work life and, as a result, some were ill-equipped to manage life outside of employment. These participants might then reject the smartphone, preferring to retain their old feature phone (*garakei*) as a physical connection to their previous social worlds. However, for others, both men and women, the smartphone gave them greater control over their part-time work in the years after retirement. Shift-organising apps meant that awkward phone calls with employers, in which it was hard to say 'no', could be avoided.

In Dar al-Hawa most of the older women have never worked in paid roles and consequently do not 'retire'. But many find smartphones play a major role when they become grandmothers. At this point increasingly their family interactions, such as allowing the grandchildren to watch YouTube on their smartphones, also changes their relationship to these devices. Most of this crafting is a gentle blending of old and new: the iconic photograph of the grandchild in the living room is complemented by the daily sharing of what the grandchild did that day. Is the smartphone part of us or is it something separate? As one participant from the

Irish fieldsite shyly admitted, now that she has a step-counter she also walks more because 'I like to impress my app'.

How does this crafting operate? At the point of purchase, the smartphone seems to be pure machine. All examples of a Samsung Galaxy Note in a shop are identical. Once purchased, however, it is up to the individual to decide which pre-installed apps they will not use or what they will do with built-in functions, such as the torch or Bluetooth. They then proceed to download new apps which they may order according to their importance, creating a foreground and periphery across several screens. In the background are rarely used or novelty apps, perhaps one that recognises birdsong when out for a walk by day and another that can identify star constellations when walking at night.

The next level involves tweaking those apps. Owners may change settings to receive only the news that they are interested in or to establish their home location as the basis for map use. The next layer of personalisation, the creation and selection of content, is usually the most significant. Many users of smartphones may have a reading app, but the app might contain anything from the *Harry Potter* series to the plays of Shakespeare. Their photographs may be selfies or PowerPoint slides. It is this choice of content across many of these apps that gives the best overall sense of that person: their tastes, values and interests. If they are from Dar al-Hawa, they might want their smartphone content to reflect the values of Islam.

Chapters 4 and 5 shifted attention from the culture of apps to the ethnography of everyday life, where the primary orientation is often towards tasks rather than apps. Many of the older people whom we met as they were learning to use smartphones were mainly concerned with sequences of actions that they either knew or were prepared to learn. The smartphone was reduced to its role in performing that task through this sequence of actions. Everything else that a smartphone might have been able to do had become irrelevant: it was simply ignored. As in other artisanal crafting, the work may involve subtraction as well as addition. The result of such crafting is an intimate relationship to smartphones that has become highly expressive of the individual, as described in chapter 6. The effect of this close relationship can be observed in events like those described in chapter 7, when Nura from Dar al-Hawa temporarily lost her connection to WhatsApp. Her momentary sense of shock and despondency came from an appreciation that WhatsApp was now integral to the infrastructure of her life. She felt pain as if the loss was visceral. This is what is meant by intimacy.

This approach to smartphone personalisation post-purchase, how-ever, by no means consigns the smartphone and app companies to irrele-vance. Firstly, they must be credited with creating all these capacities in the first place, as these make this subsequent crafting possible. It is the designers, often in conjunction with heavy user-testing, who create the conditions for all these subsequent transformations. Unlike most other goods that companies create and people consume, the smartphone has been implanted with its own capacity for extending intimacy towards the user. Apps are increasingly designed to learn through interaction. This is what is meant in technical terminology by 'smart' – the ability of the device to learn. Location is not just the capacity of GPS to tell us where we are; it may include predictions based on where we have been or shown an interest in going to. Voice-activated assistants become more accurate as devices learn our voice. In these ways machine-learning algorithms enable apps to learn from us, from our environment and from each other, as our social media piggybacks on our locational information or our search histories. This is in turn a two-way process through which the companies remain involved as collectors and processors of data; they also remain invested in the continued refinement of apps so that these can be more attuned to their role within the process, for purposes such as personalised advertising.

Yet too much emphasis on this 'smart' element of the smartphone is misleading. When inspecting a smartphone in detail, the way in which a person adapts their device to their particular needs has far greater impact on their experience than the way the device adapts to them through its algorithms. The corporations may have poured their resources into trying to get users hooked onto phone assistants, but many of our participants regarded the Samsung Galaxy 'assistant' Bixby as just an infuriating nuis-ance because you cannot get rid of it. Another phone assistant, Alexa, is rarely more than a voice-activated radio. Perhaps in the future the AI-driven chatbot may find a niche in relation to friendship, or even as our therapist,[17] but so far the impact is limited.

The result of this crafting is an intimate blend that goes beyond anthropomorphism. In his excellent book *Smarter than you Think*, Clive Thompson[18] documents the way in which human beings have become more intelligent by incorporating such devices. Not having to memorise 'facts', but instead memorising the way we use the smartphone to find facts is making us cleverer. Thompson shows why the best chess player is neither a person nor a computer, but the two working together. The best analogy for the rise of smartphones is the invention of printing and then of the book. This achievement consigned many previous memory functions, which had relied purely on cognitive memory, to the 'hard rectangle' of the book, which we can consider the precursor to the hard disk. Few people would have a problem with the suggestion that books made humanity cleverer because people were prepared to cede so much of memory to the written word.[19] Many of the research participants in this volume see the smartphone as a literal aide-memoire, keeping track of their notes. Others, such as Fernanda from Bento, use brain training apps, in part because of a huge anxiety about dementia. Almost all the portraits in this volume evoke a wider holism that transcends the distinction of person and machine when it comes to *how* we think.

Many examples in this volume show that what is being crafted is, equally, a person – a point made particularly clearly in the case of Eduardo from Bento, discussed in chapter 6. He was quite explicit about using his smartphone to help envisage what his new life would become now that he was a retiree. The outcome of such processes leads to stories such as that of Mario in NoLo. His smartphone has come to replicate many of his core interests, ranging from horticulture to the organisation of the local community allotments. Another example is Matis, originally from Lithuania and now living in Ireland. His smartphone is dominated, as he is himself, by a passion for car repair. Sometimes it is obvious that it is the smartphone crafting the person, as much as the other way around.

Crafting here may at first look like faking, for example making someone look much younger than they actually are, as implied by this meme from Yaoundé (Fig. 9.4). But perhaps it is the meme that gives us a false impression? Many of our research participants respond by telling us repeatedly that they feel much younger than their appearance suggests. As noted in chapter 6, they see their own outer appearance as fake and the smartphone as a device that can show them as they truly are. They were in any case already taking many measures to improve their external appearance. So which is the 'true' person: the unadorned body, the body as we have prepared it for public display through cosmetics and clothing or the photo crafted to present us as we imagine ourselves to be? It is not up to us as authors to provide an answer. Instead we respect that people in Yaoundé might give a very different response from people in Shanghai. There can be no absolute claim that the smartphone does, or does not, create fake images of inauthentic people. The evidence from this volume is rather that in some societies the results are regarded as a fake person, while in others this is seen as crafting a clearer expression of the real person.

The conclusion of chapter 8 quoted Goffman in repudiation of the idea that face to face is a 'natural' encounter. In China people can say through smartphones what cannot be said face to face. People refer to all kinds of emojis and stickers as '*biaoqing*', which literally means 'facial

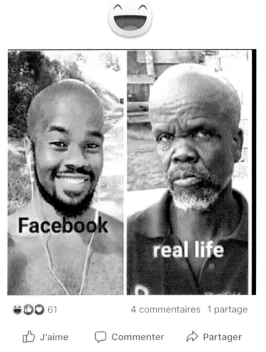

Figure 9.4 Meme circulating on social media in Yaoundé. Screengrab by Patrick Awondo.

expression' in Chinese. People frequently apply *biaoqing* to tackle situations that would otherwise be regarded as awkward or embarrassing. For example Mr Hong, a retired civil servant, has more than 100 *biaoqing* saved on his WeChat. This provides a repository of *biaoqing* for common emotions or social gestures, such as saying sorry. He skilfully appropriates different *biaoqing*, depending on specific contexts and recipients. For instance, Mr Hong might make a gentle apology to friends, when unable to attend a get-together, though a clasped hands, manga-style sticker with embedded text reading 'sorry, sorry'. Or he might tease his grandson with a head-dropping cat, also saying sorry. When Mr Hong had to turn down a friend's request for help, he replied with a crying cartoon figure saying 'Your Majesty, I really can't make it' to balance out the toughness of a rejection. As a result, he notes on one occasion:

> Sometimes I really wish I could also use these *biaoqing* in face-to-face conversation, that will make life much easier.

The problem is that in face-to-face conversation Mr Hong maintains his grave and dignified bearing, something that he has preserved for decades as a respectable male adult. When using WeChat, however, his repository of *biaoqing* can bring forth multiple layers of emotions and social skills which have no existence in the world of offline communication. In turn, we can expect that constraints and etiquette will develop around these forms of conversation over time.

A final point is that the concept of Beyond Anthropomorphism should not overly romanticise the ideal of being human. If the smartphone assumes our humanity, then equally it can express our inhumanity. Smartphones are easily used for stalking or bullying, or as instruments of power. These practices lie behind several of the accusations recorded in chapter 2, in which smartphones are viewed as standing for increasing inhumanity, responsible for making people more superficial or anti-social. As noted above, however, the Transportal Home does not need to imply domestic bliss. The home has often been a place of bullying, conflicts of power, abuse and inequality. Arguing that smartphones become like people does not necessarily imply they are good or moral. That depends rather on the person whom they have come to resemble.

The Relational Smartphone

One problem arising from both the concepts of the Transportal Home and Beyond Anthropomorphism is that they lend themselves to an emphasis upon the individual and their device. Yet one of the main contributions of this volume is that every chapter contains evidence opposing such a reduction to the individual. The smartphone has become central to all manner of relationships and groups – not just as the medium of their communication, but also as something that now partially constitutes those relationships or groups or, increasingly, networks.[20]

It was already apparent from the discussion of Social Ecology in chapter 3 that people cannot be considered as isolated individuals when it comes to the smartphone. In Lusozi many people share their phones with relatives and friends. In Shanghai some older people may not bother downloading an app; since everything is done with their partner, it makes no difference whose phone has that app. Perhaps still more important is the impact of Perpetual Opportunism, discussed in chapter 5. This concept has transformed relationships into a far more constant conversation. Now you do not have to wait until you see your friend or relative. At the moment something comes up that you feel they might want to know about, out comes LINE, WeChat or WhatsApp. Sharing is all day and every day. Elderly people who used to have a red button for emergencies may now feel secure through their smartphone, knowing that even after a fall they can still contact someone quickly and easily. These observations lead to the bigger question of whether smartphones are actually changing, rather than just facilitating, these relationships and groups.

The strongest argument for change would be the suggestion that smartphones may be reversing one of the fundamental trajectories of the family – a trajectory that has been emerging over centuries in some regions or over recent decades in others. This is in essence the shift from the extended family to the nuclear family, which is very evident in the change in traditions of house design, as houses in many regions were once built to accommodate much larger families.[21] In most affluent regions, modern apartments and other dwellings are almost always created with the assumption of occupancy by a nuclear family, or, increasingly, single occupancy, as found in Japan. Yet, as we go deeper into the more intimate uses of smartphones and apps such as WhatsApp, it seems that this long-term trend is in some measure being reversed. An example comes from the discussion of Screen Ecology, exemplified by a household in Shanghai. Here it was not simply the proliferation of screens that existed around the home, but rather that these were used to bring the extended family back into constant communication. The result was that, at least to a degree, their extended family once again shared the home with the older couple who continued to live there.

The clearest evidence of this change came from Bento. Here cousins, aunts, uncles and more distant relatives, often only ever seen at Christmas, weddings and funerals, have become a much more constant presence within the Transportal Home. The formal conversation suited for special occasions is replaced by the informality of everyday chat. Perhaps it is precisely because the extended family is not actually living in the same house that they are very welcome as residents of the Transportal Home. These relationships are close, but not too close. Within the smartphone there is always the possibility of communication, without the pressure of sharing the same physical space. If you need a chat and one person is not there, there is usually someone else to speak to in the Transportal Home. But you can also use the smartphone to tune out from the proximate sociality. On a crowded commuter train in Japan, each individual's smartphone in effect places them in a private yet social bubble, while being physically surrounded by strangers. Overall, the evidence from several fieldsites suggests that the smartphone provides a medium regarded as balancing periods when sociality has been either too intense or insufficient.

Smartphones may also be playing a role in what seems to be a more fluid relationship between kinship and friendship. In China smartphones may confirm an observation that this is the first time most people have developed any kind of significant relationship to strangers.[22] It is significant because it may *only* be possible to tell all your secrets and worst fears to someone who does not know who you are. In Ireland retirees use smartphones to organise their round of meetings with friends in cafés and to plan more regular, group-based meetings, such as book clubs. This expansion of family and friendship, in turn, leads to the increasing use of smartphones within the community. In Lusozi and Yaoundé WhatsApp might help to organise the rotating credit groups that provide financial support. In Dar al-Hawa sharing photographs meant that people unable to come on an outing were included to some degree. In NoLo Facebook was the key site used to arrange and advertise a mass convening in a public space one Saturday afternoon in May 2018. Here people stood side by side, holding hands to form a human chain (*catena umana*) 4 km in length. This gesture was made to celebrate the unity present in their community and to challenge the negative perception of the neighbourhood as an 'immigrant ghetto'. The reporter Zita Dazzi, describing the event for the *La Repubblica* online newspaper from its Milan site, wrote of a 'Human Chain against racism in Via Padova: "We are citizens, not illegal immigrants"'.

In NoLo Facebook had become the main site for local community news, community history and community photography. It is the place where people express willingness to offer their time or a helping hand to one another. One research participant in NoLo found it remarkable how if someone gets sick or needs help, for instance, and posts this to the group, there will be an average of between 20 and 30 responses each time from people willing to offer assistance, from buying some basic groceries to picking up medicines. More recently the NoLo community has also extended to Instagram, receiving attention and support from the Mayor of Milan.

The situation is similar in Ireland. Here Facebook is used to organise charity walks, to provide a posting site for local sports groups, to facilitate community development on a new housing estate and to form the public site for the most important community activities, such as the Tidy Towns competition. In Cuan, one of the Dublin fieldsites, 2,300 people (out of a population of 10,000) very quickly joined a new Facebook group called 'Cuan against Covid-19'. The Palestinians of Dar al-Hawa no longer have to worry about being out of earshot of any mosque, as their smartphone can now issue the call to prayer. Facebook has also accrued a significant role in providing religious information, such as events associated with breaking the fast during the holy month of Ramadan.

Having discussed the smartphone in relation to individuals, relationships, groups and communities, chapter 6 ended on a still more expansive note. The conclusion reflected upon how smartphones express cultural values that are naturally prominent in a study whose primary methodology is comparative ethnography. For example, many of the grandmothers Shireen knew in NoLo highlight the prominent place of 'Nonna' (grandmother) in contemporary family care and communication. As in so many other cases, the smartphone aligns with traditional ideas and idealisations of the Nonna, but also plays a role in expanding and changing social roles in line with contemporary circumstances. For example, smartphones facilitate the Nonna in playing an active role in childcare and providing practical support in urban family contexts such as in Milan.

For the Peruvians in Santiago and the people of Dar al-Hawa, smartphones are instruments of religion which may create very clear norms. Equally in Japan a common consensus often exists around what constitutes appropriate and inappropriate smartphone usage. For example, talking on the phone when travelling on public transport will attract stern glares from fellow passengers, while taking a photo of a restaurant meal before eating it is almost a prerequisite for the meal to be appreciated fully. The smartphone can also be part of the emergence of new societal norms. This was the case in Cameroon, where smartphones are becoming established as an instrument in the development of a new public sphere of political debate for a newly emergent middle class.

The Relational Smartphone creates new problems even as it resolves others. Our research revealed many examples of problematic intergenerational relations. Fundamental to chapter 7 was a digital divide between older people and younger ones in terms of the former's struggle to learn how to employ their smartphones properly. In Santiago, and across many of the other fieldsites, young people come across as shockingly impatient. They tend to pick up the smartphone and make a cursory attempt to show how something is done, but then refuse to repeat this on the next visit since the older person has 'already been shown' how the phone works. In Ireland younger people may claim that the smartphone should be 'intuitive', thereby implying that an older person who struggles to use them is rather stupid, although even a cursory inspection shows smartphones to be anything but intuitive. In Yaoundé and elsewhere this condescension reflects a much wider overturning of traditional roles of seniority and authority, as the wisdom that once reflected long-term experience is trumped by knowledge of the latest gadget.

Often older people only obtain their first phones as hand-me-downs from young people, reversing the traditional trajectory. They then find they need to delete offensive material, which appears again when the youngsters insist on borrowing back the phones. Young people can also be cruelly dismissive of the loss experienced by older people who have spent decades cultivating knowledge now made redundant by smartphones. Recall the woman who was brilliant at finding locations, having worked for years making deliveries for a florist. Who cares about those hard-won skills when there is now Google Maps?

This is a two-way process: social media has been around long enough that it may be changed in turn by these intergenerational tensions. One of the main reasons young people in many of our fieldsites avoid platforms such as Facebook is because they have become colonised by their parents' generation. When your mother, or even your grand-mother, is on Facebook, it is much better to express yourself away from their gaze, first on Instagram and then, when that also becomes colonised, by moving on to TikTok. Some of the most significant changes in social media and the usage of particular platforms have nothing to do with corporate control, nor the affordances of the platform. They simply reflect tensions in intergenerational relationships among users.

Contradiction and ambivalence

The first substantive chapter of this book was not about what people do with smartphones, but what they say about them. The overwhelming conclusion was that people are generally ambivalent. Older people talk incessantly about the harm smartphones cause young people, claiming they are 'addicted to their screens'; they have become anti-social, disconnected from the real world and superficial as a result. Regarding their own usage, older people in NoLo say they are wasting their time, or smartphones are causing 'too much confusion' (*troppo confusione*). In Japan people complain about the pressure to respond to messages imme-diately as an intrusion and an addition to the everyday social pressures to interact, always in the correct manner. In Ireland they often refer to the endless WhatsApp pings that now have to be acknowledged.

Yet at the same time people wax lyrical about the wonderful things they can do with specific apps within the smartphone. Couples no longer shout at each other in the car when the non-driver has lost their place in the street atlas they are holding; now they can swear at their GPS instead. Grandparents feel blessed that the grandchildren in Australia can be their incredibly cute selves on webcam. An older woman with dodgy knees does not have to wait in the rain, since an app tells her when the next bus will arrive. Everywhere the smartphone is simultaneously a blessing and a curse. Older Chileans in Santiago often say something highly negative and positive about smartphones in the very same sentence. It is constantly 'this…', but also 'that…'. If people, wherever they live, contradict themselves in the same sentence and what they say has little bearing on what they do, one might assume they are either hypocritical or ignorant. However, the evidence throughout this book suggests that ambivalence may be the only reasonable response to a phenomenon that is itself a mass of contradictions. Every chapter in this book has provided evidence for the simultaneous good and harm that are both the consequences of smartphones.

After our fieldwork was completed, the team discussed general findings from the project. One of these was given the title 'the fine line between care and surveillance'. We could not have predicted it, but this was shortly before this finding became a critical component in the response to the Covid-19 pandemic. At the time of writing, the smartphone has become an even more important element of a moral discourse of the kind described in chapter 2. This is because, writing in 2020 before any vaccine, the main candidate for suppressing the Covid-19 virus in many regions, and especially in East Asia, is an intensive use of tracking individual movements, based on a combination of smartphone data and interviews. This development has revealed to everyone the potential of smartphones, which, with their unprecedented intimacy, can also be vehicles of personal surveillance – Big Brother is exposed in our handbags and pockets. Yet at the same time effective use of track and trace is seen as the main reason for relative success in limiting the consequences of this virus in some regions. The smartphone thus appears to be both a potential saviour and the harbinger of dystopia.

Surveillance

There are three primary considerations within these developments. The first is to examine the issue of surveillance and the second to examine the nature of this care. Most important, however, is the third: to consider the implications of these events for the balance between care and surveillance. Surveillance has a dual presence in this book. On the one hand, it is a core topic within the explicit discourse of participants. When he taught smartphone use to older Chileans in Santiago, Alfonso could observe people simply refusing to use GPS because they regarded a device that recorded everywhere they had been as an unbearable intrusion. Similarly, these wary users picked up on targeted advertising on the internet as evidence that Google knew too much about them. Surveillance by smartphone is no secret. One of the most hyped features of the device is its supposed ability to learn from its user, such that it can pre-empt their enquiries and actions. Every time the smartphone succeeds in this mission, it simultaneously confronts its user with direct evidence of how much it now knows about them.

This surveillance is evident – but it is the tip of an iceberg of much wider surveillance that lies beneath dark waters. The same personal data about the user from which the smartphone is trying to learn may also be channelled back to corporations to become part of a vast aggregate field of Big Data. This, in turn, is the fuel that drives artificial intelligence (AI) – a process that soon spirals off into worlds we cannot see and of which we often have little understanding. This may be the most important externality of all.[23]

There are two main versions of these fears. The first was exemplified by a book, *The Age of Surveillance Capitalism* by Shoshana Zuboff.[24] She argued that the core driving force of capitalism, the search for profit, has led to an extraordinary ability to extract information about us and the relentless development of strategies that seek to employ that data in controlling our lives. If ever we find time to read the 'terms and conditions' we are required to sign in order to use a new smartphone app, we are shocked. They seem to require access to all sorts of irrelevant data held within our smartphone. Why should an entertainment app need to know our location or gain access to our social media? Companies compete to extract every minuscule detail about us that they can. Personal data is said to be the new oil.

If this feels intrusive or even sinister, that is because it is. As Zuboff puts it, companies such as Google now assume the right to 'claim all human experience as free raw material for translation into behavioural data'.[25] Tech companies spend huge sums on lobbying to prevent any curtailment either of this data extraction per se or its subsequent use in the cause of profitability. We have shifted from being mass consumers to being massively consumed, subject to continual experiment and analysis to further the corporate penetration of our lives. From this perspective the smartphone again feels like a spy in our pockets, ideally located to survey everything we do and say and are.

The second version of this critique is directed at the state rather than commercial surveillance. This was the background for the experience of people in Dar al-Hawa, who have lived most of their lives with concrete evidence of the surveillance powers of the Israeli state. For most other people, the initial revelation of the extent of state surveillance came through the whistleblowing of Edward Snowden. He used his position as a state employee to reveal the extent to which the United States could be equally rapacious in its hoovering up of private data. Snowden's revelations were then followed by the Cambridge Analytica scandal, which indicated that data about individuals might be used to alter the outcome of democratic elections through targeting messages.

Both of these critical observations seem to be supported by a third exemplar of the contemporary threat of surveillance. A recent review by John Lanchester in *The London Review of Books* was based on two new publications, *The Great Firewall of China* by James Griffiths and *We Have Been Harmonised* by Kai Strittmatter.[26] These books argue that all such fears have already been realised within contemporary China. Strittmatter suggests that digital surveillance has become the most effective instrument we have ever known for creating a totalitarian state. While the content of WhatsApp remains encrypted, every detail and private content of WeChat is available to the Chinese state. There are no restrictions on the state's use of technologies such as facial recognition. The state openly proclaims these powers, so that citizens are well aware of the degree to which they can be observed. Once again, these surveillance capacities of smartphones were made still more evident during the Covid-19 pandemic, when the Chinese state added features to people's smartphones ensuring their compliance with lockdown. In Israel surveillance systems that would have been familiar to the people of Dar al-Hawa were now extended to the entire population as a form of health control.[27]

Care

If surveillance seems to be the curse of smartphones, then care appears at first to be their blessing. There has been an explosion of propaganda in China based on the evidence that, although the Covid-19 pandemic started there and other countries had more warning, China ended up with 3 deaths per million while most European countries experienced a mortality rate more than one hundred times that figure. The Chinese state, which used the smartphone as a key instrument in the suppression of the virus, used their success as evidence that it cares more about its citizens than Western democracies. The fact that the equivalent figure in democratic Taiwan was 0.3 deaths per million,[28] one-tenth of the number in mainland China, was, of course, completely ignored.

The previous chapters provide multiple instances of the use of smartphones as instruments of care. These included the way that people in Lusozi help to care for older relatives who remain living in their home villages, often through sending mobile money. As noted above, in Shanghai visual elements within smartphone communication can break down traditions of distance and formality in family relationships between the generations. Conversely, a research participant in Kyoto preferred not to send cute stickers to her daughter because she might think them childish, yet she would happily send them to her own friends.

There has been considerable prior research on the use of digital technologies to facilitate care at a distance, whether with reference to systems within a state[29] or to the increasing need for care across continents, for example migrant diasporas.[30]

Our ethnographic evidence in this volume suggests that we have now moved from care at a distance to 'Care Transcending Distance' (Fig. 9.5). Sometimes in Bento Marília simply did not know which grandchild a research participant was referring to: the one who lived

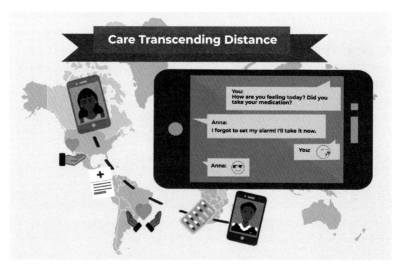

Figure 9.5 Illustration of the concept of Care Transcending Distance. Created by Georgiana Murariu.

in Bento or the one in New York. The way in which she communicated with her grandchildren through WhatsApp was just the same. Indeed, in one case a woman living alone in São Paulo, whose daughter lives in France, noted that 'My friends said she was abandoning me, but she is much closer than most of my friends' daughters who live in São Paulo'.

Care and surveillance are clearly two sides of the same coin. In Dublin many of the research participants had been involved in caring for parents, often in their nineties, which meant there was a high incidence of dementia. While those parents remained in their own homes, surveillance came to dominate many of their care practices. Across many fieldsites people quickly developed WhatsApp groups to support the frail elderly; one of these groups' prime objectives was to share the burden of monitoring the vulnerable. In Yaoundé retired people are closely monitored by their children 'to keep a better eye on them' (*avoir un oeil sur eux*), either by having them live with them or through very active family WhatsApp groups in which one or two children often play a sentinel role.

There is then no simple demarcation between care as benign and surveillance as malign. Surveillance may often be evidence of real and constant concern – including, during the Covid-19 pandemic, appropriate care by the welfare state. Equally there is a growing literature in anthropology regarding the dark side of care, especially with regard to migration issues.[31] Sometimes the way in which people expressed care included surveillance of the person they paid to be the carer. Working under surveillance was often the experience of rural migrants in China who had come to Shanghai to look after elderly people. Their employers had picked up on the potential of smartphones to spy on their paid carers. In Japan, being skilled in social surveillance is understood as important for maintaining good social relations. It can also be exhausting and, if incorrectly practised, can lead to social ostracism. Understanding the practice of care, including through social surveillance in the digital age, is critical in the context of Japan's turn towards technology to cope with an ageing population and a decreasing health and care workforce.

In all of these cases, smartphones can also become tools to constrain or negotiate autonomy. Much of the concern in Japan focuses on finding ways to respect the autonomy and dignity of older people under these conditions of constant surveillance.[32] Some older people are aware that refusing to have a smartphone can become a means of ensuring that digital contact does not take the place of physical contact, or make the familiar landline phone redundant. Yet in several fieldsites smartphone-based surveillance is what reassures adult children that it is still all right for their elderly parents to retain some autonomy by remaining in their own home. In such instances, constant surveillance secures the parents' continued autonomy.

Ideology, privacy and the fine line between care and surveillance

This discussion of surveillance, care and the relationship between the two sets the scene for the subsequent developments that have resulted from the Covid-19 pandemic. For the critics, the resulting emphasis upon smartphones as care solutions lets not only surveillance, but also the smartphone, off the hook. Evgeny Morozov[33] published an article in *The Guardian* under the title 'The tech "solutions" for coronavirus take the surveillance state to the next level'. He argued that 'the good cop in this drama is the ideology of solutionism, which has transcended its origins in Silicon Valley and now shapes the thinking of our ruling elites'. The key to solutionism is the smartphone, exemplified by China's colour-coded smartphone health-rating programme. The next day *The Economist* published an article under the title of 'A global microscope made of phones'[34] to discuss the collaboration between Apple and Google to create a contact-tracing app, as well as the role of governments in such ventures.

Yet subsequent events seem more like a vindication of the arguments of this volume. For the next stage was not straightforward adoption of these smartphone potentials, but rather incredible heterogeneity in the response to technological possibility. The reason is that deployment of track and trace foregrounds the question of balance between care and surveillance. This turns the deployment of technology into a moral issue – which in turn exposes the underlying ideologies that created the heterogeneity in regional responses. We thus also need to factor such underlying ideologies into any exploration of the 'fine line' concept.

The balance between care and surveillance is an ancient dilemma. A definition of God in most religions would be the omniscient being that sees everything and cares for everyone. It is the crux of parenting. Governments too have always traded knowledge for care. The use of smartphones in response to the Covid-19 pandemic in China had its precedents in what is called the Social Credit System. Under this system, any behaviour the state regards as anti-social – or anti-state – can result in citizens losing the ability to book flights or high-speed train journeys, as well as making them the subject of many other curtailments.

What has been lacking is a sense of how the Chinese population feel about this system, which consequently became a component of Xinyuan's research. She found three reasons to explain why, in most cases, the Social Credit System has been quite a popular development. The first reason lay in the overall shift from an agrarian economy, where people generally knew at least the reputation of pretty much everyone they dealt with on a personal level, and trust was thus based on social relations (*guanxi*). Agrarian life has been replaced by an urban society where people know almost nothing about those with whom they are forced to interact, and they believe this has led to a massive rise in fraud and deception. For many, the end (combating fraud) justifies the means of state surveillance.[35] The second reason was that people believed the Social Credit System was merely catching up with the idea of a poor credit rating in the West. The third factor was a belief that the Social Credit System tallied with traditional cosmology, based on the Daoist idea that the sky (*tian*) sees all our behaviour and a good fate depends upon good deeds.[36] There is thus a complex underlying ideology and history that must be taken into account to understand the situation in China. The Communist Party's claims of being a paternalistic carer, a role that is expressed through surveillance, are reminiscent of centuries of Imperial rule.

Ideology is just as important for understanding the response in regions such as Europe and the US. Here the principal objection to surveillance is the Western premise of privacy. The obsession with individual privacy may well appear equally extreme to a visitor from East Asia. In his book *The Comfort of People*, Danny argued that the single biggest cause of harm to hospice patients, other than their illness, was the insistence upon strict confidentiality. The result was that different members of the care teams for these dying patients failed to keep each other informed. When members of our team give talks about how we hope to use our research to improve people's welfare, the most common question asked is not how it might benefit the welfare of that population, but whether our proposals might intrude on privacy.

The reason these concerns with privacy can be termed ideological is because they are regarded as axiomatic. For most people in Europe and the US, assumptions about privacy are felt to be simply 'natural'. As ideology, they extend the fundamental ideology of historical liberalism. This belief in the prioritisation of the individual as the foundation of ethics led to the ideal of the individual as the source of basic human rights. This system of 'neo' liberalism includes the core belief that individuals have an intrinsic human right to control all information about themselves.[37] The stance on privacy derived from 'neo' liberalism is very different from, for example, socialist ideologies which assumed that if the state can enhance social welfare by collecting information about individuals, this automatically supersedes individual rights.

The US and Europe are perhaps the staunchest proponents of this 'neo' liberal privacy, but they take very different forms. In Europe, privacy is protected through bureaucratic regulation such as General Data Protection Regulation (GDPR). By contrast, the way in which privacy rights have developed in the US seems more in keeping with the neo-liberalism of the political economy, where such rights are part of an ideology associated with the freedoms of individuals and choice that is also used to legitimate contemporary capitalism. These rights are then set directly against forms of state intervention, including 'snooping', rather than becoming an instrument of state bureaucracy.[38]

Once we acknowledge the importance of 'taken for granted' ideologies, then the heterogeneity of response to the Covid-19 tracking apps makes far more sense. It was surely predictable that the US saw demonstrations protesting against the government's curtailment of individual liberty, and that Republican voters might be least expected to comply with smartphone-based surveillance. Nor is it surprising that people in countries such as South Korea, with its prior tradition of individuals being shamed on social media for any kind of inappropriate behaviour,[39] proved most compliant with a system of public surveillance at the expense of individual privacy. If this surveillance exposed people's extra-marital affairs,[40] it was justified by the greater good of supressing the virus. South Korea showed just how central the smartphone could become, outside of any authoritarian regime. Not only an instrument of contract tracing, the device was also used for remote sensing, for example of the pulse of oxygen saturation. In addition, smartphones became the means by which the state kept people constantly informed of the presence of contagion in each specific locality, through direct texting.

Since the Covid-19 pandemic occurred quite soon after the end of our fieldwork, the team was still in contact with our research participants. We could therefore watch how events unfolded in ways that reflected on our characterisation of these fieldsites as we were writing this volume. For example, in Japan the state tried to distance itself from any official 'track and trace' technology by using social media as a tool for monitoring. But people tended to be cautious about completing the surveys that were sent by the government through LINE honestly, as shown in the illustration below (Fig. 9.6). When the government offered some financial compensation (of around £1,000) to each citizen for losses incurred as a result of the virus, they asked people to sign up online using their

In Japan, the Coronavirus has brought into focus the way that care is often reliant on various forms of monitoring. Many Japanese people are not comfortable with the idea of surveillance by the government, hence the decision to use LINE as an informal method to track the virus instead of a dedicated app like in neighbouring South Korea or China.

Figure 9.6 Illustration of responses to issues of care and surveillance by Laura Haapio-Kirk, based on interviews with research participants.

'my number' identification (マイナンバー), which digitally connects all their social security records with other forms of data. However, many people expressed concern that this system was in breach of their privacy, allowing the government far too much insight into private information such as their bank balance and health records. As it turned out, signing up online for the compensation proved difficult because of an inefficiently designed website, resulting in people around the country queuing up for hours at town halls to reset their 'my number' passwords in person.[41] For many people, this episode epitomised wider concerns with ineffective digital infrastructures in Japan.

In Ireland, by contrast, many of our research participants had themselves worked for the state in health, education and the civil service. They saw it as their duty as a citizen to become the instruments of surveillance in support of the state. In this meme circulated by a Dublin participant (Fig. 9.7), they point fun at the sheer extent of their adoption of surveillance during a time when people were only supposed to be in the presence of others from the same household.

This culminated in the release during July 2020 of the state's smartphone app designed for 'track and trace'. Ireland seems to have had one of the most positive responses of any country where there was no enforcement behind downloading the app. Within two days one million downloads had taken place, around one-quarter of the eligible population.[42]

In conclusion, the underlying factor is once again the huge power and prevalence of smartphones, whether in respect to surveillance or care. But this is only ever the start of a process. Equally important are the ways in which underlying cultural values determine how such technologies relate to local normative ideals. This is what *The Global Smartphone*

Figure 9.7 Meme circulating in Dublin. Screenshot by Daniel Miller.

means – the smartphone is a device for making global diversity manifest rather than suppressing it.

Two further implications will be mentioned briefly, one regarding the policy implications of this volume and the second regarding future research. First, with respect to policy, we can clearly only learn from states where there was at least the possibility of reflecting cultural values and popular sentiment. There is little to be gleaned from authoritarian regimes that imposed solutions; we can only voice our support for the subsequently oppressed populations. But while both South Korea and Sweden are relatively consensus-led populations, they had very different responses to the pandemic. We have just illustrated a similar contrast between Japan and Ireland.

What is significant is that our observations about the fine line between care and surveillance derived from our research prior to the pandemic. This suggests that ordinary people are already highly experienced in dealing with the relevant issue. Balancing care and surveillance is central to the dilemma of looking after older people's health while respecting their autonomy and dignity. It is at least as important as the way in which parents negotiate their relationship to teenage children.[43] Surely every parent sees the monitoring of their child's use of their smartphone as an example of care, while every teenager views exactly the same behaviour as an example of surveillance. This whole volume is replete with examples of the consequent ambivalence people feel about smartphones.

In short, we are all relevant experts, and this is the expertise that ought to be drawn upon when faced with public crises such as the Covid-19 pandemic. Most of the core dilemmas the virus posed to governments were intractable and contradictory moral choices. They were about the rights of older people as against younger ones, of education as against health, or of individuals as against the collective. These dilemmas are then subject to cultural relativism, as each population will require its own internal negotiation over their 'least bad' option and the appropriate role for smartphones. The evidence of this project is that populations not only have the right to be consulted, but that they also have the qualifications – as a result of the considerable experience most people have accumulated in trying to balance care and surveillance in their everyday lives. This is the time to insist that populations are allowed, through consulting, to have a say in deciding the appropriate balance.

The second point is the importance of continual research. When the Covid-19 pandemic developed, we helped to initiate[44] a kind of 'citizen science' response by setting up a website called *anthrocovid.com*. This provided a space for anthropologists to post about events as they were developing on the ground, using their access to local populations usually developed through prior fieldwork. Many similar ventures also took place, especially in the field of medical anthropology.[45] These anthropological investigations are complemented by other disciplines researching different perspectives and evidence. A brief survey of some of this literature was given in chapter 1. The speed of response is illustrated by Deborah Lupton, who quickly published a survey of relevant literature, as also by Evgeny Morozov's compilation of sources.[46] The story of smartphones has only just begun. We hope our volume has shown how important it is to develop further research into their consequences for humanity.[47] We hope also that this volume will play its role in stimulating further research across the disciplines.

Conclusion: 'smart from below'

These observations regarding the Covid-19 pandemic – the fine line between care and surveillance, the expertise of ordinary citizens and the need for further collaborative research – are all points which return this volume to the ethos with which it began: the justification for the project's premise of 'smart from below'. We all need to observe, listen to and learn from the way in which people use smartphones in everyday life. So much of the initial response to Covid-19 took the form of top-down technological solutionism. Governments selected the smartphone apps for contact tracing and then informed populations of how they should comply. There is plenty of science behind the development of apps and making them effective, but the balance between care and surveillance is a political decision. Politics should imply consultation, where this is appropriate. The last section has argued that we have always negotiated this balance in our lives; for populations to embrace these technologies effectively, governments would do far better to respect that diffused experience.

The entire book is underpinned by a similar ethos. If the results of this study look different from mainstream discussion of smartphones, it is not because our research team are trying to make smartphones look either 'good' or 'bad'. It is because our research is based on empathetic respect for the resourcefulness and crafting of ordinary people. Had the Covid-19 pandemic never arisen, there was just as strong a case to be made for 'smart from below' perspectives, based on our studies of mHealth. Our research found that it was the creative use of ubiquitous apps such as WhatsApp[48] that makes a real impression on health management, as yet far more consequential than top-down bespoke health apps. A third example was the finding that it is generally not the S.M.A.R.T. machine learning from usage that makes smartphones 'smart', but rather the subsequent adaptations and creation of content by users.

For this research it has also been important not to rely simply on questionnaires and surveys about the relationship between an individual person and their smartphone. The holistic approach of our ethnographies has focused just as much on the impact of couples, groups, networks and wider cultural values. This also accounts for the style in which this volume has been written. The use of portraits throughout reflects a humanistic concern not to traduce what is special about individuals. The analytical generalisations around concepts such as the Transportal Home and Beyond Anthropomorphism, which form the substance of this final chapter, may apply to several of the populations we studied, but rarely all. Theory and conceptualisation in turn can lead to highly abstract and overgeneralised characterisation, detached from the messy and diverse worlds that emerge from ethnographic observation. Countering this trend is another reason for constantly re-immersing general conclusions in stories about people. A generalisation is made about men in Dublin, but then there is Eamon; a generalisation is made about women in Dar al-Hawa, but then there is Nura.

Nor should our approach gainsay the contribution of the smartphone designers and developers and corporations – something that we have tried to acknowledge from time to time, but did not directly research. We also recognise the influence of our research methodology, resulting in a book focused on what was apparent from ethnography. Designers and developers were unlikely to be present in these ethnographic fieldsites, a point that is also true of many other significant contributions which can only be acknowledged as externalities. These are reasons why we suggest this book should be used in tandem with research by other disciplines engaged in the study of other contexts relevant to understanding smartphones, in ways that were not available to our ethnographic approach. Some of these are discussed in the literature review of the first chapter.

Part of our 'smart from below' ethos has been to give empathetic respect to populations that are less likely to be represented in studies of smartphones, through concentrating on a demographic that is neither young nor elderly, but is here mostly just termed 'older'. The logic for selecting the fieldsites was not that we considered them in any way special. It was rather simply to represent the diversity of the world that justifies calling this volume *The Global Smartphone*.

One of the joys of studying the smartphone is the access it gives to the myriad ways in which *people* are smart. They are not necessarily good. The book is replete with examples of how the smartphone reflects our inhumanity as well as our humanity. But foregrounding the way in which ordinary people have contributed their creativity to making what smartphones turn out to be might just help humanity to recover a little of its self-esteem – especially in the face of intimidating new capacities and new technologies, and the overwhelmingly powerful corporations and states behind them. It is ordinary people who have turned the capacity of smartphones from being just 'clever-smart' into the capacity to be 'sensitive-smart'. It is thanks to them that every smartphone is unique. The smartphone's potential to go beyond anthropomorphism may have been created by corporations, but any subsequent humanity or inhumanity discernible within the smartphone – the primary evidence we have presented here – was created by the people you have met in this volume.

Notes

1. Greschke 2012. See also Morley 2000.
2. It should be obvious, given that this is a book about diversity, that this and similar statements in this chapter come with the caveat that there will be many exceptions.
3. Cairncross 1997.
4. Jackson 1995.
5. These work at different temporal rates and different scales at local and global levels. See Eriksen 2016.
6. Alison 2014.
7. Augé 2008.
8. Douglas 1991, 306.
9. Several papers in de Souza e Silva 2014 argue that the phone is often used to link us with places in new ways.
10. Zuboff 2019, 6.
11. For a precedent see Boullier 2002. Portals as such tend to be present more in science fiction or children's books, such as *The Lion, the Witch and the Wardrobe* by C. S. Lewis or Phillip Pullman's *The Subtle Knife*, the portkey in the *Harry Potter* series or the *doraemon* in Japanese cartoons. It is an example of how a technology has allowed humanity to 'attain' something that was previously more of a fantasy and ideal. This is discussed more abstractly as a Theory of Attainment in the book *Webcam*. See Miller and Sinanan 2014.
12. These arguments build on previous studies as well as this one, where we can see people using these new media to try and create a sense of co-presence while living in different countries. See for example Madianou and Miller 2012 and Miller and Sinanan 2014.
13. For example Hjorth et al. 2021.
14. Russell 2017.
15. This is becoming less true today as we start to think of robots more as automated factory workers or within the realm of medical surgery. Neither of these instances imply that kind of anthropomorphism. See for example Hockstein et al. 2007.
16. An exception would be Lusozi, where most people were still working.
17. As predicted in Sherry Turkle 1984.
18. Thompson 2013.
19. Ong 1982.

20. Rainie and Wellman 2014.
21. Waterson 2014.
22. McDonald 2016.
23. A variety of examples of digital surveillance are discussed in Lupton 2015. There are many relevant papers in the journal *Surveillance and Society*.
24. Zuboff 2019.
25. Zuboff 2019, 14.
26. Lanchester 2019.
27. Bateman 2020.
28. Worldometers.info 2020. Data compiled from official sources around the world. Last accessed 1 October 2020.
29. Pols 2012 and Oudshoorn 2011.
30. Wilding and Baldassar 2018. See also Lutz 2018. See also Baldassar et al. 2017 and Baldassar et al. 2016.
31. For example Ticktin 2011, also the special issue of *Ethnos* dedicated to 'Care in Asia'. See Johnson and Lindquist 2020. For a sense of the overall entanglements around care and surveillance see also Schwennesen 2019.
32. Kavedžija 2019.
33. Morozov 2020.
34. *The Economist*, 16 April 2020.
35. Wang 2019a.
36. Wang 2019b.
37. See Rossler 2005, 22: 'The theory of privacy with which we will be dealing in what follows is grounded within a particular political and philosophical framework. Namely liberalism.' See De Bruin 2010.
38. It should be noted that this particular argument about 'neo'-liberal privacy is primarily a stance held by Danny, based in part on his research: for example, the harm caused to hospice patients by the emphasis upon confidentiality (Miller 2017b, 41–50) and the problems faced by the voluntary sector caused by the European GDPR regulations. There would not be consensus on such issues among the authors, several of whom would put far more emphasis on the dangers of surveillance rather than the dangers of privacy.
39. Such as the 2005 Dog Poop Girl incident. See Henig 2005.
40. BBC News 2020, 5 March 2020.
41. Asahi Digital 2020.
42. McGrath 2020.
43. A common topic in the ongoing blogposts of 'Parenting for a Digital Future' run by Professor Sonia Livingstone, Dr Alicia Blum-Ross, Kate Gilchrist and Paige Mustain. Last accessed 1 October 2020. This can be read here: https://blogs.lse.ac.uk/parenting4digitalfuture.
44. Alongside Haidy Geismar and Hannah Knox of the Centre for Digital Anthropology at UCL. The website can be accessed at https://anthrocovid.com/.
45. For example our colleagues at UCL in Medical Anthropology, who blog here: https://www.ucl.ac.uk/anthropology/study/graduate-taught/biosocial-medical-anthropology-msc/medical-anthropology-blog-posts as well as other collaborative blogs see e.g. Somatosphere.net 2020.
46. See the coronavirus readings from Morozov's 'The Syllabus' at https://the-syllabus.com/coronavirus-readings/. See also Lupton 2020.
47. The relationship between care and surveillance, particularly in the context of families and households, has also been recently discussed in depth in Hjorth et al. 2020. One of their main examples of what they term 'friendly surveillance' includes evidence based on their own fieldwork in Shanghai, which they too relate to the development of care at a distance. See Hjorth et al. 2020, 65–73.
48. Duque 2020.

Appendix: methodology and content

The context

The first chapter of this volume began with an acknowledgement that the smartphone bears little relation to earlier mobile phones, having expanded its capabilities to encompass an extraordinary range of uses. By the end of this book, it should be hard to imagine any significant area of life which does not now involve smartphones, at least potentially. Fortunately ethnography, the primary method of anthropology, is tailored exactly to match the problem this poses for researching the smartphone. Within this project ethnography is based on 'holistic contextualisation', which means that everything we study is in turn the context for everything else we study. For example, in order to understand the family, we may view the concept of gender as its context. Then, in order to understand how people conceptualise gender, we might examine the family as its context. Rather than use hypotheses, ethnographers admit that they simply do not know in advance what will turn out to be relevant to the topic they are studying. They respond by trying to include observations that cover a wide range of aspects of everyday life.

Although we call holistic contextualisation a method, it is also quite simply a reflection of the reality of people's lives. No people exist only in relation to their family, nor to their work, nor to their online activity, nor to their politics, nor to their eating habits. They exist in relation to all these things simultaneously. In real life we all practise holistic contextualisation and ethnography acknowledges this. Such recognition dovetails with the other definition of ethnography: a method that studies people within their normal life circumstances rather than within a more artificial setting, for example a laboratory or focus group.

The ideal of being holistic does not stop at the borders of our fieldsites. The reasons why people behave as they do may also be influenced by commercial forces, government regulations, the weather or other factors. So holistic contextualisation not only defines ethnography but also transcends it. Here 'holistic' means including whatever turns out to be relevant to understanding the experiences of our research participants, whether it is possible to observe this within the ethnography or not. This is why, at times, this volume may have drawn upon materials derived from history, the media or the wider political economy, as in the conclusions to chapter 9. However, the main emphasis within *The Global Smartphone* is on our own original findings, based on our primary ethnographic observations.

While the ideal is to consider all aspects of our research participants' lives, there is bound to be a focus that will privilege some activities over others. This was determined by the original application to the European Research Council, which funded the project. The application specified a three-pronged approach to ageing, smartphones and mHealth, and our subsequent research was bound to follow those commitments. As a result, every chapter is focused upon older people, although rarely in isolation – after all, their smartphones connected them to family and friends. Living in one place for 16 months, researchers naturally made friends across all age groups too, including people of their own age.

The concept of 'older people' admittedly sounds rather vague. Originally we saw our focus as the middle aged, who typically see themselves as neither young nor elderly. But the diversity of our fieldsites confronted us with highly variable experiences, ranging from Japan, where some participants may not feel 'elderly' at 80, to Uganda, where people may be considered old at 40, depending on their lifestyle. One of the arguments within this volume was that the smartphone has played its own role in changing people's perceptions of age. As shown in chapter 7, people who had difficulties in using the smartphone often felt that this placed them in the 'old' category, while being proficient was often a reason for thinking of themselves as relatively young. Discussions of intergenerational relations recur throughout the volume, simply because it turned out that the smartphone has become deeply involved in those relations. Far more information about the project's findings with respect to ageing may be found in the individual monographs. All of these are being published under the title *Ageing with Smartphones in…* (the respective fieldsites).

The third element of this project, in addition to smartphones and ageing, was mHealth; this accounts for the focus on health issues in, for example, chapter 4. As noted in chapter 1, our ambitions for this part of the project were rather different from the studies of ageing or smartphones because they were also orientated to more practical concerns. The aim was to develop studies or interventions that would ideally have directly beneficial consequences for the welfare of populations in the regions where we worked. As with our work on ageing and smartphones, however, our understanding of the topic changed substantially from our initial expectations. In brief, as the research proceeded the main trajectory of these studies took the project away from an initial focus on conventional mHealth, understood as the production of bespoke health apps made for smartphones. We found that mostly these apps had as yet made relatively little impact upon these ethnographic populations. The focus therefore changed to an emphasis upon the way in which people already use everyday, general purpose apps such as WeChat, WhatsApp or YouTube for health purposes. The results of these studies will be published elsewhere,[1] but they have significantly influenced the approach of this volume. For example, the discussion of health dominates chapter 4 and the first third of chapter 8.

Ethnography

Ethnography, as noted above, is the primary method of research for anthropologists. Its goal is to study people as they go about their everyday lives – something that there is no one way of achieving. It is more important to be flexible and change methods as one learns about each specific population, and indeed each individual research participant. In one place friendship might be created by going to parties; in another through attending religious ceremonies. For a sense of how the findings in this book emerged, it may be helpful to imagine ethnography as a circle with four segments, each of which blurs into the others (Fig. A.1).

The first segment consists of what is usually understood to be the heart of ethnography – participant observation. The team spent time with their research participants by directly sharing their experiences. Both Charlotte and Patrick joined rotating savings schemes with their frequent meetings. Pauline went on regular walks with a group in Thornhill. Alfonso became heavily involved in the religious activities of Peruvians living in Santiago. Laura regularly accompanied her participants on girls' night dinners and volunteered at a health check, while Shireen joined choirs and sewing groups. Some participation resulted from our own initiatives. Alfonso, Danny, Marília, Maya and Pauline all became involved in teaching smartphone use to older people, while Xinyuan helped to develop exhibitions for and with her neighbourhood. Laila was an active participant in one of the women's groups in al-Quds.

For a researcher, most days are spent in such participant observation. It is the most immersive, most heightened involvement in research.

ETHNOGRAPHY

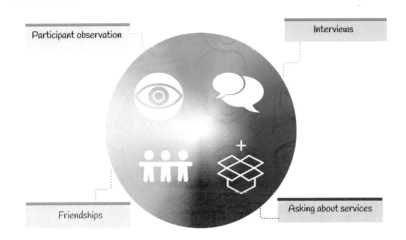

Participant observation

Interviews

Friendships

Asking about services

Figure A.1 Infographic showing ethnography as a circle with elements blurring into each another. Created by Xinyuan Wang.

As a project in digital anthropology, participant observation now extends to our direct involvement in online worlds such as social media platforms. This is another domain within which we can casually observe research participants' modes of communication, noting, for example, the popularity of stickers in LINE chats in Japan.

The second segment consisted of interviews. For each of the three key research areas (ageing, smartphone and health), the researcher agreed to interview and record at least 25 people. These were not perceived as a formal sample, but used to gain some sense of what was common to most of our research participants and therefore suitable for generalisation. These interviews were open-ended and informal. Interviews are helpful to record how people discuss these topics in their own words; they have allowed this volume to include many examples of the ways in which they expressed themselves. They also help the ethnographer to assess what seems typical of a population, and to explore what seem the more quirky preferences of a particular individual. These interviews were especially important in contributing to chapter 2, which is a discussion of discourse – i.e. what people say about smartphones.

One set of interviews were particularly important for the content of this book; they are summarised in chapter 4. During these interviews we asked people to go through every single app of their smartphone in turn and discuss their usage. We might then ask for other details, such as how many phone calls had they made in the previous week, what WhatsApp groups they belonged to and how many of these groups consisted just of family members. As noted in chapter 2, older people might dismiss their own use of their smartphones in general discussion, claiming they only used it for texting and voice calls. By going through every single app on their phones, however, a different picture emerged. It often turned out that these same research participants were using around 25 or 30 different apps and functions, the discussion of each of which could lead to stories and examples that might not have been directly observed.

However, over 16 months, the interviews may be far less important than the three-hour conversation you had with someone when walking in the countryside or the gossip you heard over a cup of coffee. Those encounters may lead to the third segment, which consists of friendships. It would be strange to live in a place for 16 months without making friends. These friends are by no means 'fake' or just instrumental. Friendships developed during fieldwork are often long lasting, continuing far beyond the period of fieldwork. Many of the key insights that inform this book are from friends that researchers made during fieldwork; so are some more everyday ones (Fig. A.2). We write about friends with their permission and following discussions about ethics and anonymity. It is part of friendship to make clear to those involved that we are living in that place as professional ethnographers and we are trying to learn about daily life there, which means inevitably also learning from them.

Figure A.2 Danny soon learned not to turn up at someone's house without a brack, a type of fruit loaf popular in Ireland. Photo by Daniel Miller.

We explain to our research participants how such knowledge may be used in our publications and for education purposes, ensuring that no one is unaware of this. Over time people become reassured as to our discretion in not conveying gossip; they often find that once trust develops, it is quite cathartic to have someone to talk to who is neither a relative nor part of an established social network. Danny remains friends today with some of the Trinidadians he first met during fieldwork in Trinidad in the 1980s. They started out as research participants but have now become good friends. He plans to return to Trinidad fairly soon, in which case they will become research participants again.

Thanks to social media, this continuity tends to be much stronger today. The foundation of friendship lies in trust, which is also the foundation for ethnography. It is what makes these studies a collaboration – not just with research participants, but also between the authors. Almost everyone is fascinated by smartphones. The research participants who have become our friends may be just as interested in trying to understand and explain what they do with smartphones as we are. Commonly, an anthropologist will talk about their initial insights and analysis, then ask for comments as to whether these seem plausible and true to the experiences of these friends. But we are also prepared to contradict our research participants when, for example, their practices seem very different from what they claim them to be.

The final segment comprises discussions with people who provide services in the local fieldsite. They might include staff in phone repair shops, health workers, hairdressers or individuals who work in public places such as bars. They may drive a taxi, work for the police or provide counselling or psychotherapy; they may be politicians or cleaners. People who occupy these positions gain access to experiences or observations that can provide valuable additional information. This in turn helps the ethnographer to gain a better understanding of the context.

The reason the centre of the circle shown in Fig. A1 is blurred is because these are not separate segments. The same individual may appear within all four. All of them depend on the fundamental commitment of spending a significant amount of time in our respective fieldsites. A period of 16 months ensures that the research is not merely anecdotal, but secured through seeking patterns and repetitions of behaviour. It is essential for the establishment of trust as this, in turn, is what allows people to feel comfortable talking about what they *really* feel – as opposed to what they think they are supposed to feel or 'should' say to a researcher. The ideal of holistic contextualisation depends upon this same long-term commitment. It takes time to sense what it is like to live in a neighbourhood, to feel the rhythms of daily life, to explore both places where there is a community and those where there is isolation and loneliness. It is within the latter that we find people whom one would simply not encounter or hear about until one has lived at the fieldsite for many months.

This research derives from the establishment of a programme in digital anthropology at University College London[2] – a programme intended to acknowledge the increasing importance of online activities in our lives. However, digital anthropology is not the same as online ethnography. Most of the fieldwork for this project consisted of traditional offline ethnography. The online components of the ethnography tended to develop more organically from our offline presence. As people increasingly use WeChat or Facebook in their interactions with other people, then researchers will tend to become involved with them online as well as offline. As they become part of the local community, the researchers participate in the public face of online worlds that have become an important resource for communities. Fieldwork for this project mostly finished in June 2019, with the exception being that conducted in al-Quds, since the researchers there also have other employment. But these days, as ethnographies develop friendships and participate in social media, there is bound to be continuity – something especially important in this instance because of the events surrounding the Covid-19 pandemic. Naturally all the authors of the volume were concerned about the welfare of their participants and remained in touch with them over this period. As a result a number of observations contained in this volume cover these more recent events, rather than ending with the formal closure of fieldwork.

While the use of the smartphone is the research subject, it is also becoming an important research tool. Smartphones have allowed this team to be connected constantly with the people they worked among, in much the same way as those people connected with each other.[3] Just as with our research participants, the smartphone enabled us to remain in contact with friends, partners and relatives back home. New digital technologies also allowed us to work collaboratively and comparatively in a way that might not have been possible previously. Throughout the fieldwork, the researchers wrote 5,000-word reports every month; these were then read by all other members of the team and collectively discussed over Skype every month. These meetings also included team discussion concerning what to focus on in the following month. More constant interaction took place through WhatsApp and email. Through discussions, arguments, drinking and laughter, we have shared our stories from the field and found comparative elements naturally through conversation on and offline.

Comparison and generalisation

The project consisted of 11 researchers who carried out 10 ethnographies in 9 countries. A very brief introduction to the researchers and their fieldsites can be found in this short film (Fig. A.3).

The ability to compare and contrast one's findings with those of other researchers within this team was a considerable asset. After a few months, the ethnographer is bound to start taking for granted the way in which people locally use their smartphones. During discussion, they come to realise that what they have been observing as a logical use of WhatsApp turns out in another fieldsite to be something people do more through YouTube. People in one fieldsite contrast with those in another over the degree to which they accept that people have changed their appearance when posting online. Being constantly confronted with evidence that the population in another fieldsite does things differently reminds each anthropologist that they need to explain *why* things are done in the manner that they have observed – that their population is no more 'natural' than any other population.

As noted several times within this volume, we are careful in delineating the units that are being compared. One fieldsite in China does not stand for 'the Chinese'. Nor is every low-income, middle-aged man

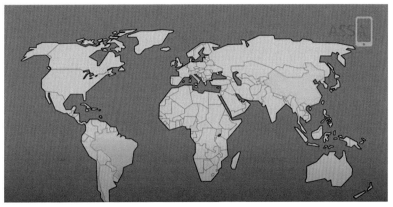

Figure A.3   Film: *Who we are*. Available at http://bit.ly/assawhoweare.

from Santiago the same. Usually people are bewilderingly different as individuals – yet this book is replete with generalisations. We have also noted that one of the main points of long-term ethnography is to observe repetition and pattern in order to assess what kind of things might be regarded as typical, which is not the same as stereotyping. Stating that people in the Shanghai fieldsite like to use QR codes more than people in the Irish fieldsite is a generalisation based on our observations. It does *not* imply that this preference is an inherent property of being Chinese. There may well be an individual Shanghai participant who hates QR codes, and 20 years from now people in the Irish fieldsite may like QR codes more than those in Shanghai. Everything we observe is a result of people growing up within the norms and expectations of their society as they develop over time. Their behaviour would have been different if socialised in another region. A generalisation is not a stereotype because it is not essentialist. Nothing in this volume relates to behaviour that might be considered innate or an essential property of any person of any particular category of humanity.

Shireen worked in a neighbourhood of Milan that includes many people originally from other countries. These people, among them individuals from Egypt, Peru and the Philippines, as well as the Hazara from Afghanistan, had come to Italy at different times and in varying circumstances. In dealing with topics such as citizenship and identity, Shireen might have organised her findings as a contrast between Italians and migrants. Instead, she focused on the wider diversity of experience in this urban setting, recognising that many Italians in Milan were themselves migrants from other parts of Italy. Her approach highlights how important it is to understand the social, legal and political categorisation of people, a process that involved documenting experiences of exclusion and inclusion. A final issue of generalisation is that this volume is multi-authored. When a statement appears in this book, it cannot be assumed that every author agrees with that statement or that every one of the fieldsites accords with that claim. It would, however, be a very tedious book if that caveat was included with every sentence.

All the team carried out more quantitative surveys than appear in this volume. We prefer to see our quantitative findings as complementary to our core qualitative evidence. Surveys may help us to understand how typical something is, but we do not want to privilege that which can be counted over factors not amenable to quantification. An example of this is the discussion of how many apps people use in chapter 4. The authority for the volume derives far more from the scholarship of 16-month immersive participant observation. At the other end of the methodological spectrum, the use of stories and vignettes may give the equally misleading impression that the fieldwork was anecdotal. Anecdotes can be collected in a two-week visit. The point of staying for a period of 16 months is to observe patterns of everyday behaviour over time, enabling the anthropologist to be clear whether a case study is reasonably typical or eccentric – and why.

Ethics

A research project that includes 10 simultaneous ethnographies across 9 countries involved the implementation of a wide variety of ethical standards. Some of these are based on compliance with a range of requirements established by ethics committees including the European Research Council and University College London, as well as following the specific policies of institutional and national ethics committees in each of the fieldsites. These policies ensure that participants are fully informed as to the research and dissemination aims of the project, including the use of consent forms and data protection. Central to the ethics of our project is anonymity: people should not be recognised unless, as in the case of some of the films, they have elected to be so. Various methods are used to assist anonymisation. Names are changed and in some cases pseudonyms used for places. Details about people not relevant to the points being made may also have been changed for this reason. Working online adds a further dimension. In some cases people are comfortable with you observing their postings on social media, but do not expect you to post. In other places people are only comfortable if you post in the same manner as themselves.

For anthropologists, ethics extend well beyond the requirements of compliance established by ethics committees. In our team, the basic dictum was simply to ensure that no harm ever came to someone through participating in our research. Preventing harm requires us to be sensitive to local and personal ideas of appropriate behaviour, for example around sharing information. Sometimes the ethnographer also faces difficult decisions about how to respond to health emergencies or financial difficulties, which can have a long-term impact on how people view them or evaluate the nature of their friendship. In addition, they may pay attention to what the anthropologist Didier Fassin describes as 'situated ethics'.[4] The place of the ethnographer includes people's feelings about their nationality or gender. The ASSA team include both researchers from the population being studied and researchers from very different populations than those of their fieldsite.[5] In Ireland, for example, Pauline worked as an Irish anthropologist; Danny, being British, could be seen as a representative of the erstwhile colonial regime. In some fieldsites research assistants were employed. Charlotte, a British researcher, worked in Kampala with research participants originating from various regions in Uganda and speaking many different languages. The project was facilitated by a co-researcher and her family. They grew up in the area and are well known and respected among the community.

Laila is an active member of a community in al-Quds. Patrick was born in Cameroon, but had been living abroad for 10 years prior to fieldwork, although he returned annually. The profound changes that have taken place in Cameroon over that decade create an ambiguity in his sense of being an 'insider'. Such an ambiguity is reflected in the attitude of local people, some of whom may see him as a *mbenguist* – a local term for Cameroonians of the Western diaspora. As the Cameroonian anthropologist Francis B. Nyamnjoh argues,[6] ethnography is an ongoing dialogue, a 'collaboration of voices'.[7] Issues of 'connectivity' are likely to come to the fore when the topic of study is smartphones, the main device today through which people connect.

Dissemination

Academic books may be written in a variety of styles depending upon their intended audience. In this case, our assumption has been that a fuller understanding of the use and consequences of smartphones would be a topic of interest to pretty much anyone anywhere. For that reason, the book is written for a much wider audience than most books in social science. Although the discussions in chapter 9 have been termed 'theory', we have tried to write them entirely in colloquial language, accessible to someone who has completed secondary school or is just starting university. The books in this series are all published as open-access books. They can thus be downloaded for free, so there is no cost involved for the reader. As far as budget allows, we are investing in translations to ensure that the populations we researched also have free access to the results. We have also been blogging throughout fieldwork and while writing this book, in addition to making short films as a way to communicate our research findings in an accessible manner. All of this can be accessed from the project's website.[8]

As noted in the main text, the style of this volume was deliberately constructed using short portraits of individuals. These portraits are intended to convey our commitment to a humanism that respects the unique character of all the individuals we met as research participants. Acknowledging them as individuals balances our need to make analytical and theoretical generalisations and abstractions within an academic volume. In addition to text, the book includes photographs and infographics. To convey the multi-sensorial atmosphere of fieldsites, however, we strongly recommend also looking at some of the short films recorded as part of the fieldwork.[9] Every author in this volume will feel frustrated that the excerpts that appear here from their fieldsites are so brief and outside of the wider context. They would therefore hope that readers who want to engage with and understand those wider contexts might now be tempted also to read their monographs.

Notes

1. For example Duque 2020.
2. Horst and Miller 2012. Digital Anthropology is also the field in which several of the team were trained.
3. See de Bruijn et al. 2009, 15. The volume's introduction includes an excerpt from *Married But Available, A Novel by Francis B. Nyamnjoh*. In this story Nyamnjoh's fictional visiting European researcher in Africa, 'Lily Loveless', loses her phone, and at the same time her networks, relationships, identity and sense of security and normality. This story is used to depict the mobile phone as both ethnographic tool and subject matter, linking people to their social relations and thus exposing them.
4. Fassin 2008.
5. For further discussion of this situation of insiders and outsiders see Griffith 1998 and Merton 1972.
6. Nyamnjoh 2012.
7. Clifford 1986.
8. Our project's blog can be found at https://www.ucl.ac.uk/anthropology/assa/. Our project's website can be found at https://www.ucl.ac.uk/anthropology/assa/.
9. These films are generally a few minutes long and can be viewed on our project's YouTube channel at https://www.youtube.com/channel/UC8gpt3_urYwiNuoB83PVJlg. For films specifically about methodology see https://www.youtube.com/playlist?list=PLm6rBY2z_0_gCJCxU5ninztHVIP_ZewZn.

Bibliography

Abacus News (Part of SMCP). 2019. 'Podcasts are booming in China and Ximalaya FM leads the charge'. 30 August 2019. Accessed 1 October 2020. https://www.abacusnews.com/digital-life/podcasts-are-booming-china-and-ximalaya-fm-leads-charge/article/3025066.

Accessa, S. P. 2018. 'RG 033 – Resultados – POnline 2017'. Accessed 1 October 2020. http://www.acessasp.sp.gov.br/wp-content/uploads/2019/01/ponline-2017.pdf.

Agar, Jon. 2013. *Constant Touch: A global history of the mobile phone*. London: Icon.

Ahlin, Tanja. 2018a. 'Frequent callers: "Good care" with ICTs in Indian transnational families'. *Medical Anthropology* 39 (1): 69–82. https://doi.org/10.1080/01459740.2018.1532424.

Ahlin, Tanja. 2018b. 'Only near is dear? Doing elderly care with everyday ICTs in Indian trans-national families: Elderly care with ICTs in Indian families'. *Medical Anthropology Quarterly* 32 (1): 85–102. https://doi.org/10.1111/maq.12404.

Ahmed, Sara. 2004. 'Affective economies'. *Social Text* 22 (2): 117–39. https://doi.org/10.1215/01642472-22-2_79-117.

Akimoto, A. 2013. 'Looking at 2013's Japanese social-media scene'. *The Japan Times*, 17 December 2013. Accessed 1 October 2020. https://www.japantimes.co.jp/life/2013/12/17/digital/looking-at-2013s-japanese-social-media-scene-3/#.Xl4ycaj7Q2w.

Al Jazeera. 2017. 'Cameroon shuts down internet in English-speaking areas'. *Al Jazeera*, 26 January 2017. Accessed 1 October 2020. https://www.aljazeera.com/news/2017/01/cameroon-anglophone-areas-suffer-internet-blackout-170125174215077.html.

Albarrán-Torres, C. and G. Goggin. 2017. 'Mobile betting apps – Odds on the social'. In *Smartphone Cultures*, edited by J. Vincent and L. Haddon, 25–37. London: Routledge.

Al-Heeti, Abrar. 2019. 'Facebook lost 15 million US users in the past two years, report says – CNET'. CNET. 6 March 2019. Accessed 1 October 2020. https://www.cnet.com/news/facebook-lost-15-million-us-users-in-the-past-two-years-report-says/.

Allison, Anne. 2014. *Precarious Japan*. Durham, NC: Duke University Press.

Andall, Jacqueline. 2002. 'Second-generation attitude? African-Italians in Milan'. *Journal of Ethnic and Migration Studies* 28 (3): 389–407. https://doi.org/10.1080/13691830220146518.

Anderson, M. and A. Perrin. 2017. 'Tech adoption climbs among older adults'. PEW Research Center. 17 May 2017. Accessed 1 October 2020. https://www.pewresearch.org/internet/2017/05/17/tech-adoption-climbs-among-older-adults/.

Andjelic, J. 2020. 'WhatsApp statistics: Revenue, usage, and history (updated May 2020)'. Fortunly. May 2020. Accessed 1 October 2020. https://fortunly.com/statistics/whatsapp-statistics/#gref.

Anthrocovid.com. 2020. 'Collecting COVID-19 | anthropological responses'. Anthrocovid.com. 2020. Accessed 1 October 2020. http://anthrocovid.com/.

Antonsich, M., S. Camilotti, L. Mari, S. Pasta, V. Pecorelli, R. Petrillo and S. Pozzi. 2020. 'New Italians: The re-making of the nation in the age of migration'. Research website. New Italians. 2020. Accessed 1 October 2020. http://newitalians.eu/en/.

Apple Inc. 2020. 'Buy iPhone 11 Pro'. Apple website. 2020. Accessed 1 October 2020. https://www.apple.com/us-hed/shop/buy-iphone/iphone-11-pro.

Apple Inc. 2020. 'Preparing apps for review'. Apple Developer. Accessed 1 October 2020. https://developer.apple.com/app-store/review/.

Archambault, J. 2017. *Mobile Secrets: Youth, intimacy, and the politics of pretense in Mozambique*. Chicago: University of Chicago Press.

Ardener, Shirley. 1964. 'The comparative study of rotating credit associations'. *The Journal of the Royal Anthropological Institute of Great Britain and Ireland* 94 (2): 201. https://doi.org/10.2307/2844382.

Asahi Digital. 2020. '10万円給付、窓口に人が殺到 総務相「改善が必要」'. Asahi Digital, 12 May 2020. Accessed 1 October 2020. https://www.asahi.com/articles/ASN5D3K6YN5DULFA00C.html.

Augé, Marc. 2008. *Non-Places: Introduction to an anthropology of supermodernity*, 2nd English language ed. London; New York: Verso.

Baldassar, Loretta, Mihaela Nedelcu, Laura Merla and Raelene Wilding. 2016. 'ICT-based co-presence in transnational families and communities: Challenging the premise of face-to-face proximity in sustaining relationships'. *Global Networks* 16 (2): 133–44. https://doi.org/10.1111/glob.12108.

Baldassar, Loretta, Raeline Wilding, Paolo Boccagni and Laura Merla. 2017. 'Aging in place in a mobile world: New media and older people's support networks'. *Transnational Social Review* 7 (1): 2–9. https://doi.org/10.1080/21931674.2016.1277864.

Barry, Christopher T., Hannah Doucette, Della C. Loflin, Nicole Rivera-Hudson and Lacey L. Herrington. 2017. '"Let me take a selfie": Associations between self-photography, narcissism, and self-esteem'. *Psychology of Popular Media Culture* 6 (1): 48–60. https://doi.org/10.1037/ppm0000089.

Bateman, Tom. 2020. 'Coronavirus: Israel turns surveillance tools on itself'. *BBC News*, 11 May 2020. Accessed 1 October 2020. https://www.bbc.com/news/world-middle-east-52579475.

Baym, N. 2010. *Personal Connections in the Digital Age*. Cambridge: Polity.

BBC News. 2007. 'Apple's "magical" iPhone unveiled'. 9 January 2007. Accessed 1 October 2020. http://news.bbc.co.uk/1/hi/technology/6246063.stm.

BBC News. 2014. 'Facebook to buy messaging app WhatsApp for $19bn'. 20 February 2014. Accessed 1 October 2020. https://www.bbc.co.uk/news/business-26266689.

BBC News. 2016. 'WhatsApp is now free (and there still won't be adverts)'. 18 January 2016. Accessed 1 October 2020. http://www.bbc.co.uk/newsbeat/article/35345731/whatsapp-is-now-free-and-there-still-wont-be-adverts.

BBC News. 2020. 'Coronavirus privacy: Are South Korea's alerts too revealing?' 5 March 2020. Accessed 1 October 2020. https://www.bbc.co.uk/news/world-asia-51733145.

Bell, C. and J. Lyall. 2005. '"I was here": Pixelated evidence'. In *The Media and the Tourist Imagination: Converging cultures*, edited by D. Crouch, R. Jackson, and F. Thompson. London: Penguin Books.

Benedict, R. 1946. *The Chrysanthemum and the Sword: Patterns of Japanese culture*. Boston, MA: Houghton Mifflin.

Bernal, Victoria. 2014. *Nation as Network: Diaspora, cyberspace, and citizenship*. Chicago: University of Chicago Press.

Bhardwaj, P. 2018. 'Tencent's business is about as big as Facebook's thanks to its stronghold in China'. *Business Insider*. 16 May 2018. Accessed 1 October 2020. https://www.businessinsider.com/tencent-compare-facebook-revenue-charts-2018-5?r=US&IR=T.

Bikoko, A. B. 2017. 'Cameroun: Le téléphone portable, au-delà de la valeur d'usage, la mort'. *Mediaterre*, 26 July 2017. Accessed 1 October 2020. https://www.mediaterre.org/climat/actu,20170726042927,6.html.

Bogost, Ian. 2020. 'Every place is the same now'. News website. *The Atlantic*. 16 January 2020. Accessed 1 October 2020. https://www.theatlantic.com/technology/archive/2020/01/smartphone-has-ruined-space/605077/.

Bolter, Jay David and Richard Grusin. 2003. *Remediation: Understanding new media*. 6th edition. Cambridge, MA: MIT Press.

Boullier, D. 2002. 'Objets communicants, avez-vous donc une âme ? Enjeux anthropologiques'. *Les Cahiers Du Numérique* 3 (4): 45–60.

Boumans, J. 2005. 'Paid content: From free to fee'. In *E-Content Technologies and Perspectives for the European Market*, edited by P. A. Bruck, Z. Karssen, A. Buchholz and A. Zerfass, 55–75. Berlin; Heidelberg: Springer. https://doi.org/10.1007/3-540-26387-X_3.

Bourdieu, Pierre. 1977. *Outline of a Theory of Practice*. Cambridge: Cambridge University Press.

boyd, danah and Kate Crawford. 2012. 'Critical questions for Big Data: Provocations for a cultural, technological, and scholarly phenomenon'. *Information, Communication & Society* 15 (5): 662–79. https://doi.org/10.1080/1369118X.2012.678878.

boyd, danah. 2014. *It's Complicated: The social lives of networked teens*. New Haven, CT: Yale University Press.

Boyd, Joshua. 2019. 'The history of Facebook: From BASIC to global giant'. *Brandwatch Blog* (blog). 25 January 2019. Accessed 1 October 2020. https://www.brandwatch.com/blog/history-of-facebook/.

Boylan, Dan. 2018. 'Ugandans riot after President imposes social media tax to fight "fake news" and gossip'. *The Washington Times*, 15 July 2018. Accessed 1 October 2020. https://www.washingtontimes.com/news/2018/jul/15/yoweri-museveni-uganda-president-fights-fake-news-/.

Bruijn, M. de, F. Nyamnjoh, and I. Brinkman, eds. 2009. *Mobile phones: The new talking drums of everyday Africa*. Bamenda, Cameroon: Langaa Publishers.

Bruns, Axel. 2019. *Are Filter Bubbles Real?* Cambridge, UK; Medford, MA: Polity.

Bruns, Axel, Gunn Enli, E. Skogerbo, Anders Olof Larsson and C. Christensen. 2018. *The Routledge Companion to Social Media and Politics*. New York; London: Routledge.

Brunton, F. 2018. 'WeChat: Messaging apps and new social currency transaction tools'. In *Appified: Culture in the age of apps*, edited by Jeremy Wade Morris and Sarah Murray, 179–87. Ann Arbor, MI: University of Michigan Press.

Buganda.com site. 2020. 'The clans of Buganda'. Buganda.com. Accessed 1 October 2020. http://www.buganda.com/ebika.htm.

Bunz, Mercedes and Graham Meikle. 2017. *The Internet of Things*. Cambridge, UK; Malden, MA, USA: Polity.

Burgess, Adam. 2004. *Cellular Phones, Public Fears, and a Culture of Precaution*. New York: Cambridge University Press.

Burke, Hilda. 2019. *The Phone Addiction Workbook: How to identify smartphone dependency, stop compulsive behavior and develop a healthy relationship with your devices*. Berkeley, CA: Ulysses Press.

Burrell, Jenna. 2010. 'Evaluating shared access: Social equality and the circulation of mobile phones in rural Uganda'. *Journal of Computer-Mediated Communication* 15 (2): 230–50. https://doi.org/10.1111/j.1083-6101.2010.01518.x.

Burrell, Jenna. 2012. *Invisible Users: Youth in the internet cafes of urban Ghana*. Cambridge, MA: MIT Press.

Bushey, R. 2014. 'How Japan's most popular messaging app emerged from the 2011 earthquake'. *Business Insider*. 12 January 2014. Accessed 1 October 2020. https://www.businessinsider.com/history-of-line-japan-app-2014-1?r=US&IR=T.

Cadwalladr, Carol and Emma Graham-Harrison. 2018. 'Revealed: 50 million Facebook profiles harvested for Cambridge Analytica in major data breach'. *The Guardian*, 17 March 2018. Accessed 1 October 2020. https://www.theguardian.com/news/2018/mar/17/cambridge-analytica-facebook-influence-us-election.

Cairncross, Frances. 1997. *The Death of Distance: How the communications revolution will change our lives*. Boston, MA: Harvard Business School Press.

Carrier, Mark. 2018. *From Smartphones to Social Media: How technology affects our brains and behavior*. Santa Barbara, CA: Greenwood, an imprint of ABC-CLIO, LLC.

Carroll, R. 2020. 'Why Ireland's data centre boom is complicating climate efforts'. *The Irish Times*, 6 January 2020. Accessed 1 October 2020. https://www.irishtimes.com/business/technology/why-ireland-s-data-centre-boom-is-complicating-climate-efforts-1.4131768.

Cecilia. 2014. 'WeChat dominates APAC mobile messaging in Q3 2014'. China Internet Watch. 27 November 2014. Accessed 1 October 2020. https://www.chinainternetwatch.com/10939/wechat-dominates-apac-mobile-messaging-q3-2014/.

Chambers, D. 2014. *Social Media and Personal Relationships*. London: Palgrave Macmillan.

Chatzimilioudis, Georgios, Andreas Konstantinidis, Christos Laoudias and Demetrios Zeinalipour-Yazti. 2012. 'Crowdsourcing with smartphones'. *IEEE Internet Computing* 16 (5): 36–44. https://doi.org/10.1109/MIC.2012.70.

Chen, X. and P. H. Ang. 2011. 'The internet police in China: Regulation, scope and myths'. In *Online Society in China: Creating, celebrating, and instrumentalising the online carnival*, edited by D. K. Herold and P. Marolt, 52–64. Abingdon, Oxon; New York: Routledge.

Chen, Yujie, Zhifei Mao and Jack Linchuan Qiu. 2018. *Super-Sticky WeChat and Chinese Society*. United Kingdom: Emerald Publishing.

Cheng, Yinghong. 2009. *Creating the 'New Man': From Enlightenment ideals to socialist realities*. Honolulu: University of Hawai'i Press.

Clark, Lynn Schofield. 2013. *The Parent App: Understanding families in the digital age*. Oxford; New York: Oxford University Press.

Clements, Alan. 2014. *Computer Organization & Architecture: Themes and variations*. Stamford, CT: Cengage Learning.

Clifford, J. 1986. 'Introduction: Partial truths'. In *Writing Culture: The poetics and politics of ethnography*, edited by J. Clifford and G. E. Marcus, 1–26. Berkeley, CA: University of California Press.

Clough Marinaro, I. and J. Walston. 2010. 'Italy's "second generations": The sons and daughters of migrants'. *Bulletin of Italian Politics* 2 (1): 5–19.

Coleman, E. Gabriella. 2013. *Coding Freedom: The ethics and aesthetics of hacking*. Princeton, NJ: Princeton University Press.

Coleman, E. Gabriella. 2014. *Hacker, Hoaxer, Whistleblower, Spy: The many faces of Anonymous*. London; New York: Verso.

Costa, Elisabetta. 2018. 'Affordances-in-Practice: An ethnographic critique of social media logic and context collapse'. *New Media & Society* 20 (10): 3641–56. https://doi.org/10.1177/1461444818756290.

Couldry, Nick, Sonia Livingstone and Tim Markham. 2007. 'Connection or disconnection?: Tracking the mediated public sphere in everyday life'. In *Media and Public Spheres*, edited by Richard Butsch, 28–42. Basingstoke, UK: Palgrave Macmillan.

Couldry, Nick and Ulises Ali Mejias. 2019. *The Costs of Connection: How data is colonizing human life and appropriating it for capitalism*. Stanford, CA: Stanford University Press.

Counterpoint. 2019. 'India smartphone market share: By quarter'. *Counterpoint Research* (blog). 27 November 2019. Accessed 30 September 2020. https://www.counterpointresearch.com/india-smartphone-share/.

Court of Justice of the European Union. 2014. 'Judgment in Joined Cases C-293/12 and C-594/12: Digital rights Ireland and Seitlinger and others. The Court of Justice declares the data retention directive to be invalid'. 8 April 2014. Accessed 25 May 2020. https://curia.europa.eu/jcms/upload/docs/application/pdf/2014-04/cp140054en.pdf.

Cronin, Michael. 2013. *Translation in the Digital Age*, 1st ed. London: Routledge. https://doi.org/10.4324/9780203073599.

Cruz, Edgar Gómez and Ramaswami Harindranath. 2020. 'WhatsApp as "technology of life": Reframing research agendas'. *First Monday* 25 (12). https://doi.org/10.5210/fm.v25i12.10405.

Daniels, Inge. 2015. 'Feeling at home in contemporary Japan: Space, atmosphere and intimacy'. *Emotion, Space and Society* 15 (May): 47–55. https://doi.org/10.1016/j.emospa.2014.11.003.

DataSenado. 2019. 'Redes sociais, notícias falsas e privacidade de dados na internet'. Accessed 30 September 2020. https://www12.senado.leg.br/institucional/datasenado/arquivos/mais-de-80-dos-brasileiros-acreditam-que-redes-sociais-influenciam-muito-a-opiniao-das-pessoas.

Dazzi, Zita. 2018. 'Catena umana contro il razzismo in via Padova: "Siamo cittadini, non clandestini"'. *La Repubblica*, 5 May 2018. Accessed 30 September 2020. https://milano.repubblica.it/cronaca/2018/05/05/news/catena_umana_via_padova-195600267/.

De Bruin, B. 2010. 'The liberal value of privacy'. *Law and Philosophy* 29 (5): 505–34.

De Pasquale, C., C. Sciacca and Z. Hichy. 2017. 'Italian validation of smartphone addiction scale short version for adolescent and young adults (SAS-SV)'. *Psychology* 8 (10): 1513–18. https://doi.org/10.4236/psych.2017.810100.

Deloitte. 2016. 'Game of phones: Deloitte's mobile consumer survey. The Africa cut 2015/2016'. Accessed 30 September 2020. https://www2.deloitte.com/content/dam/Deloitte/za/Documents/technology-media-telecommunications/ZA_Deloitte-Mobile-consumer-survey-Africa-300816.pdf.

Denworth, L. 2019. 'Social media has not destroyed a generation'. *Scientific American*, November 2019. Accessed 30 September 2020. https://www.scientificamerican.com/article/social-media-has-not-destroyed-a-generation/.

Deursen, Alexander J. A. M. van, Colin L. Bolle, Sabrina M. Hegner and Piet A. M. Kommers. 2015. 'Modeling habitual and addictive smartphone behavior'. *Computers in Human Behavior* 45 (April): 411–20. https://doi.org/10.1016/j.chb.2014.12.039.

Dijck, José van. 2007. *Mediated Memories in the Digital Age*. Stanford, CA: Stanford University Press.

Dijk, Jan A. G. M. van. 2006. 'Digital divide research, achievements and shortcomings'. *Poetics* 34 (4–5): 221–35. https://doi.org/10.1016/j.poetic.2006.05.004.

Dijk, Jan A. van and Alexander van Deursen. 2014. *Digital Skills: Unlocking the information society*. New York, NY: Palgrave Macmillan.

Doi, Takeo. 1985. *Anatomy of Self: The individual versus society*. Japan: Kodansha.

Donner, Jonathan. 2015. *After Access: Inclusion, development, and a more mobile internet*. Cambridge, MA: MIT Press.

Donner, Jonathan and Patricia Mechael. 2013. *MHealth in Practice: Mobile technology for health promotion in the developing world*. Bloomsbury Academic. https://doi.org/10.5040/9781780932798.

Doron, Assa and Robin Jeffrey. 2013. *The Great Indian Phone Book: How the cheap cell phone changes business, politics, and daily life*. Cambridge, MA: Harvard University Press.

Douglas, M. 1991. 'The idea of home: A kind of space'. *Social Research* 58 (1): 287–307.

Drazin, Adam and David Frohlich. 2007. 'Good intentions: Remembering through framing photographs in English homes'. *Ethnos* 72 (1): 51–76. https://doi.org/10.1080/00141840701219536.

Drozdiak, N. 2016. 'WhatsApp to drop subscription fee'. *The Wall Street Journal*, 18 January 2016. Accessed 1 October 2020. https://www.wsj.com/articles/whatsapp-to-drop-subscription-fee-1453115467.

Duque Pereira, Marília. 2018. 'Seriam os dados sublimes?' *Novos Olhares* 7 (2): 38–52. https://doi.org/10.11606/issn.2238-7714.no.2018.149040.

Duque, Marília. 2020. *Learning from WhatsApp: Best practices for health. Communication protocols for hospitals and medical clinics*. London: ASSA.

Duque, Marília and A. Lima. 2019. ' "Share on the Whats": How WhatsApp is turning São Paulo into a smart city for older people'. The Global South Conference in São Paulo, Brazil.

Edwards, Elaine. 2018. 'Department seeks tender to monitor social media for "keywords" '. *The Irish Times*, 27 August 2018. Accessed 30 September 2020. https://www.irishtimes.com/news/social-affairs/department-seeks-tender-to-monitor-social-media-for-keywords-1.3608275.

Eede, Yoni van den. 2019. *The Beauty of Detours: A Batesonian philosophy of technology*. Albany, NY: State University of New York.

Elhai, Jon D., Haibo Yang, Jianwen Fang, Xuejun Bai and Brian J. Hall. 2020. 'Depression and anxiety symptoms are related to problematic smartphone use severity in Chinese young adults: Fear of missing out as a mediator'. *Addictive Behaviors* 101 (February): 105962. https://doi.org/10.1016/j.addbeh.2019.04.020.

Encyclopaedia Iranica Online. 2020. 'HAZĀRA Iv. Hazāragi Dialect'. In *Encyclopaedia Iranica Online*. Accessed 1 October 2020. http://www.iranicaonline.org/articles/hazara-4#.

Eriksen, Thomas Hylland. 2016. *Overheating: An anthropology of accelerated change*. London: Pluto Press.

European Commission. 2020. 'eHealth Network'. European Commission website. 2020. http://ec.europa.eu/health/ehealth/policy/network/index_en.htm.

Fan, Zhang. 2018. 'People's daily commentator observes: "Learning is the best retirement" '. *The People's Daily*, 15 November 2018. http://opinion.people.com.cn/n1/2018/1115/c1003-30401293.html.

Fassin, Didier. 2008. 'L'éthique, au-delà de la règle: Réflexions autour d'une enquête ethnographique sur les pratiques de soins en Afrique du Sud'. *Sociétés contemporaines* 71 (3): 117. https://doi.org/10.3917/soco.071.0117.

Favero, Paolo S. H. 2018. *The Present Image: Visible stories in a digital habitat*. London: Palgrave Macmillan.

Feigenbaum, E. 2003. *Chinese Techno-Warriors: National security and strategic competition from the Nuclear Age to the Information Age*. Stanford, CA: Stanford University Press.

Fiegerman, Seth. 2013. 'WhatsApp tops 250 million active users'. *Mashable*. 21 June 2013. Accessed 1 October 2020. https://mashable.com/2013/06/21/whatsapp-250-million-users/?europe=true.

Fischer, Claude. 1992. *America Calling: A social history of the telephone to 1940*. Berkeley and Los Angeles: University of California Press.

Fortunati, Leopoldina. 2002. 'Italy: Stereotypes, true and false'. In *Perpetual Contact,* edited by J. E. Katz and M. Aakhus, 42–62. Cambridge: Cambridge University Press.

Fortunati, Leopoldina. 2013. 'The mobile phone between fashion and design'. *Mobile Media & Communication* 1 (1): 102–9. https://doi.org/10.1177/2050157912459497.

Fortunati, Leopoldina, James E. Katz and Raimonda Ricini. 2003. *Mediating the Human Body*. Mahwah, NJ: Lawrence Erlbaum Associates, Inc.

Foster, R., and H. Horst, eds. 2018. *The Moral Economy of Mobile Phones: Pacific Islands perspectives*. Acton, Australia: Australian National University Press.

Fox, Kate. 2014. *Watching the English: The hidden rules of English behavior*. Revised and updated. Boston, MA: Nicholas Brealey Publishing.

Frey, Nancy. 1998. *Pilgrim Stories: On and off the road to Santiago*. Berkeley, CA: University of California Press.

Frey, Nancy. 2017. 'The Smart Camino: Pilgrimage in the internet age'. Presented at the Annual General Meeting of the London Confraternity of St James, St Alban's Centre, London, 28 January 2017. Accessed 1 October 2020. https://www.walkingtopresence.com/home/research/text-pilgrimage-in-the-internet-age.

Friedberg, Anne. 2006. *The Virtual Window: From Alberti to Microsoft*, 1st paperback ed. Cambridge, MA: MIT Press.

Frith, Jordan. 2015. *Smartphones as Locative Media*. Cambridge, UK: Malden, MA: Polity.

Fu, Xiaolan, Zhongjuan Sun and Pervez N. Ghauri. 2018. 'Reverse knowledge acquisition in emerging market MNEs: The experiences of Huawei and ZTE', *Journal of Business Research* 93 (December): 202–15. https://doi.org/10.1016/j.jbusres.2018.04.022.

Gadgets Now. 2019. '10 biggest smartphone companies of the world | Gadgets Now'. February 2019. Accessed 1 October 2020. https://www.gadgetsnow.com/slideshows/10-biggest-smartphone-companies-of-the-world/photolist/68097589.cms.

Garnham, N. 1986. 'The media and the public sphere'. In *Communicating Politics*, 44–53. Leicester: Leicester University Press.

Garsten, Christina. 1994. *Apple World: Core and periphery in a transnational organizational culture: A study of Apple Computer Inc*. Stockholm: Almqvist & Wiksell International.

Giordano, Cristiana. 2014. *Migrants in Translation: Caring and the logics of difference in contemporary Italy*. Berkeley, CA: University of California Press.

'Giovani Musulmani d'Italia GMI'. 2020. Facebook group page, 2020. Accessed 1 October 2020. https://www.facebook.com/GiovaniMusulmanidItaliaGMI/.

Goffman, Erving. 1971. *Relations in Public: Microstudies of the public order*. New York: Basic Books.

Goffman, Erving. 1972. *Frame Analysis*. New York: Harper and Row.

Gombrich, E. H. 1984. *The Sense of Order: A study in the psychology of decorative art*, 2nd ed. London: Phaidon Press.

Gómez Cruz, Edgar and Eric T. Meyer. 2012. 'Creation and control in the photographic process: iPhones and the emerging fifth moment of photography'. *Photographies* 5 (2): 203–21. https://doi.org/10.1080/17540763.2012.702123.

Gómez Cruz, Edgar and Asko Lehmuskallio, eds. 2016. *Digital Photography and Everyday Life: Empirical studies on material visual practices*. London; New York: Routledge.

Gopinath, Sumanth S. and Jason Stanyek, eds. 2014. *The Oxford Handbook of Mobile Music Studies, Volume 2*. Oxford: Oxford University Press.

Governo do Brazil (Government of Brazil). 2020. 'Governo trabalha para digitalizar todos serviços públicos'. Gov.br. Official government website for Brazil. 13 July 2020. Accessed 20 September 2020. https://www.gov.br/pt-br/noticias/financas-impostos-e-gestao-publica/2020/07/governo-trabalha-para-digitalizar-todos-servicos-publicos.

Governo Federal (Brazilian federal government). 2020. 'Desenvolvimento social'. Ministério Da Cidadania (Brazil). 2020. Accessed 1 October 2020. http://desenvolvimentosocial.gov.br/auxilio-emergencial/auxilio-emergencial-de-600.

Graham, Mark and William Dutton, eds. 2019. *Society and the Internet: How networks of information and communication are changing*, 2nd edition. Oxford: Oxford University Press.

Gray, Mary L. and Siddharth Suri. 2019. *Ghost Work: How to stop Silicon Valley from building a new global underclass*. Boston, MA: Houghton Mifflin Harcourt.

Graziani, Tomas. 2019. 'WeChat Official Account: A simple guide'. Walk the Chat. 11 December 2019. Accessed 1 October 2020. https://walkthechat.com/wechat-official-account-simple-guide/#wechat-official-acct.

Green, Nicola and Leslie Haddon. 2009. *Mobile Communications: An introduction to new media*, English ed. Oxford; New York: Berg.

Greenwald, Glenn. 2014. *No Place to Hide: Edward Snowden, the NSA and the surveillance state*. London: Hamilton.

Greschke, Heike Mónika. 2012. *Is There a Home in Cyberspace? The internet in migrants' everyday life and the emergence of global communities*. Abingdon, Oxon; New York: Routledge.

Griffith, Alison I. 1998. 'Insider / outsider: Epistemological privilege and mothering work'. *Human Studies* 21 (4): 361–76. https://doi.org/10.1023/A:1005421211078.

Grupo Casa. 2012. 'Waze arrives officially in Brazil'. 22 June 2012. Accessed 1 October 2020. http://grupocasa.com.br/waze-arrives-officially-in-brazil/.

Guess, Andrew, Jonathan Nagler and Joshua Tucker. 2019. 'Less than you think: Prevalence and predictors of fake news dissemination on Facebook'. *Science Advances* 5 (1): eaau4586. Accessed 1 October 2020. https://doi.org/10.1126/sciadv.aau4586.

Gupta, S. and I. Dhillon. 2014. 'Can Xiaomi shake the global smartphone industry with an innovative "services-based business model?"' *Journal of Management & Research* 8 (3/4): 2177–97.

Habermas, J. 1989. *The Structural Transformation of the Public Sphere*. Cambridge: Polity.

Halavais, Alexander M. Campbell. 2017. *Search Engine Society*. Cambridge, UK; Medford, MA: Polity.

Haynes, Nell. 2016. *Social Media in Northern Chile*. London: UCL Press.

Headspace. 2020. 'Mindfulness for your everyday life'. Headspace app website. 2020. Accessed 1 October 2020. https://www.headspace.com/.

Hell-Valle, J. and A. Storm-Mathisen, eds. 2020. *Media Practices and Changing African Socialities*. London: Berghahn.

Hendry, J. 1995. *Wrapping Culture: Politeness, presentation, and power in Japan and other societies*. Oxford: Oxford University Press.

Henig, Samantha. 2005. 'The tale of dog poop girl is not so funny after all'. *Columbia Journalism Review*, 7 July 2005. https://archives.cjr.org/behind_the_news/the_tale_of_dog_poop_girl_is_n.php.

Henrique, Alfredo. 2019. 'Cidade de São Paulo tem 13 celulares roubados por hora' ('Thirteen mobile phones were stolen every hour in São Paulo'), *Folha de São Paulo*, 7 June 2019. Accessed 1 October 2020. https://agora.folha.uol.com.br/sao-paulo/2019/06/cidade-de-sao-paulo-tem-13-celulares-roubados-por-hora.shtml.

Hingle, Melanie, Mimi Nichter, Melanie Medeiros and Samantha Grace. 2013. 'Texting for health: The use of participatory methods to develop healthy lifestyle messages for teens'. *Journal of Nutrition Education and Behavior* 45 (1): 12–19. https://doi.org/10.1016/j.jneb.2012.05.001.

Hirshauga, O. and H. Sheizaf. 2017. 'Targeted prevention: The new method of dealing with terrorism is exposed'. *Haaretz*, 26 May 2017. Accessed 1 October 2020. https://www.haaretz.co.il/magazine/.premium-MAGAZINE-1.4124379.

Hjorth, L., K. Ohashi, J. Sinanan, H. Horst, Sarah Pink, F. Kato and B. Zhou. 2020. *Digital Media Practices in Households*. Amsterdam: Amsterdam University Press.

Hobbis, Geoffrey. 2020. *The Digitizing Family: An ethnography of Melanesian smartphone*. Cham, Switzerland: Springer Nature Switzerland AG.

Hockstein, N. G., C. G. Gourin, R. A. Faust and D. J. Terris. 2007. 'A history of robots: From science fiction to surgical robotics'. *Journal of Robotic Surgery* 1 (2): 113–18. https://doi.org/10.1007/s11701-007-0021-2.

Holroyd, K. 2017. 'The digital Galapagos: Japan's digital media and digital content economy'. *Japan Studies Association Journal* 15 (1): 41–65.

Horst, Heather A. 2013. 'The infrastructures of mobile media: Towards a future research agenda'. *Mobile Media & Communication* 1 (1): 147–52. https://doi.org/10.1177/2050157912464490.

Horst, Horst, Heather A. and Daniel Miller, eds. 2012. *Digital Anthropology*, English ed. London; New York: Berg.

Huang, Zheping. 2019. 'China's most popular app is a propaganda tool teaching Xi Jinping thought'. *South China Morning Post*, 14 February 2019. Accessed 1 October 2020. https://www.scmp.com/tech/apps-social/article/2186037/chinas-most-popular-app-propaganda-tool-teaching-xi-jinping-thought.

Hughes, Christopher and Gudrun Whacker. 2003. *China and the Internet: Politics of the digital leap forward*. London; New York: Routledge.

Humphreys, Lee. 2018. *The Qualified Self: Social media and the accounting of everyday life*. Cambridge, MA: MIT Press.

IEEE. 2020. *IEEE Internet of Things Journal*. 2020. https://ieeexplore.ieee.org/xpl/RecentIssue.jsp?punumber=6488907.

Instituto Nacional De Estadisticas (INE) and Departamento de Extranjeria y Migracion (DEM). 2019. 'Estimación de personas extranjeras residentes en Chile 31 de Diciembre 2018'. Santiago,

Chile: Estadísticas Migratorias. Accessed 1 October 2020. https://www.extranjeria.gob.cl/media/2019/04/Presentaci%C3%B3n-Extranjeros-Residentes-en-Chile.-31-Diciembre-2018.pdf.

Iqbal, Mansoor. 2019. 'WhatsApp revenue and usage statistics (2019)'. *Business of Apps*, 19 February 2019. https://www.businessofapps.com/data/whatsapp-statistics/.

Iqbal, Mansoor. 2020. 'Line revenue and usage statistics (2020)'. *Business of Apps*, 28 April 2020. Accessed 1 October 2020. https://www.businessofapps.com/data/line-statistics/.

Israel's Ministry of Social Equality. 2020. 'Headquarters for the National Digital Israel Initiative, Ministry of Social Equality'. Israeli government website. 2020. Accessed 1 October 2020. https://www.gov.il/en/departments/digital_israel.

Istepanian, R. S. H., S. Laxminarayan and C. Pattichis, eds. 2006. *M-Health: Emerging mobile health systems*. New York: Springer.

Ito, Mizuko. 2005. 'Mobile phones, Japanese youth, and the re-placement of social contact'. In *Mobile Communications*, 131–48. London: Springer-Verlag. https://doi.org/10.1007/1-84628-248-9_9.

Itō, Mizuko, Daisuke Okabe and Misa Matsuda, eds. 2005. *Personal, Portable, Pedestrian: Mobile phones in Japanese life*. Cambridge, MA: MIT Press.

Jackson, Michael. 1995. *At Home in the World*. Durham, NC: Duke University Press.

Jao, N. 2018. 'A clone of a failed mobile game has just gone viral on WeChat'. *Technode*, 9 January 2018. Accessed 1 October 2020. https://technode.com/2018/01/09/wechat-viral-game/.

Jia, Kai, Martin Kenney and John Zysman. 2018. 'Global competitors? Mapping the internationalization strategies of Chinese digital platform firms'. In *International Business in the Information and Digital Age*, edited by Rob van Tulder, Alain Verbeke and Lucia Piscitello, 187–215. Progress in International Business Research series, Vol. 13, chap. 8. https://doi.org/10.1108/S1745-886220180000013009.

Jia, Lianrui and Dwayne Winseck. 2018. 'The political economy of Chinese internet companies: Financialization, concentration, and capitalization'. *International Communication Gazette* 80 (1): 30–59. https://doi.org/10.1177/1748048517742783.

Jiang, M. 2012. 'Internet companies in China: Dancing between the party line and the bottom line'. *Asie Visions* 47 (January). https://ssrn.com/abstract=1998976.

Johnson, M. and J. Lindquist. 2020. 'Care in Asia'. *Ethnos* 85 (2): 195–399.

Jorgensen, D. 2018. 'Toby and "the mobile system": Apocalypse and salvation in Papua New Guinea's wireless network'. In *The Moral Economy of Mobile Phones: Pacific Islands perspectives*, edited by R. Foster and H. Horst, 53–73. Acton, Australia: Australian National University Press.

Jovicic, Suzana. Under review. 'Scrolling and the in-between spaces of boredom: Youths on the periphery of Vienna'.

Jurgenson, N. 2019. *The Social Photo: On photography and social media*. London; New York: Verso.

Katz, James Everett and Mark A. Aakhus, eds. 2002. *Perpetual Contact: Mobile communication, private talk, public performance*. Cambridge; New York: Cambridge University Press. https://doi.org/10.1017/CBO9780511489471.

Kavedžija, Iza. 2019. *Making Meaningful Lives: Tales from an aging Japan*. Philadelphia: University of Pennsylvania Press.

Keane, Michael. 2020. 'Civilization, China and digital technology'. Open access. E-International Relations, 1 February 2020. Accessed 1 October 2020. https://www.e-ir.info/2020/02/01/civilization-china-and-digital-technology/.

Kedmey, D. 2014. 'Facebook's new tool lets you tell your friends you're safe during an emergency'. *TIME Magazine*, 16 October 2014. Accessed 1 October 2020. https://time.com/3513016/facebook-safety-check/.

Kelty, Christopher M. 2008. *Two Bits: The cultural significance of free software*. Durham, NC: Duke University Press.

Kemp, Simon. 2020. 'Digital trends 2020: Every single stat you need to know about the internet'. *The Next Web*, 30 January 2020. Accessed 1 October 2020. https://thenextweb.com/growthquarters/2020/01/30/digital-trends-2020-every-single-stat-you-need-to-know-about-the-internet/.

Kim, S. D. 2002. 'Korea: Personal meanings'. In *Perpetual Contact: Mobile communication, private talk, public performance*, edited by J. Katz and M. Aakhus, 63–79. Cambridge: Cambridge University Press.

Kirkpatrick, David. 2010. *The Facebook Effect: The real inside story of Mark Zuckerberg and the world's fastest-growing company*. New York: Simon and Schuster.

Kodama, M., ed. 2015. *Collaborative Innovation: Developing health support ecosystems*, Vol. 39. New York; London: Routledge.

Kress, Gunther R. 2003. *Literacy in the New Media Age*. London; New York: Routledge.

Kriedte, Peter, Hans Medick and Jürgen Schlumbohm. 1981. *Industrialization before Industrialization: Rural industry in the genesis of capitalism*. Cambridge; New York: Cambridge University Press.

Ku, Yi-Cheng, Yi-an Lin and Zhijun Yan. 2017. "Factors driving mobile app users to pay for freemium services". Paper presented at 21st Pacific Asia Conference on Information Systems (PACIS 2017): Langkawi, Malaysia, 16–20 July 2017. Accessed 1 October 2020. https://pdfs.semanticscholar.org/1414/42501c8130fb480e4958a300bd295482d26d.pdf.

Kumar, V. 2014. 'Making "freemium" work'. *Harvard Business Review*, May 2014. https://hbr.org/2014/05/making-freemium-work.

Kurniawan, Sri. 2006. 'An exploratory study of how older women use mobile phones'. In *UbiComp 2006: Ubiquitous computing*, ed. by Paul Dourish and Adrian Friday, 4206:105–22. Berlin; Heidleberg: Springer. https://doi.org/10.1007/11853565_7.

Kusimba, Sibel, Yang Yang and Nitesh Chawla. 2016. 'Hearthholds of mobile money in Western Kenya'. *Economic Anthropology* 3 (2): 266–79. https://doi.org/10.1002/sea2.12055.

Kyodo News Agency. 2019a. '613,000 in Japan aged 40 to 64 are recluses, says first government survey of *hikikomori*', 29 March 2019. Accessed 30 September 2020. https://www.japantimes.co.jp/news/2019/03/29/national/613000-japan-aged-40-64-recluses-says-first-government-survey-hikikomori/#.Xl6UCKj7Q2w.

Kyodo News Agency. 2019b. 'Japan enacts bill aimed at lowering mobile phone fees'. *Japan Times*, 10 May 2019. Accessed 30 September 2020. https://www.japantimes.co.jp/news/2019/05/10/business/corporate-business/japan-enacts-bill-aimed-lowering-mobile-phone-fees/#.Xr6LymhKg2x.

Lanchester, J. 2019. 'Document number nine'. *London Review of Books*, 10 October 2019. Accessed 30 September 2020. https://www.lrb.co.uk/the-paper/v41/n19/john-lanchester/document-number-nine.

Lasch, Christopher. 1979. *The Culture of Narcissism: American life in an age of diminishing expectations*. New York: Norton & Company.

Lavado, T. 2019. 'Facebook lança rival do tinder no Brasil'. *Globo*, 20 April 2019. Accessed 30 September 2020. https://g1.globo.com/economia/tecnologia/noticia/2019/04/30/facebook-lanca-rival-do-tinder-no-brasil.ghtml.

Leswing, Kif. 2019. 'Inside Apple's team that greenlights iPhone apps for the App Store'. CNBC, 21 June 2019. Accessed 30 September 2020. https://www.cnbc.com/2019/06/21/how-apples-app-review-process-for-the-app-store-works.html.

Leung, Rock, Charlotte Tang, Shathel Haddad, Joanna Mcgrenere, Peter Graf and Vilia Ingriany. 2012. 'How older adults learn to use mobile devices: Survey and field investigations'. *ACM Transactions on Accessible Computing* 4 (3): 1–33. https://doi.org/10.1145/2399193.2399195.

Li, Shancang, Li D. Xu and Imed Romdhani. 2017. *Securing the Internet of Things*. Cambridge, MA: Syngress.

Li Sun, Sunny, Hao Chen and Erin G. Pleggenkuhle-Miles. 2010. 'Moving upward in global value chains: The innovations of mobile phone developers in China'. Edited by Robert Tiong. *Chinese Management Studies* 4 (4): 305–21. https://doi.org/10.1108/17506141011094118.

Licoppe, C. and Heurtin, J.-P. 2002. 'France: Preserving the image'. In *Perpetual Contact*, edited by J. Katz and M. Aakhus, 94–109. Cambridge: Cambridge University Press.

Lim, S. S. 2020. *Transcendent Parenting: Raising children in the digital age*. Oxford: Oxford University Press.

Linecorp 2019. 'LINE Announces Custom Stickers – Create Your Own Stickers in Minutes Using Popular LINE Characters'. Linecorp website, 11 April 2019. Accessed 8 January 2021. https://linecorp.com/en/pr/news/en/2019/2666.html.

Ling, Richard Seyler. 2004. *The Mobile Connection: The cell phone's impact on society*. San Francisco, CA: Morgan Kaufmann.

Ling, Richard Seyler. 2012. *Taken for Grantedness: The embedding of mobile communication into society*. Cambridge, MA: MIT Press.

Ling, Richard Seyler. and Birgitte Yttri. 2002. 'Hyper-coordination via mobile phones in Norway'. In *Perpetual Contact*, edited by J. Katz and M. Aakhus, 170–92. Cambridge: Cambridge University Press.

Lipset, D. 2018. 'A handset dangling in a doorway: Mobile phone sharing in a rural sepik village (Papua New Guinea)'. In *The Moral Economy of Mobile Phones: Pacific Islands perspectives*, edited by R. Foster and H. Horst, 19–38. Acton, Australia: Australian National University Press.

Liu, Xuefeng, Yuying Xie and Mangui Wu. 2015. 'How latecomers innovate through technology modularization. Evidence from China's Shanzhai industry'. *Innovation* 17 (2): 266–80. https://doi.org/10.1080/14479338.2015.1039636.

Livingstone, Sonia. 2009. *Children and the Internet*. Cambridge: Polity.

Livingstone, Sonia. 2019. 'Parenting in the digital age'. TED Talk presented at the TED Summit 2019, July 2019. Accessed 1 October 2020. https://www.ted.com/talks/sonia_livingstone_parenting_in_the_digital_age.

Livingstone, Sonia M. and Julian Sefton-Green. 2016. *The Class: Living and learning in the digital age*. New York: New York University Press.

Livingstone, Sonia, Alicia Blum Ross, Kate Gilchrist and Paige Mustain. 2020. 'Welcome to our blog'. *Parenting 4 Digital Future Blog (LSE) – A blog about growing up in a digital world*. 2020. Accessed 1 October 2020. https://blogs.lse.ac.uk/parenting4digitalfuture/.

Long, Susan O. 2012. 'Bodies, technologies, and aging in Japan: Thinking about old people and their silver products'. *Journal of Cross Cultural Gerontology* 27 (2): 119–37. https://doi.org/10.1007/s10823-012-9164-3.

Lui, Natalie. 2019. 'WeChat mini programs: The complete guide for business'. *Dragonsocial* (a commercial website). 19 June 2019. Accessed 1 October 2020. https://www.dragonsocial.net/blog/wechat-mini-programs/.

Luo, Chris. 2014. 'China's latest internet sensation: Young man's hand-drawn guide to WeChat for his parents'. *South China Morning Post*, 26 February 2014. Accessed 30 September 2020. https://www.scmp.com/news/china-insider/article/1435568/sons-hand-drawn-guide-wechat-parents-goes-down-storm-chinese.

Lupton, Deborah. 2015. *Digital Sociology*. Abingdon, Oxon; New York: Routledge.

Lupton, Deborah. 2020. 'Topical mapping of academic publications on social aspects of Covid-19'. 2020. Accessed 30 September 2020. https://simplysociology.files.wordpress.com/2020/07/lupton-map-of-social-research-on-covid-19-july-2020-3.pdf.

Lury, Celia. 1997. *Prosthetic Culture: Photography, memory and identity*. Abingdon, Oxon; New York: Routledge.

Lutz, Helma. 2018. 'Care migration: The connectivity between care chains, care circulation and transnational social inequality'. *Current Sociology* 66 (4): 577–89. https://doi.org/10.1177/0011392118765213.

MacKenzie, Donald A. and Judy Wajcman, eds. 1999. *The Social Shaping of Technology*, 2nd ed. Buckingham, UK; Philadelphia, PA: Open University Press.

Madianou, Mirca. 2015. 'Digital inequality and second-order disasters: Social media in the typhoon Haiyan recovery'. *Social Media + Society* 1 (2): 205630511560338. https://doi.org/10.1177/2056305115603386.

Madianou, Mirca and Daniel Miller. 2012. *Migration and New Media: Transnational families and polymedia*. Abingdon, Oxon; New York: Routledge.

Maistre, Xavier de and Stephen Sartarelli. 1994. *Voyage around My Room: Selected works of Xavier DeMaistre*. New York, NY: New Directions.

Margetts, Helen, Peter John, Scott A. Hale and Taha Yasseri. 2016. *Political Turbulence: How social media shape collective action*. Princeton, NJ: Princeton University Press.

Marwick, Alice E. and danah boyd. 2010. 'I tweet honestly, I tweet passionately: Twitter users, context collapse, and the imagined audience'. *New Media & Society* 13 (1): 114–33. https://doi.org/10.1177/1461444810365313.

Maurer, Bill. 2012. 'Mobile money: Communication, consumption and change in the payments space'. *Journal of Development Studies* 48 (5): 589–604. https://doi.org/10.1080/00220388.2011.621944.

Maxwell, Richard and Toby Miller. 2012. *Greening the Media*. New York: Oxford University Press.

Maxwell, Richard and Toby Miller. 2020. *How Green Is Your Smartphone?* Cambridge, UK; Medford, MA: Polity.

McCulloch, Gretchen. 2019. *Because Internet: Understanding how language is changing*. London: Harvill Secker.

McDonald, Tom. 2016. *Social Media in Rural China: Social networks and moral frameworks*. London: UCL Press.

McGrath, Dominic. 2020. 'Why was the Covid-19 app so successful in Ireland?' *The Journal.Ie*, 11 July 2020. Accessed 1 October 2020. https://www.thejournal.ie/covid-19-app-ireland-success-5146093-Jul2020/.

Mcintosh, Janet. 2010. 'Mobile phones and Mipoho's prophecy: The powers and dangers of flying language'. *American Ethnologist* 37 (2): 337–53. Accessed 1 October 2020. https://doi.org/10.1111/j.1548-1425.2010.01259.x.

McNamee, Roger. 2019. *Zucked: Waking up to the Facebook catastrophe*. New York: Penguin Press.

'Mensaje Presidencial de S.E. el Presidente de la República, Sebastián Piñera Echenique, en su Cuenta Pública ante el Congreso Nacional'. 2018. 1 June 2018. Accessed 1 October 2020. https://prensa.presidencia.cl/lfi-content/uploads/2018/06/jun012018arm-cuenta-publica-presidencial_3.pdf.

Merola, Francesco. 2018. 'Italiani, sempre più smartphone-mania: Il 61% li usa a letto, Il 34% a tavola'. *La Repubblica*, 26 June 2018. Accessed 1 October 2020. https://www.repubblica.it/tecnologia/2018/06/26/news/dipendenza_degli_italiani_ad_internet-200069807/.

Merton, Robert K. 1972. 'Insiders and outsiders: A chapter in the sociology of knowledge'. *American Journal of Sociology* 78 (1): 9–47. https://doi.org/10.1086/225294.

Miller, Daniel. 1987. *Material Culture and Mass Consumption*. Oxford: Blackwell.

Miller, Daniel. 1995. 'Style and ontology in Trinidad'. In *Consumption and Identity*, edited by J. Friedman, 71–96. Chur, Switzerland: Harwood Academic.

Miller, Daniel. 1997. *Capitalism: An ethnographic approach*. Oxford, UK; Washington, D.C: Berg.

Miller, Daniel, ed. 2009. *Anthropology and the Individual: A material culture perspective*. Oxford; New York: Berg.

Miller, Daniel. 2011. *Tales from Facebook*. Cambridge, UK; Malden, MA: Polity.

Miller, Daniel. 2013. 'What will we learn from the fall of Facebook?' *UCL Blogs – Global social media impact study* (university blog). 24 November 2013.

Miller, Daniel. 2015. 'Photography in the age of Snapchat'. *Anthropology & Photography*, Vol.1. Royal Anthropological Institute. https://www.therai.org.uk/images/stories/photography/AnthandPhotoVol1.pdf.

Miller, Daniel. 2016. *Social Media in an English Village*. London: UCL Press.

Miller, Daniel. 2017a. 'The ideology of friendship in the era of Facebook'. *HAU: Journal of Ethnographic Theory* 7 (1): 377–95. https://doi.org/10.14318/hau7.1.025.

Miller, Daniel. 2017b. *The Comfort of People*. Cambridge, UK; Medford, MA: Polity.

Miller, Daniel and D. Slater. 2000. *The Internet: An ethnographic approach*. Oxford: Berg.

Miller, Daniel and Jolynna Sinanan. 2014. *Webcam*. Cambridge: Polity.

Miller, Daniel, Elisabetta Costa, Nell Haynes, Tom McDonald, Razvan Nicolescu, Jolynna Sinanan, Juliano Spyer, Shriram Venkatraman and Xinyuan Wang. 2016. *How the World Changed Social Media*. London: UCL Press.

Miller, Daniel and Jolynna Sinanan. 2017. *Visualising Facebook: A comparative perspective*. London: UCL Press.

Mirzoeff, Nicholas. 2015. *How to See the World: A Pelican introduction*. London: Penguin UK.

Mitchel, W. 1992. *The Reconfigured Eye: Visual truth in the post-photographic era*. Cambridge, MA: MIT Press.

Mobile Internet Statistics 2020. Accessed 3 December 2020. https://www.finder.com/uk/mobile-internet-statistics#:~:text=Quick%20overview,up%20from%2066%25%20in%202018.

Mohan, Babu. 2019. 'Google now takes three days to approve new play store apps'. *Android Central* (blog). 20 August 2019. Accessed 1 October 2020. https://www.androidcentral.com/google-now-takes-three-days-approve-new-play-store-apps.

Monnerat, A. 2019. 'Idosos compartilham sete vezes mais noticias falsas do que jovens no Facebook, diz Pesquisa'. *O Estadão*, 11 January 2019. Accessed 1 October 2020. https://politica.estadao.com.br/blogs/estadao-verifica/idosos-compartilham-sete-vezes-mais-noticias-falsas-do-que-usuarios-mais-jovens-no-facebook-diz-pesquisa/.

Moore, G. 1991. *Crossing the Chasm: Marketing and selling high-tech goods to mainstream customers*. New York: Harper Business.

Morley, David. 2000. *Home Territories: Media, mobility and identity*. London; New York: Routledge.

Morosanu Firth, S. Rintel and A. Sellen. 2020. 'Everyday time travel: Future nostalgia, multitemporality, and temporal mobility with smartphones'. In *Beyond Chrono(dys)topia: Making time for digital lives*, edited by Anne Kaun, C. Pentzold and C. Lohmeier. London: Rowman & Littlefield.

Morozov, Evgeny. 2012. *The Net Delusion: How not to liberate the world*. London: Penguin Books.

Morozov, Evgeny. 2013. *To Save Everything, Click Here: Technology, solutionism and the urge to fix problems that don't exist*. London: Allen Lane.

Morozov, Evgeny. 2020. 'The tech "solutions" for coronavirus take the surveillance state to the next level'. *The Guardian*, 15 April 2020. Accessed 1 October 2020. https://www.theguardian.com/commentisfree/2020/apr/15/tech-coronavirus-surveilance-state-digital-disrupt.

Morris, Jeremy Wade and Sarah Murray, eds. 2018. *Appified: Culture in the age of apps*. Ann Arbor, MI: University of Michigan Press.

Morris, Jeremy. 2018. 'Is It Tuesday? Novelty apps and digital solutionism'. In *Appified: Culture in the age of apps*, edited by Jeremy Wade Morris and Sarah Murray, 91–103. Ann Arbor, MI: University of Michigan Press.

Mugerwa, Yasiin and Tom Malaba. 2018. 'Museveni slaps taxes on social media users'. *The Daily Monitor*, 1 April 2018. Accessed 1 October 2020. https://www.monitor.co.ug/News/National/Museveni-taxes-social-media-users-Twitter-Skype/688334-4366608-oilivjz/index.htm.

Mumbere, Daniel. 2018. 'Digital in 2018: Africa's internet users increase by 20%'. *Africa News*, 6 February 2018. Accessed 1 October 2020. https://www.africanews.com/2018/02/06/digital-in-2018-africa-s-internet-users-increase-by-20-percent/.

Murray, Susan. 2008. 'Digital images, photo-sharing, and our shifting notions of everyday aesthetics'. *Journal of Visual Culture* 7 (2): 147–63. https://doi.org/10.1177/1470412908091935.

Namatovu, Esther and Oystein Saebo. 2015. 'Motivation and consequences of internet and mobile phone usage among the urban poor in Kampala, Uganda'. In *2015 48th Hawaii International Conference on System Sciences*, 4335–44. HI, USA: IEEE. https://doi.org/10.1109/HICSS.2015.519.

National Information Technology Authority (NITA). 2018. 'National Information Technology Survey 2017/18 Report'. Accessed 1 October 2020. https://www.nita.go.ug/sites/default/files/publications/National%20IT%20Survey%20April%2010th.pdf.

Naughton, J. 2000. *A Brief History of the Future: The origins of the internet*. London: Phoenix (Orion Books).

Nicolescu, Razvan. 2016. *Social Media in South Italy*. London: UCL Press.

Nissenbaum, Helen Fay. 2010. *Privacy in Context: Technology, policy, and the integrity of social life*. Stanford, CA: Stanford Law Books.

Norman, Jeremy M., ed. 2005. *From Gutenberg to the Internet: A sourcebook on the history of information technology*. Novato, CA: Historyofscience.com.

Nyamnjoh, Francis B. 2012. '"Potted plants in greenhouses": A critical reflection on the resilience of colonial education in Africa'. *Journal of Asian and African Studies* 47 (2): 129–54. https://doi.org/10.1177/0021909611417240.

O Estado de S. Paulo. 2017. 'Roubos de celular atingem metade das ruas de São Paulo'. *O Estado de S. Pãulo*, 30 September 2017. Accessed 1 October 2020. https://sao-paulo.estadao.com.br/noticias/geral,roubos-de-celular-atingem-metade-das-ruas-de-sao-paulo,70002022457.

O Globo. 2018. 'Golpes na internet: Veja as fraudes mais comuns e como se proteger'. *O Globo*, 2018. Accessed 1 October 2020. https://oglobo.globo.com/economia/defesa-do-consumidor/golpes-na-internet-veja-as-fraudes-mais-comuns-como-se-proteger-22485183.

Ong, W. 1982. *Orality and Literacy: The technologizing of the word*. London: Methuen.

Otaegui, Alfonso. 2019. 'Older adults in Chile as digital immigrants: Facing the "digital transformation" towards a paperless world'. *UCL ASSA Blog* (academic blog). 22 April 2019. Accessed on 1 October 2020. https://blogs.ucl.ac.uk/assa/2019/04/22/older-adults-in-chile-as-digital-immigrants-facing-the-digital-transformation-towards-a-paperless-world/.

Oudshoorn, Nelly. 2011. *Telecare Technologies and the Transformation of Healthcare*. Houndmills, Basingstoke, UK; New York: Palgrave Macmillan.

Papacharissi, Zizi. 2010. *A Networked Self: Identity, community, and culture on social network sites*. London: Taylor and Francis.

Papacharissi, Zizi. 2018. *A Networked Self and Love*. London: Taylor and Francis.

Pariser, Eli. 2012. *The Filter Bubble: What the internet is hiding from you*. London: Penguin Books.

Parulis-Cook, S. 2019. 'Survey: WeChat mini-program use for travel'. *DragonTrail Interactive* (marketing website). 19 February 2019. Accessed 1 October 2020. https://dragontrail.com/resources/blog/wechat-mini-program-travel-survey.

Patil, Adwait. 2016. 'Tracking down India's $4 smartphone'. *The Verge*. 2016. Accessed 1 October 2020. https://www.theverge.com/2016/3/18/11260488/india-ringing-bells-4-dollar-smartphone-controversy.

Peters, Benjamin. 2016. *How Not to Network a Nation: The uneasy history of the Soviet internet*. Cambridge, MA: MIT Press.

Petsas, Thanasis, Antonis Papadogiannakis, Michalis Polychronakis, Evangelos P. Markatos and Thomas Karagiannis. 2013. 'Rise of the Planet of the Apps: A systematic study of the mobile app ecosystem'. In *Proceedings of the 2013 Conference on Internet Measurement Conference – IMC '13*, 277–90. Barcelona, Spain: ACM Press. https://doi.org/10.1145/2504730.2504749.

Pinney, Christopher. 2012. 'Seven theses on photography'. *Thesis Eleven* 113 (1): 141–56. https://doi.org/10.1177/0725513612457864.

Plantin, Jean-Christophe and Gabriele de Seta. 2019. 'WeChat as infrastructure: The techno-nationalist shaping of Chinese digital platforms'. *Chinese Journal of Communication* 12 (3): 257–73. https://doi.org/10.1080/17544750.2019.1572633.

Pols, Jeanette. 2012. *Care at a Distance: On the closeness of technology*. Amsterdam: Amsterdam University Press.

Postill, John. 2011. *Localizing the Internet: An anthropological account*. Anthropology of Media, vol. 5. New York: Berghahn Books.

Postill, John. 2018. *The Rise of Nerd Politics: Digital activism and political change*. London: Pluto Press.

Prefeitura de São Paulo (São Paulo City Hall). 2013. 'LEI Nº 15.937 DE 23 DE DEZEMBRO DE 2013'. Prefeitura de São Paulo. http://legislacao.prefeitura.sp.gov.br/leis/lei-15937-de-23-de-dezembro-de-2013.

Prendergast, D. 2019. 'Ethnography, technology design and the future of "ageing in place"'. HRB Grant Holder's Conference, Athlone, Ireland. 2019. Accessed 1 October 2020. https://www.youtube.com/watch?v=5sSWrz5Dkig&list=PL5egX8ZzHdSyM4FCC9vJ5v1fTcTlOW5ZG&index=4.

Price, Catherine. 2018. *How to Break up with Your Phone*. London: Trapeze.

Pulse News KR. 2019. 'Naver takes telemedicine business to Japan through JV with M3', *Pulse News KR*, 16 January 2019. Accessed 1 October 2020. https://pulsenews.co.kr/view.php?year=2019&no=33579#:~:text=South%20Korean%20internet%20giant%20Naver,platform%20firm%20M3%20in%20Tokyo.

Pype, Katrien. 2015. 'Remediations of Congolese urban dance music in Kinshasa'. *Journal of African Media Studies* 7 (1): 25–36.

Pype, Katrien. 2016. 'Blackberry girls and Jesus's brides'. *Journal of Religion in Africa* 46 (4): 390–416. https://doi.org/10.1163/15700666-12341106.

Pype, Katrien. 2017. 'Smartness from Below: Variations on technology and creativity in contemporary Kinshasa'. In *What Do Science, Technology, and Innovation Mean from Africa?*, 97–115. Cambridge, MA: The MIT Press.

Rainie, Lee and B. Wellman. 2014. *Networked: The new social operating system*. Cambridge, MA: MIT Press.

Reuters Institute and OII. n.d. 'Reuters Institute digital news report 2019'. Accessed 1 October 2020. https://reutersinstitute.politics.ox.ac.uk/sites/default/files/inline-files/DNR_2019_FINAL.pdf.

Roberts, Sarah T. 2019. *Behind the Screen: Content moderation in the shadows of social media*. New Haven, CT: Yale University Press.

Rossler, Beate. 2005. *The Value of Privacy*. Cambridge: Polity.

RTÉ Radio 1. 2020. 'News at One', 15 January 2020. Accessed 1 October 2020. www.rte.ie/radio/radioplayer/html5/#/radio1/11140162.

Russell, Ben. 2017. *Robots: The 500-year quest to make machines human*. London: Scala Arts & Heritage Publishers Ltd.

Russell, John. 2019. 'Chat app line injects $182m into its mobile payment business'. *TechCrunch*, 4 February 2019. Accessed 1 October 2020. https://techcrunch.com/2019/02/04/line-pay/.

Samat, Sameer. 2019. 'Improving the update process with your feedback'. *Android Developers Blog*, 15 April 2019. Accessed 1 October 2020. https://android-developers.googleblog.com/2019/04/improving-update-process-with-your.html.

Sarvas, Risto and David M. Frohlich. 2011. *From Snapshots to Social Media: The changing picture of domestic photography*. London; New York: Springer.

Scancarello, G. 2020. *#Addicted: Viaggio dentro le manipolazioni della tecnologia*. Milano: Hoepli.

Schafer, M. 2015. 'Digital public sphere'. In *The International Encyclopaedia of Political Communication*, edited by Gianpietro Mazzoleni, K. Barnhurst, K. Ikedia, R. Maia and H. Wessler, 322–28. London: Wiley Blackwell.

Schaffer, Rebecca, Kristine Kuczynski and Debra Skinner. 2008. 'Producing genetic knowledge and citizenship through the internet: Mothers, pediatric genetics, and cybermedicine'. *Sociology of Health & Illness* 30 (1): 145–59. https://doi.org/10.1111/j.1467-9566.2007.01042.x.

Schwennessen, Nete. 2019. 'Surveillance entanglements: Digital data flows and ageing bodies in motion in the Danish welfare state'. *Anthropology & Aging* 40 (2): 10–22.

Serger, Sylvia Schwaag and Magnus Breidne. 2007. 'China's fifteen-year plan for science and technology: An assessment'. *Asia Policy*, no. 4: 135–64. https://doi.org/10.1353/asp.2007.0013.

Servidio, R. 2019. 'Self-control and problematic smartphone use among Italian university students: The mediating role of the fear of missing out and of smartphone use patterns'. *Current Psychology*, July 2019. https://doi.org/10.1007/s12144-019-00373-z.

Sheng, Wei. 2020. 'WeChat mini programs: The future is e-Commerce'. *TechNode*, 15 January 2020. Accessed 1 October 2020. https://technode.com/2020/01/15/wechat-mini-programs-the-future-is-e-commerce/.

Shifman, Limor. 2013. *Memes in Digital Culture*. Cambridge, MA: MIT Press.

Shim, Yongwoon and Dong-Hee Shin. 2016. 'Neo-techno nationalism: The case of China's handset industry'. *Telecommunications Policy* 40 (2–3): 197–209. https://doi.org/10.1016/j.telpol.2015.09.006.

Shirky, Clay. 2008. *Here Comes Everybody*. London: Allen Lane.

Shirky, Clay. 2015. *Little Rice: Smartphones, Xiaomi, and the Chinese Dream*. New York: Columbia Global Reports.

Shu, C. 2015. 'The secret language of line stickers'. *TechCrunch*, 10 July 2015. https://techcrunch.com/2015/07/10/creepy-cute-line/.

Shuken, Ryan. n.d. 'Growth hacking an audio sharing platform with Tian Sun, Vice President of Business Intelligence Center at Ximalaya App'. *China Star Pulse*. Accessed 1 October 2020. https://chinastartuppulse.simplecast.com/episodes/growth-hacking-an-audio-sharing-platform-tian-sun-ximalaya.

Silverstone, R. and D. Morley, eds. 1992. *Consuming Technology*. London; New York: Routledge.

Simmel, George. 1968. *The Conflict in Modern Culture and Other Essays*. New York: New York Teachers' College Press.

Simoni, Emilio. 2019. 'Carta do diretor'. *PSafe*, 2019. Accessed 1 October 2020. https://www.psafe.com/dfndr-lab/relatorio-da-seguranca-digital-2018/.

Sina Technology Comprehensive (Sina Corp). 2019. 'People's daily overseas edition: Involving the elderly in the internet needs multiple efforts'. Sina Technology Comprehensive (Sina Corp), 22 February 2019. Accessed 1 October 2020. https://tech.sina.cn/i/gn/2019-02-22/detail-ihqfskcp7412236.d.html?from=wap.

Singh, R. 2015. 'Older people and constant contact media'. In *Aging and the Digital Life Course*, edited by David Prendergast and Chiara Garattini, 1st ed., 63–83. New York, Oxford: Berghahn Books. Retrieved 2 October 2020, from https://www.jstor.org/stable/j.ctt9qdb6b.

Slater, D., K. Nishimura and L. Kindstrand. 2012. 'Social media, information, and political activism in Japan's 3.11 Crisis'. *The Asia-Pacific Journal* 1, 10 (24). Accessed 1 October 2020. https://apjjf.org/2012/10/24/David-H.-Slater/3762/article.html.

Smith, Craig. 2020. '65 amazing LINE statistics and facts'. DMR – Business Statistics. 20 February 2020. Accessed 1 October 2020. https://expandedramblings.com/index.php/line-statistics/.

Social Street. 2020. 'Social Street: Dal virtuale al reale al virtuoso'. 2020. Accessed 1 October 2020. http://www.socialstreet.it/.

Solon, Olivia. 2018. 'Teens are abandoning Facebook in dramatic numbers, study finds'. *The Guardian*, 1 June 2018. Accessed 1 October 2020. https://www.theguardian.com/technology/2018/jun/01/facebook-teens-leaving-instagram-snapchat-study-user-numbers.

Somatosphere.net. 2020. 'Medical anthropology weekly: COVID-19', 2020. Accessed 1 October 2020. http://somatosphere.net/medical-anthropology-weekly-covid-19/.

Sorokowski, P., A. Sorokowska, A. Oleszkiewicz, T. Frackowiak, A. Huk and K. Pisanski. 2015. 'Selfie posting behaviours are associated with narcissism among men'. *Personality and Individual Differences* 85: 123–27.

Sousa Pinto, A. E. de. 2018. 'Uso do celular prolonga saúde mental de idosos'. *Folha de São Paulo*, May 2018. Accessed 1 October 2020. https://www1.folha.uol.com.br/cotidiano/2019/05/uso-do-celular-prolonga-saude-mental-de-idosos.shtml.

Souza e Silva, Adriana de. 2014. *Mobility and Locative Media: Mobile communication in hybrid spaces*. London; New York: Routledge. https://doi.org/10.4324/9781315772226.

Spadafora, A. 2018. 'Tablet device sales struggle again'. 2 November 2018. Accessed 1 October 2020. https://www.techradar.com/news/tablet-device-sales-struggle.

Spyer, Juliano. 2017. *Social Media in Emergent Brazil: How the internet affects social change*. London: UCL Press.

Srnicek, Nick. 2017. *Platform Capitalism*. Cambridge, UK; Malden, MA: Polity.

Standage, Tom. 2013. *Writing on the Wall: Social media – the first 2,000 years*. London: Bloomsbury.

Stark, Luke and Kate Crawford. 2015. 'The conservatism of emoji: Work, affect and communication'. *Social Media + Society* 1 (2): 205630511560485. https://doi.org/10.1177/2056305115604853.

Statista. 2019. 'Number of smartphone users by country as of September 2019 (in millions)'. *Statista*, September 2019. Accessed 30 September 2020. https://www.statista.com/statistics/748053/worldwide-top-countries-smartphone-users/.

Statista. 2020. 'Number of monthly active WeChat users from 2nd Quarter 2011 to 1st Quarter 2020'. *Statista*, 20 May 2020. https://www.statista.com/statistics/255778/number-of-active-wechat-messenger-accounts/.

Steinberg, Marc. 2020. 'LINE as super app: Platformization in East Asia'. *Social Media + Society* 6 (2): 205630512093328. https://doi.org/10.1177/2056305120933285.

Subsecretaria de Telecomunicaciones (Subsecretary of Telecommunications, Chile). 2019. 'Conexiones 4G se disparan 35% en 2018 y abre expectativas de cara al despliegue de 5G'. Chilean government website, subtel.Gob.Cl, 10 April 2019. Accessed 1 October 2020. https://www.subtel.gob.cl/conexiones-4g-se-disparan-35-en-2018-y-abre-expectativas-de-cara-al-despliegue-de-5g/.

Sumpter, David. 2018. *Outnumbered: From Facebook and Google to fake news and filter-bubbles – the algorithms that control our lives*. London: Bloomsbury Sigma.

Sutton, Theodora. 2017. 'Disconnect to reconnect: The food/technology metaphor in digital detoxing'. *First Monday*, June 2017. https://doi.org/10.5210/fm.v22i6.7561.

Sutton, Theodora. 2020. 'Digital harm and addiction: An anthropological view'. *Anthropology Today* 36 (1): 17–22. https://doi.org/10.1111/1467-8322.12553.

Sweeny, Alastair. 2009. *BlackBerry Planet: The story of research in motion and the little device that took the world by storm*. Mississauga, Ont: John Wiley & Sons Canada.

Tagal, J. 2008. 'The mosaic browser democratises the world wide web, 1993'. *Financial Times*, 5 July 2008. Accessed 1 October 2020. https://www.ft.com/content/2126bb5c-4ffc-11dd-a851-000077b07658.

Taub Center. 2017. 'מרכז טאוב זכרם בואט יארק למחקר מדינות ישראל'. לארשיב תיברעה היסולכואה תואירב (2017), .םילשורי 'The health of the Arab Israeli population'. Accessed 1 October 2020. http://taubcenter.org.il/wp-content/files_mf/healthofthearabisraelipopulationheb.pdf.

Tenhunen, S. 2018. *A Village Goes Mobile: Telephony, mediation, and social change in rural India*. Oxford: Oxford University Press.

The Economist. 2019. 'A global timepass economy – How the pursuit of leisure drives internet use'. *The Economist*, 8 June 2019.

The Economist. 2020a. 'A global microscope made of phones'. *The Economist*, 16 April 2020.

The Economist. 2020b. 'England's contact-tracing system (finally) gets parochial'. *The Economist*, 'Fighting Covid-19' section, 15 August 2020.

The Economist. 2020c. 'How centralisation impeded Britain's Covid-19 response'. *The Economist*, 18 July 2020.

The Guardian [editorial]. 2013. 'Civil liberties: Surveillance and the state'. *The Guardian [Editorial]*, 16 June 2013. Accessed 30 September 2020. https://www.theguardian.com/commentisfree/2013/jun/16/civil-liberties-surveillance-state-editorial.

The Local (no author). 2019. 'Italian government unveils plan to tackle smartphone addiction'. *The Local (IT)*, 22 July 2019. Accessed 30 September 2020. https://www.thelocal.it/20190722/italian-government-unveils-plan-to-tackle-smartphone-addiction.

The Telegraph. 2019. 'Quarter of mobile phone users make less than five calls a month, Ofcom figures show'. *The Telegraph*, 10 October 2019. Accessed 1 October 2020. https://www.telegraph.co.uk/news/2019/10/09/quarter-mobile-phone-users-make-less-five-calls-month-ofcom/.

Thompson, Clive. 2013. *Smarter than You Think: How technology is changing our minds for the better*. New York: Penguin Books.

Thumala, Daniela. 2017. 'Imágenes sociales del envejecimiento'. Lecture/course material presented at the 'Cómo envejecemos: una mirada transdisciplinaria', Universidad Abierta, Universidad de Chile.

Ticktin, Miriam Iris. 2011. *Casualties of Care: Immigration and the politics of humanitarianism in France*. Berkeley, CA: University of California Press.

Tiongson, James. 2015. 'Mobile app marketing insights: How consumers really find and use your apps'. Think with Google. 2015. Accessed 1 October 2020. https://www.thinkwithgoogle.com/consumer-insights/mobile-app-marketing-insights/.

Travezuk, Thomas. 2018. 'Brasil soma quase 26 mil tentativas de golpes virtuais por dia'. *R7*, 29 July 2018. Accessed 1 October 2020. https://noticias.r7.com/economia/brasil-soma-quase-26-mil-tentativas-de-golpes-virtuais-por-dia-29072018.

Turkle, Sherry. 1984. *The Second Self: Computers and the human spirit*. Cambridge, MA: MIT Press.

UCL Anthropology. 2020. 'Medical anthropology blog posts'. *UCL Medical Anthropology Blog Posts*, 2020. Accessed 1 October 2020. https://www.ucl.ac.uk/anthropology/study/graduate-taught/biosocial-medical-anthropology-msc/medical-anthropology-blog-posts.

Venkatraman, S. 2017. *Social Media in South India*. London: UCL Press.

Vertesi, Janet. 2014. 'Seamful spaces: Heterogeneous infrastructures in interaction'. *Science, Technology, & Human Values* 39 (2): 264–84. https://doi.org/10.1177/0162243913516012.

Vieira, N. 2019. 'Idosos: Um público cada vez mais adepto à tecnologia'. *CanalTech*, 17 November 2019. Accessed 1 October 2020. https://canaltech.com.br/comportamento/idosos-um-publico-cada-vez-mais-adepto-a-tecnologia-154977/.

Villalobos, A. 2017. 'Conceptos básicos acerca del autocuidado'. Lecture/course material presented at the 'Cómo envejecemos: una mirada transdisciplinaria', Universidad Abierta, Universidad de Chile.

de Vries, M. Under review. 'The voice of silence: Practices of participation among East Jerusalem Palestinians'.

Wallis, Cara. 2013. *Technomobility in China: Young migrant women and mobile phones*. New York: New York University Press.

Walton, S. 2016. 'Photographic truth in motion – The case of Iranian photoblogs'. *Anthropology & Photography* 4. Accessed 30 September 2020. http://www.therai.org.uk/images/stories/photography/AnthandPhotoVol4.pdf.

Wang, H. 2014. 'Machine for a long revolution: Computer as the nexus of technology and class politics in China 1955–1984'. PhD thesis. Hong Kong: The Chinese University of Hong Kong.

Wang, Xinyuan. 2016. *Social Media in Industrial China*. London: UCL Press.

Wang, Xinyuan. 2019a. 'Hundreds of Chinese citizens told me what they thought about the controversial social credit system'. *The Conversation*, 17 December 2019. Accessed 1 October 2020. https://theconversation.com/hundreds-of-chinese-citizens-told-me-what-they-thought-about-the-controversial-social-credit-system-127467.

Wang, X. 2019b. 'China's social credit system: The Chinese citizens perspective'. *UCL ASSA blog*. 9 December 2019. Accessed 1 October 2020. https://blogs.ucl.ac.uk/assa/2019/12/09/chinas-social-credit-system-the-chinese-citizens-perspective/.

Ward, Mark. 2009. 'Celebrating 40 years of the net'. BBC News, 29 October 2009. Accessed 1 October 2020. http://news.bbc.co.uk/1/hi/technology/8331253.stm.

Wardlow, H. 2018. 'HIV, phone friends and affective technology in Papua New Guinea'. In *The Moral Economy of Mobile Phones: Pacific Islands perspectives*, edited by R. Foster and H. Horst, 39–52. Acton, Australia: Australian National University Press.

Waterson, Roxana. 2014. *The Living House: An anthropology of architecture in South East Asia*. North Clarendon, VT: Tuttle Publishing.

WeAreSocial. 2018. 'Digital 2018: Cameroon'. Accessed 1 October 2020. https://datareportal.com/reports/digital-2018-cameroon.

WeAreSocial. 2020. 'Digital 2020: Cameroon'. Accessed 1 October 2020. https://datareportal.com/reports/digital-2020-cameroon.

Web Foundation. 2020. 'Sir Tim Berners-Lee invented the world wide web in 1989'. Web Foundation website, 2020. Accessed 1 October 2020. https://webfoundation.org/about/vision/history-of-the-web/.

Weiser, Eric B. 2015. '#Me: Narcissism and its facets as predictors of selfie-posting frequency'. *Personality and Individual Differences* 86 (November): 477–81. https://doi.org/10.1016/j.paid.2015.07.007.

Wilding, Raelene and Loretta Baldassar. 2018. 'Ageing, migration and new media: The significance of transnational care'. *Journal of Sociology* 54 (2): 226–35. https://doi.org/10.1177/1440783318766168.

Wilken, R., G. Goggin and Heather A. Horst, eds. 2019. *Location Technologies in International Context*. Abingdon, Oxon; New York: Routledge.

Williams, L. and C. Smith. 2005. 'QSEMSM: Quantitative scalability evaluation method'. Paper presented at Int. CMG (International Computer Measurement Group) conference, Orlando, Florida, 2005. PerfX and Performance Engineering Services. Accessed 1 October 2020. https://pdfs.semanticscholar.org/1ba0/8541f2cf3723d1af109c0ef08e2e12f46c74.pdf?_ga=2.77758556.952171762.1582645803-397802861.1582645803.

Wired Magazine. 2019. 'Oggi la tecnologia non ha età'. *Wired Italy*, 18 January 2019. Accessed 1 October 2020. https://www.wired.it/attualita/tech/2019/01/18/tecnologia-amplifon-eta/.

Worldometers.info. n.d. 'Covid-19 Coronavirus pandemic'. *Worldometers.info*. Accessed 1 October 2020. https://www.worldometers.info/coronavirus/.

Woyke, Elizabeth. 2014. *The Smartphone: Anatomy of an industry*. New York: The New Press.

Wright, J. 2019. 'The new frontier of robotics in the lives of elders: Perspectives from Japan and Europe'. In *The Cultural Context of Aging: Worldwide perspectives*, edited by J. Sokolovsky, 4th ed. Westport, CT: Praeger.

Wu, Jyh-Jeng, Chien Shu-Hua and Liu Kang-Ping. 2017. 'Why should I pay? Exploring the determinants influencing smartphone users' intentions to download paid app'. *Telematics and Informatics* 34 (5): 645–54. https://doi.org/10.1016/j.tele.2016.12.003.

Xiang, Biao. 2007. *Global 'Body Shopping': An Indian labor system in the information technology industry*. Princeton, NJ: Princeton University Press.

Xinhua. 2019. 'Chinese smartphone brand transsion most popular in Africa in Q2: IDC Study – Xinhua | English.News.Cn', 2019. Accessed 1 October 2020. http://www.xinhuanet.com/english/2019-08/29/c_138345934.htm.

Yalla Italia Twitter Account. 2020. 'Yalla Italia Twitter Account' (social media account), 2020. Accessed 1 October 2020. https://twitter.com/yallaitalia.

Yong, V. and Saito, Y. 2012. 'National long-term care insurance policy in Japan a decade after implementation: Some lessons for aging countries'. *Ageing International* 37: 271–84. https://doi.org/10.1007/s12126-011-9109-0; https://link.springer.com/article/10.1007/s12126-011-9109-0.

Zhao, X. 2018. 'Deals | Offering middle-aged users with a content generation tool, post editing app Meipian Banks $6.6m'. 3 January 2018. Accessed 1 October 2020. https://kr-asia.com/offering-middle-aged-users-with-its-content-generation-tool-post-editing-app-meipian-banks-6-6m.

Zuboff, Shoshana. 2019. *The Age of Surveillance Capitalism: The fight for a human future at the new frontier of power*. London: Profile Books.

Index